New Trends in Nonlinear Dynamics

Walter Lacarbonara • Balakumar Balachandran
Jun Ma • J. A. Tenreiro Machado • Gabor Stepan
Editors

New Trends in Nonlinear Dynamics

Proceedings of the First International Nonlinear Dynamics Conference (NODYCON 2019), Volume III

Springer

Editors
Walter Lacarbonara
Department of Structural and Geotechnical Engineering
Sapienza University of Rome
Rome, Italy

Jun Ma
Department of Physics
Lanzhou University of Technology
Lanzhou, Gansu, China

Gabor Stepan
Department of Applied Mechanics
Budapest University of Technology and Economics
Budapest, Hungary

Balakumar Balachandran
Department of Mechanical Engineering
University of Maryland
College Park, MD, USA

J. A. Tenreiro Machado
Department of Electrical Engineering
Polytechnic of Porto - School of Engineering (ISEP)
Porto, Portugal

ISBN 978-3-030-34723-9 ISBN 978-3-030-34724-6 (eBook)
https://doi.org/10.1007/978-3-030-34724-6

This Springer imprint is published by the registered company Springer Nature Switzerland AG.
The registered company address is: Gewerbestrasse 11, 6330 Cham, Switzerland

Preface

This volume is part of three volumes collecting the *Proceedings of the First International Nonlinear Dynamics Conference (NODYCON 2019)* held in Rome, February 17–20, 2019. NODYCON was launched to foster the tradition of the conference series originally established by Prof. Ali H. Nayfeh in 1986 at Virginia Polytechnic Institute and State University (Virginia Tech), Blacksburg, VA, USA, as the Nonlinear Vibrations, Stability, and Dynamics of Structures Conference. With the passing in 2017 of Prof. Nayfeh, who was also the founder of the Springer journal *Nonlinear Dynamics* in 1990, NODYCON 2019 was organized as a collective tribute of the community to Prof. Nayfeh for being one of the most influential leaders of nonlinear dynamics. NODYCON 2019 was also established to look to and dream about the future. The call for papers attracted contributions dealing with established nonlinear dynamics research topics as well as with the latest trends and developments. At the same time, to reflect the rich spectrum of topics covered by the journal *Nonlinear Dynamics*, the call included diverse and multidisciplinary topics, to mention a few, multi-scale dynamics, experimental dynamics, dynamics of structures/industrial machines/equipment/facilities, dynamics of adaptive, multifunctional, metamaterial structures, dynamics of composite/nanocomposite structures, reduced-order modeling, nonsmooth dynamics, fractional-order system dynamics, nonlinear interactions and parametric vibrations, computational techniques, nonlinear system identification, dynamics of NEMS/MEMS/nanomaterials, multibody dynamics, fluid/structure interaction, influence of nonlinearities on vibration control systems, human–machine interaction, nonlinear wave propagation in discrete and continuous media, chaotic map-based cryptography, ecosystem dynamics, social media dynamics, complexity in engineering, and network dynamics.

For NODYCON 2019, the organizers received 450 two-page abstracts and based on 467 reviews from the Program Committee, the Steering and Advisory Committees, and external reviewers, 391 papers and 17 posters were accepted, published in the Book of Abstracts (NODYS Publications, Rome, ISBN 978-88-944229-0-0), and presented by nearly 400 participants from 68 countries. The diverse topics covered by the papers were organized along four major themes to organize the technical sessions:

(a) Concepts and methods in nonlinear dynamics
(b) Nonlinear dynamics of mechanical and structural systems
(c) Nonlinear dynamics and control
(d) Recent trends in nonlinear dynamics

The authors of a selection of approximately 60 papers were invited to publish in the Special Issue of *Nonlinear Dynamics* entitled "NODYCON 2019 First International Nonlinear Dynamics Conference." Over 200 full papers were submitted to the *Proceedings of the First International Nonlinear Dynamics Conference* (NODYCON 2019) and only 121 of them were accepted. These papers have been collected into three volumes, which are listed below together with a sub-topical organization.

Volume 1: Nonlinear Dynamics of Structures, Systems, and Devices

(a) Methods for nonlinear dynamics
(b) Bifurcations and nonsmooth systems
(c) Nonlinear phenomena in mechanical systems and structures
(d) Experimental dynamics, system identification and monitoring
(e) Fluid–structure interaction, multibody system dynamics
(f) Turning processes, rotating systems, and systems with time delays

Volume 2: Nonlinear Dynamics and Control

(g) Vibration absorbers and isolators
(h) Control of nonlinear systems
(i) Sensors and actuators
(j) Network synchronization

Volume 3: New Trends in Nonlinear Dynamics

(k) Smart materials, metamaterials, composite and nanocomposite materials, and structures
(l) MEMS/NEMS and energy harvesters
(m) Nonlinear phenomena in bio- and ecosystem dynamics
(n) Chaos in electronic systems
(o) Fractional-order systems

I wish to acknowledge the work of the Co-Editors of the NODYCON 2019 Proceedings: Prof. Balakumar Balachandran (University of Maryland, College Park, MD, USA), Prof. Jun Ma (Lanzhou University of Technology, China), Prof. J. A. Tenreiro Machado (Instituto Superior de Engenharia do Porto, Portugal), Prof. Gabor Stepan (Budapest University of Technology and Economics, Hungary).

The success of NODYCON 2019 relied primarily on the efforts, talent, energy, and enthusiasm of researchers in the field of nonlinear dynamics who wrote and submitted these papers. Special praise is also deserved for the reviewers who invested significant time in reading, examining, and assessing multiple papers, thus ensuring a high standard of quality for this conference proceedings.

Rome, Italy
August 2019

Walter Lacarbonara

Preface for Volume 3: New Trends in Nonlinear Dynamics

Volume 3 of the NODYCON 2019 Proceedings includes 33 papers on new trends in nonlinear dynamics and is organized in five parts.

The first part includes 12 papers and explores the mechanical properties and dynamical response of smart materials, metamaterials, composite and nanocomposite materials. A. Elhady, M. Basha, and E. Abdel-Rahman present an electric permittivity sensor based on the Bleustein–Gulyaev waves. The modeling suggests that under a bias of only a few volts, the sensor can be driven into a nonlinear regime where its sensitivity can be tuned to match that of aqueous solutions, thus making it suitable for biomedical applications. M. Bukhari and O. Barry examine the role of stiffness nonlinearity on a periodic one-dimensional chain with multiple local resonators. The dispersion equation for the system is derived analytically by the method of multiple scales. The nonlinearity shows enhancement in the bandgap regions, especially with increasing number of local resonators. S. Zhu, J. Li, J. Zhou, and T. Quan investigate the nonlinear dynamic response of a simply supported concave hexagonal honeycomb sandwich plate with negative Poisson's ratio. The results provide theoretical guidance towards nonlinear vibration control for the metamaterial honeycomb sandwich structures. F. Mezzani, A. Rezaei, and A. Carcaterra provide a general method to deal with nonlinear integro-differential equations by using the statistical linearization and Fredholm's approach. The elastic metamaterial is characterized by long-range nonlocal interactions besides a nonlinear short-range constitutive relationship. The analytical results are obtained to unveil the onset of unconventional propagation. F. Coppo, F. Mezzani, S. Pensalfini, and A. Carcaterra present a numerical investigation of a metamaterial in the form of a one-dimensional elastic waveguide, equipped with nonlocal, nonlinear interactions. Numerical simulations, comparing the linear vs. nonlinear models, unveil wave-stopping and backward propagation phenomena. M. Taló, B. Carboni, G. Formica, G. Lanzara, M. Snyder, and W. Lacarbonara investigate the nonlinear dynamic response of carbon nanotube (CNT) nanocomposite cantilevers, both experimentally and theoretically. Nanocomposite cantilevers made of a thermoplastic polymer and high aspect ratio CNTs are subject to a primary resonance base excitation. The CNT/polymer frictional sliding hysteresis described by a hysteretic restoring force in the context of a nonlinear beam theory is shown to be responsible for a peculiar softening-hardening frequency response trend. L. Leonetti, G. Formica, D. Magisano, M. Taló, G. Garcea, and W. Lacarbonara suggest an efficient continuation strategy based on the Riks method to describe the stable and unstable branches of the response of CNT/polymer nanocomposite shells. The equilibrium paths and the static bifurcations of CNT nanocomposites are investigated and numerically described highlighting the effects of material parameters such as the CNT orientation and weight fraction. K. Lidiya, S. Tetyana, and J. Awrejcewicz study the buckling behavior and parametric vibrations of sandwich plates with arbitrary forms and made of isotropic and functionally graded materials (FGM). To calculate the mechanical characteristics for different lamination schemes, the analytical expressions are obtained in the context of first-order shear deformation theory. F. Silva, P. Rodrigues, and P. Gonçalves explore the influence of the elastic foundation discontinuities on the nonlinear response of a functionally graded cylindrical shell with internal flowing fluid. Different discontinuities of elastic foundation and axial fluid flow velocities are considered, and the influence of these parameters on the stable branches of the resonance curves is confirmed. V. Burlayenko, T. Sadowski, and S. Dimitrova investigate double cantilever beam interlaminar fracture toughness sandwich specimens under different kinds of dynamic loading and loading rates. Cohesive finite elements are used to simulate the dynamic fracturing of the specimens. The influence of inertia on interfacial crack propagation in the specimens is evaluated as a direct outcome of the finite element analysis. A. Nabarrete, E. Araujo, J. Balthazar, and A. Tusset discuss the dynamic buckling of a sandwich plate to identify the response signals and to find the relevant frequencies and amplitudes. The nonlinear variation of the response for the out-of-plane displacements is analyzed with the continuous wavelet transform method for characterizing the observed behavior. J. Zhang, S. Chen, and L. Chen study torsional stress waves, buckling of functionally graded material cylindrical shells under torsional impact load by the symplectic method. The influence of the material gradient and the geometric parameters on buckling loads is analyzed and discussed.

The second part includes four papers dealing with model setting and estimation of micro-electromechanical systems and energy harvesters. S. Arakelian, I. Chestnov, A. Istratov, T. Khudaiberganov, and O. Butkovskiy study laser-induced nanocluster structures of different types. The problem of optical response and high temperature superconductivity, due to topological surface structures with correlated states, is considered in the frame of nonlinear dynamic modeling. The quantum mobility of electrons over different trajectories in the spatially inhomogeneous structures/nanocluster systems is presented in accordance with the path integral theory approach. K. Khorkov, D. Kochuev, R. Chkalov, V. Prokoshev, and S. Arakelian adopt a nonstationary technique for the laser-induced functional elements synthesis based on micro- and nanostructures in graphite samples. Carbon nanostructures such as graphene, nanopeaks, and crystals are obtained. The nonlinear formation mechanisms of nanostructures and microcrystals under femtosecond laser radiation for graphite in liquid nitrogen are analyzed. K. Chinnam, A. Casalotti, E. Bemporad, and G. Lanzara estimate the influence of the electrical and mechanical coupling on the dynamic response of an electrospun piezoelectric microfiber. Particular focus is given to the piezoelectric response dependence on the applied voltage. The results highlight the possibility of making materials characterized by tunable stiffness and resonance frequency. R. Mohamed, A. El-Badawy, A. Moustafa, A. Kirolos, M. Soliman, and E. Abdel-Rahman obtain an analytical model of an electromagnetic levitation energy harvester and validate it by comparing its result with FEM simulations and experimental measurements. The model is based on dipole moment approximations of magnetic fields and interaction forces. The level of agreement of both models with measurements is discussed.

The third part of this volume includes nine papers and analyzes biological and ecological systems. P. Feng, R. Wang, and Y. Wu study the role of regular waves in cortical information processing in the mammalian neocortex. A variety of spatiotemporal patterns can be induced for selecting plane waves, spiral and irregular waves, and even chaotic spatial patterns by changing the coupling strength between neurons connected by chemical synapses. The results indicate that the stability of neural network depends on the coupling function and type. E. Kaslik and R. Muresan analyze the dynamical behavior of the interplay of homeostatic regulation and coupling time delay in a pair of reciprocally coupled Wilson–Cowan networks. The occurrence of rich dynamical behavior is explored both theoretically and numerically. X. Mao, X. Zhou, T. Shi, and L. Qiao evaluate the effects of the autaptic connection on the behaviors of ring-coupled FitzHugh–Nagumo neurons. Different transmission time delays between neurons and one autaptic time delay on the self-connection are considered. Several examples reveal phenomena such as stability switches, different multi-periodic oscillation patterns, chaotic motions, and coexisting attractors. E. Kaslik, M. Neamtu, and A. Radulescu present a nonlinear model of dopamine-modulated prefrontal–limbic interactions in schizophrenia, including discrete time delays. A stability and bifurcation analysis is carried out in the neighborhood of the system positive equilibrium. The results reveal the importance of time delays in modulating dopamine reactivity. E. Kaslik, E. Kokovics, and A. Radulescu generalize the Wilson–Cowan model of excitatory and inhibitory interactions in localized neuronal populations by taking into consideration distributed time delays. The stability region in the characteristic parameter plane is determined and a comparison is given for several types of delay kernels. Important differences are also highlighted by comparing the generalized model with the original Wilson–Cowan model without time delays. T. Trifonova, S. Arakelian, D. Trifonov, S. Abrakhin, V. Koneshov, A. Nikolaev, and M. Arakelian propose a scheme to explain and predict the process of a flood and/or mudflow (debris) formation and spreading out over the river beds in mountain conditions taking into account nonlinear dynamics. The approach can enable more reasonable forecast and early warning for the natural water hazard/disaster taking into account the groundwater nonlinear flow contribution as a dominant factor under some conditions to the land surface water. H. Zhao, J. Yu, J. Cao, and W. Liao address the complexity quantification of human gait and physiologic signals. The study calculates a refined weighted-permutation entropy by assigning fewer weights to outliers and more weights to regular spiky patterns according to the normal distribution function. The human gait and ECG experimental data are analyzed by the proposed method. J. Simonović analyzes the bone cell communication dynamics using the bone cell population model by means of a system of coupled ordinary differential equations with power-law nonlinearities. The work explores several in silico experiments and provides more realistic approaches for interpreting the development of interventions for patients with bone trauma and diseases. S. George, R. Misra, and G. Ambika consider the nonlinear dynamics of RRc Lyrae variable stars, using intensity data from the Kepler space telescope. The results support the existence of two distinct subcategories of RRc Lyrae stars and indicate a link between the nonlinear and astrophysical properties of RRc Lyrae variable stars.

The fourth part includes three papers with focus on resonance phenomena, the emergence of chaos in circuits with finite variables, and the multistability property that can be induced in circuits with memristor. M. Bucolo, A. Buscarino, C. Famoso, L. Fortuna, and M. Frasca describe the existence of multi-jump resonance for the driven Chua circuit. Multi-jump resonance is investigated to characterize its onset as a function of system parameter values. A physical implementation of the driven Chua circuit in which jump and multi-jump resonances occur is presented and discussed. S. Seth studies robust chaos occurring in piecewise smooth dynamical systems. The first experimental observation of this phenomenon in a 3D electronic switching system is reported and the region of its parameter space is determined experimentally. C. Li selects a chaotic system with amplitude/frequency parameter that controls the scale and speed of oscillations without changing its basic feature of chaos. By introducing a memristor into the feedback for amplitude/frequency control, a special regime of homogenous multistability emerges, where the initial condition of the internal variable only determines the amplitude of the variables without changing the essential chaotic oscillation.

The fifth and last part includes five papers dealing with fractional-order and nonsmooth systems. Here, researchers analyze dynamical effects, such as stability, bifurcations, and complex vibrations. M. Shitikova and B. Ajarmah investigate nonlinear damped vibrations of a cylindrical shell embedded into a fractional derivative medium. They consider the case of the combinational internal resonance, resulting in modal interactions, using two different numerical methods. The damping properties of the surrounding medium are described by the fractional derivative Kelvin–Voigt model adopting the Riemann–Liouville fractional derivatives. O. Brandibur, E. Kaslik, D. Mozyrska, and M. Wyrwas discuss the stability of the Caputo-type linear fractional variable-order discrete-time equations known as biquadratic equations. Linear equations with constant coefficients and variable-order differences defined by functions with values from the interval (0, 2] are considered, and some sufficient conditions for the asymptotic stability are presented. O. Brandibur, E. Kaslik, D. Mozyrska, and M. Wyrwas find the necessary and sufficient conditions for the asymptotic stability and instability of two-dimensional linear autonomous noncommensurate systems of fractional-order Caputo difference equations. The results are applied to the fractional-order version of the Rulkov neuronal model. K. Hedrih addresses the topic of independent fractional-type modes of free oscillations and hybrid modes of forced vibrations. The paper describes the hybrid forced fractional-type vibration modes and the corresponding analytical approximate solution expressed by convolution integral. Y. Yu and Z. Wang propose two fractional-order Chua's memristive circuits. The first is a fractional-order memristive circuit with only the memristor described by a fractional-order derivative due to the memory loss which is observed experimentally. The second is a direct fractional-order generalization of integer-order Chua's memristive circuit without considering the physical background. Numerical simulations show that both models exhibit multistability and different steady states switch via grazing bifurcation.

We hope that the multifaceted state-of-the-art contributions collected in this volume will provide fruitful inspiration for new advances in the relevant challenging areas of research.

Rome, Italy — Walter Lacarbonara
College Park, MD, USA — Balakumar Balachandran
Lanzhou, China — Jun Ma
Porto, Portugal — J. A. Tenreiro Machado
Budapest, Hungary — Gabor Stepan
August 2019

Contents

Contributors

Eihab M. Abdel-Rahman Department of Systems Design Engineering, University of Waterloo, Waterloo, ON, Canada

Sergei Abrakhin Vladimir State University, Vladimir, Russia

Basem Ajarmah Research Center on Dynamics of Solids and Structures, Voronezh State Technical University, Voronezh, Russia
Al-istiqlal University, Jericho, Palestine

G. Ambika Indian Institute of Science Education and Research, Tirupati, Tirupati, India

Mileta Arakelian Yerevan State University, Yerevan, Armenia

Sergei M. Arakelian Department of Physics and Applied Mathematics, Vladimir State University, Vladimir, Russia

Jan Awrejcewicz Department of Automation, Biomechanics and Mechatronics, Lodz University of Technology, Lodz, Poland

Balakumar Balachandran Department of Mechanical Engineering, University of Maryland, College Park, MD, USA

Jose Manoel Balthazar Department of Electronic Engineering, UTFPR, Ponta Grossa, SP, Brazil

Oumar Barry Department of Mechanical Engineering, Virginia Tech, Blacksburg, VA, USA

Mohamed Basha Department of Systems Design Engineering, University of Waterloo, Waterloo, ON, Canada

Edoardo Bemporad Department of Engineering, Roma Tre University, Rome, Italy

Oana Brandibur Department of Mathematics and Computer Science, West University of Timisoara, Timisoara, Romania

Maide Bucolo Dipartimento di Ingegneria Elettrica Elettronica e Informatica, University of Catania, Catania, Italy

Mohammad Bukhari Department of Mechanical Engineering, Virginia Tech, Blacksburg, VA, USA

Vyacheslav N. Burlayenko Lublin University of Technology, Lublin, Poland
National Technical University 'KhPI', Kharkiv, Ukraine

Arturo Buscarino Dipartimento di Ingegneria Elettrica Elettronica e Informatica, University of Catania, Catania, Italy
IASI, Consiglio Nazionale delle Ricerche (CNR), Roma, Italy

Oleg Ya. Butkovskiy Department of Physics and Applied Mathematics, Vladimir State University, Vladimir, Russia

Junyi Cao Xi'an Jiaotong University, Xi'an, China

Biagio Carboni Department of Structural and Geotechnical Engineering, Sapienza University of Rome, Rome, Italy

Antonio Carcaterra Department of Mechanical and Aerospace Engineering, Sapienza University of Rome, Rome, Italy

Arnaldo Casalotti Department of Engineering, Roma Tre University, Rome, Italy

Like Chen Department of Engineering Mechanics, Lanzhou University of Technology, Lanzhou, China

Shuai Chen Faculty of Vehicle Engineering and Mechanics, Dalian University of Technology, Dalian, China

Igor Yu. Chestnov Department of Physics and Applied Mathematics, Vladimir State University, Vladimir, Russia
Institute of Natural Sciences, Westlake Institute for Advanced Study, Zhejiang Province, China

Krishna Chytanya Chinnam Department of Engineering, Roma Tre University, Rome, Italy

Ruslan Chkalov Vladimir State University, Vladimir, Russia

Francesco Coppo Department of Mechanical and Aerospace Engineering, Sapienza University of Rome, Rome, Italy

Eduardo Francisco Rocha de Araujo Department of Aerospace Engineering, Instituto Tecnológico de Aeronáutica, São José dos Campos, SP, Brazil

Svetlana D. Dimitrova National Technical University 'KhPI', Kharkiv, Ukraine

Ayman El-Badawy Department of Mechanical Engineering, Al-Azhar University, Cairo, Egypt
Mechatronics Engineering Department, German University, Cairo, Egypt

Alaa Elhady Department of Systems Design Engineering, University of Waterloo, Waterloo, ON, Canada

Carlo Famoso Dipartimento di Ingegneria Elettrica Elettronica e Informatica, University of Catania, Catania, Italy

Peihua Feng School of Aerospace Engineering, Xi'an Jiaotong University, Xi'an, China

Giovanni Formica Department of Architecture, University of Roma Tre, Rome, Italy

Luigi Fortuna Dipartimento di Ingegneria Elettrica Elettronica e Informatica, University of Catania, Catania, Italy
IASI, Consiglio Nazionale delle Ricerche (CNR), Roma, Italy

Mattia Frasca Dipartimento di Ingegneria Elettrica Elettronica e Informatica, University of Catania, Catania, Italy
IASI, Consiglio Nazionale delle Ricerche (CNR), Roma, Italy

Giovanni Garcea Department of Informatics, Electronics and System Engineering, University of Calabria, Rende, Italy

Sandip V. George Indian Institute of Science Education and Research, Pune, India
University Medical Center Groningen, Groningen, the Netherlands

Paulo B. Gonçalves Department of Civil Engineering, Pontifical Catholic University, Rio de Janeiro, Brazil

Alexander V. Istratov Department of Physics and Applied Mathematics, Vladimir State University, Vladimir, Russia

Eva Kaslik Department of Mathematics and Computer Science, West University of Timisoara, Timisoara, Romania

Kirill Khorkov Vladimir State University, Vladimir, Russia

Timur A. Khudaiberganov Department of Physics and Applied Mathematics, Vladimir State University, Vladimir, Russia

Andrew Kirolos Mechatronics Engineering Department, German University, Cairo, Egypt

Dmitriy Kochuev Vladimir State University, Vladimir, Russia

Emanuel-Attila Kokovics West University of Timişoara, Timişoara, Romania

Vyacheslav Koneshov Schmidt Institute of Physics of the Earth, RAS, Moscow, Russia

Walter Lacarbonara Department of Structural and Geotechnical Engineering, Sapienza University of Rome, Rome, Italy

Giulia Lanzara Department of Engineering, University of Roma Tre, Rome, Italy

Leonardo Leonetti Department of Informatics, Electronics and System Engineering, University of Calabria, Rende,
Italy

Chunbiao Li Jiangsu Collaborative Innovation Center of Atmospheric Environment and Equipment Technology (CICAEET), Nanjing University of Information Science and Technology, Nanjing, China
School of Electronic and Information Engineering, Nanjing University of Information Science and Technology, Nanjing, China

Jing Li College of Applied Sciences, Beijing University of Technology, Beijing, P. R. China

Wei-Hsin Liao The Chinese University of Hong Kong, Hong Kong, China

Kurpa Lidiya Department of Applied Mathematics, National Technical University "KhPI", Kharkov, Ukraine

Jun Ma Department of Physics, Lanzhou University of Technology, Lanzhou, Gansu, China

J. A. Tenreiro Machado Department of Electrical Engineering, Polytechnic of Porto - School of Engineering (ISEP), Porto, Portugal

Domenico Magisano Department of Informatics, Electronics and System Engineering, University of Calabria, Rende, Italy

Xiaochen Mao Department of Engineering Mechanics, College of Mechanics and Materials, Hohai University, Nanjing, China

Federica Mezzani Department of Mechanical and Aerospace Engineering, Sapienza University of Rome, Rome, Italy

Ranjeev Misra Inter-University Centre for Astronomy and Astrophysics, Pune, India

Ramy A. Mohamed Department of Mechanical Engineering, Al-Azhar University, Cairo, Egypt

Ahmed Moustafa Mechatronics Engineering Department, German University, Cairo, Egypt

Dorota Mozyrska Faculty of Computer Science, Bialystok University of Technology, Bialystok, Poland

Raluca Mureşan Department of Mathematics and Computer Science, West University of Timişoara, Timişoara, Romania

Airton Nabarrete Department of Aerospace Engineering, Instituto Tecnológico de Aeronáutica, São José dos Campos, SP, Brazil

Mihaela Neamţu West University of Timişoara, Timişoara, Romania
Academy of Romanian Scientists, Bucharest, Romania

Alexei Nikolaev Schmidt Institute of Physics of the Earth, RAS, Moscow, Russia

Sara Pensalfini Department of Mechanical and Aerospace Engineering, Sapienza University of Rome, Rome, Italy

Valery Prokoshev Vladimir State University, Vladimir, Russia

Lei Qiao Department of Engineering Mechanics, College of Mechanics and Materials, Hohai University, Nanjing, China

Tingting Quan College of Applied Sciences, Beijing University of Technology, Beijing, P. R. China
School of Science, Tianjin Chengjian University, Tianjin, P. R. China

Anca Rădulescu SUNY New Paltz, New Paltz, NY, USA

Amir Sajjad Rezaei Department of Mechanical and Aerospace Engineering, Sapienza University of Rome, Rome, Italy

Patrícia C. Rodrigues School of Civil and Environmental Engineering, Federal University of Goiás, Goiânia, Brazil

Tomasz Sadowski Lublin University of Technology, Lublin, Poland
National Technical University 'KhPI', Kharkiv, Ukraine

Soumyajit Seth Indian Institute of Science Education and Research Kolkata, Mohanpur, India

Tiantian Shi Department of Engineering Mechanics, College of Mechanics and Materials, Hohai University, Nanjing, China

Marina V. Shitikova Research Center on Dynamics of Solids and Structures, Voronezh State Technical University, Voronezh, Russia

Frederico M. A. Silva School of Civil and Environmental Engineering, Federal University of Goiás, Goiânia, Brazil

Julijana Simonović Biomedical Engineering Department, School of Engineering, Cardiff University, Wels, UK

Matthew Snyder Department of Engineering Mechanics, United States Air Force Academy, CO, USA

Mostafa Soliman Power Electronics Department, Electronics Research Institute, Cairo, Egypt

Gabor Stepan Department of Applied Mechanics, Budapest University of Technology and Economics, Budapest, Hungary

Katica R. (Stevanović) Hedrih Department of Mechanics, Mathematical Institute of the Serbian Academy of Science and Arts, Belgrade, Serbia

Michela Talò Department of Structural and Geotechnical Engineering, Sapienza University of Rome, Rome, Italy

Shmatko Tetyana Department of Higher Mathematics, National Technical University "KhPI", Kharkov, Ukraine

Dmitri Trifonov Vladimir State University, Vladimir, Russia

Tatiana Trifonova Lomonosov Moscow State University, Moscow, Russia

Angelo Marcelo Tusset Department of Electronic Engineering, UTFPR, Ponta Grossa, SP, Brazil

Rong Wang Xi'an University of Science Technology, College of Science, Xi'an, China

Zaihua Wang State Key Lab of Mechanics and Control of Mechanical Structures, Nanjing University of Aeronautics and Astronautics, Nanjing, China

Ying Wu State Key Laboratory for Strength and Vibration of Mechanical Structures, Shaanxi Engineering Laboratory for Vibration Control of Aerospace Structures, School of Aerospace Engineering, Xi'an Jiaotong University, Xi'an, China

Małgorzata Wyrwas Faculty of Computer Science, Bialystok University of Technology, Bialystok, Poland

Jian Yu China Ship Development and Design Center, Wuhan, China

Yajuan Yu School of Mathematics and Physics, Changzhou University, Changzhou, China

Jinghua Zhang Department of Engineering Mechanics, Lanzhou University of Technology, Lanzhou, China

Huan Zhao Xi'an Jiaotong University, Xi'an, China

Ji Zhou State Key Laboratory of New Ceramics and Fine Processing, School of Materials Science and Engineering, Tsinghua University, Beijing, P. R. China

Xiangyu Zhou Department of Engineering Mechanics, College of Mechanics and Materials, Hohai University, Nanjing, China

Shaotao Zhu College of Applied Sciences, Beijing University of Technology, Beijing, P. R. China

Part I
Smart Materials, Metamaterials, Composite and Nanocomposite Materials, and Structures

Tunable Bleustein–Gulyaev Permittivity Sensors

Alaa Elhady, Mohamed Basha, and Eihab M. Abdel-Rahman

Abstract We present a novel electric permittivity sensor based on Bleustein–Gulyaev (BG) waves. We also demonstrate a mechanism by which biasing can be used to modulate the sensitivity of permittivity sensors to match different electric permittivity ranges. We formulate the nonlinear electromechanical differential equations governing the dynamics of BG waves. Model results suggest that under a bias of only a few volts, the sensor can be driven into a nonlinear regime where its sensitivity can be tuned to match that of aqueous solutions, thereby allowing for biomedical applications.

Keywords Bleustein–Gulyaev waves · Permittivity sensors

1 Introduction

This work proposes a novel miniature permittivity sensor with a large dynamic range that can operate in the MHz range. Dielectric permittivity sensors are widely used in industrial applications, such as oil characterization [1], environmental applications such as soil testing [2] and sea water salinity testing [3], and biological applications, such as DNA and tissue discrimination [4, 5], cancer cell identification [6], and blood analysis [7, 8]. They are typically RF devices with operating frequencies set in excess of 100 GHz in order to limit the sensor size [5]. The use of high frequency signals introduces additional complexity and the attendant challenges of cost, reliability, and overall device size. Indeed, operating sensors at lower frequencies inflate their size to a few centimeters [9].

Bleustein–Gulyaev (BG) waves [10, 11] are coupled acoustic-electromagnetic waves that propagate exclusively along the surface of shear poled piezoelectric

A. Elhady (✉) · M. Basha · E. M. Abdel-Rahman
Department of Systems Design Engineering, University of Waterloo, Waterloo, ON, Canada
e-mail: aesamymo@uwaterloo.ca

W. Lacarbonara et al. (eds.), *New Trends in Nonlinear Dynamics*,
https://doi.org/10.1007/978-3-030-34724-6_1

materials at acoustic speeds. These waves have been used in viscosity sensors [12, 13] and telecommunication filters [14]. We exploit the wave's acoustic speed to develop sensors that detect disturbances to the electromagnetic field at the shorter wavelength of BG waves, thereby reducing the device size and operating frequency by several orders of magnitude. Prototypes ranging in size from a few hundred micrometers to a few millimeters are under fabrication. Further, we present a novel technique to deploy the nonlinear properties of dielectrics to tune the sensors' sensitivity on-the-fly, thereby extending its dynamic range.

2 Model

BG waves couple an electromagnetic component, made of a transverse electric field and a magnetic field perpendicular to it, with a surface acoustic shear wave. It propagates along the surface of shear poled piezoelectrics.

The wave constituent electric, magnetic, and displacement fields, Fig. 1, can be written as:

$$\mathbf{u}(x, y, t) = \begin{bmatrix} 0 \\ 0 \\ u \end{bmatrix}, \qquad \mathbf{E}(x, y, t) = \begin{bmatrix} E_x \\ E_y \\ 0 \end{bmatrix}, \qquad \mathbf{H}(x, y, t) = \begin{bmatrix} 0 \\ 0 \\ H_z \end{bmatrix}. \tag{1}$$

The wave is governed by the equation of motion for a piezoelectric element in the lower half-space

$$\nabla \cdot \boldsymbol{\sigma} = \rho \ddot{\mathbf{u}}; \qquad y > 0 \tag{2}$$

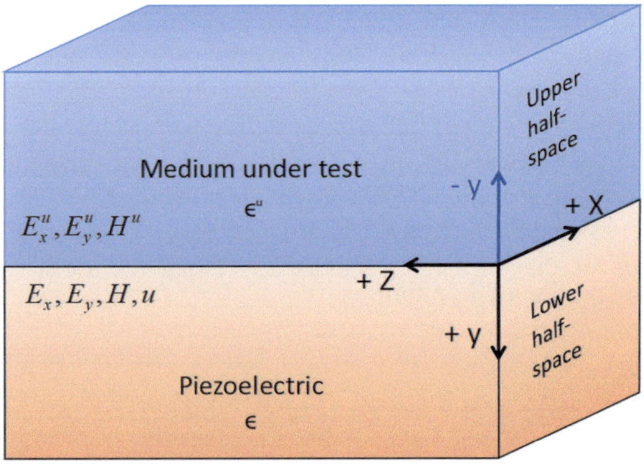

Fig. 1 Propagation of BG waves

where ρ is the piezoelectric density. Further, Maxwell's equations in the lower and upper half-spaces [10] are

$$\nabla \times \mathbf{E} = -\mu \dot{\mathbf{H}} \qquad y > 0\,, \qquad \nabla \times \mathbf{E^u} = -\mu \dot{\mathbf{H^u}} \qquad y < 0 \tag{3}$$

$$\nabla \times \mathbf{H} = \dot{\mathbf{D}} \qquad y > 0\,, \qquad \nabla \times \mathbf{H^u} = \dot{\mathbf{D^u}} \qquad y < 0. \tag{4}$$

These equations are subject to a compatibility condition on the electromagnetic and mechanical fields:

$$\nabla \cdot \mathbf{D} = 0; \qquad y > 0 \tag{5}$$

and boundary conditions at the interface plane ($y = 0$) between the two half-spaces, as well as reflector boundary conditions placed at both ends of the propagation direction (x-axis) to create a standing wave.

In the upper half-space, the relevant relations are only the linear Maxwell equations. In the lower half-space, we adopt a binomial expansion of the crystal's Gibbs free energy (PE) to capture the piezoelectric and electrostrictive material nonlinearities.

$$PE = \alpha_1 E^2 + \alpha_2 \sigma^2 + \alpha_3 \sigma E + \alpha_4 E^3 + \alpha_5 \sigma^3 + \alpha_6 \sigma^2 E + \alpha_7 E^2 \sigma + \alpha_8 \sigma^4 + \alpha_9 E^4 + \alpha_{10} E^3 \sigma + \alpha_{11} \sigma^3 E + \alpha_{12} \sigma^2 E^2 + H.O.T. \tag{6}$$

To represent the relative orders of the shear deformation and electric fields, we scale the electric field $\mathbf{E}$ at $O(\hat{\epsilon})$ and the stress tensor $\boldsymbol{\sigma}$ at $O(\hat{\epsilon}^{\,2})$, where $\hat{\epsilon}$ is a small bookkeeping parameter. Retaining terms up to order $O(\hat{\epsilon}^{\,4})$, Gibbs free energy reduces to:

$$PE = \alpha_1 E^2 + \alpha_2 \sigma^2 + \alpha_3 \sigma E + \alpha_4 E^3 + \alpha_7 \sigma E^2 + \alpha_9 E^4 + O(\hat{\epsilon}^{\,5}). \tag{7}$$

The mechanical strain and electric displacement can then be derived as:

$$S = \frac{\partial PE}{\partial \sigma} = 2\alpha_2 \sigma + \alpha_3 E + \alpha_7 E^2 \tag{8}$$

$$D = \frac{\partial PE}{\partial E} = 3\alpha_1 E + \alpha_3 \sigma + 3\alpha_4 E^2 + 2\alpha_7 \sigma E + 4\alpha_9 E^3. \tag{9}$$

Using the field definitions in Eq. (1), the material definitions for the coefficients in Eqs. (8) and (9) for $y > 0$, and rearranging we obtain:

$$\sigma_{yz} = G\frac{\partial u}{\partial y} - \xi E_y - GME_x^2 + GME_y^2 \tag{10}$$

$$\sigma_{xz} = G\frac{\partial u}{\partial x} - \xi E_x - 2GME_x E_y \tag{11}$$

$$D_x = \epsilon E_x + \xi \frac{\partial u}{\partial x} + 2GM \frac{\partial u}{\partial y} E_x + 4GM \frac{\partial u}{\partial x} E_y - 6\xi M E_x E_y$$
$$- 6GM^2 E_x E_y^2 - 2GM^2 E_x^3 \tag{12}$$

$$D_y = \epsilon E_y + \xi \frac{\partial u}{\partial y} - 4\xi M E_x^2 + 2\xi M E_y^2 - 2GM \frac{\partial u}{\partial y} E_y + 4GM \frac{\partial u}{\partial x} E_x$$
$$- 6GM^2 E_x^2 E_y - 2GM^2 E_y^3, \tag{13}$$

where G is the shear modulus, ξ is the shear piezoelectric coefficient, M is the electrostrictive coefficient, and ϵ is the zero strain electric permittivity of the piezoelectric material.

3 Permittivity Sensor

Substituting with Eqs. (10)–(13) into Eqs. (2), (3), and (4), we obtain the governing system of equations. We set $M = 0$ to obtain the linear wave equations and solve the corresponding eigenvalue problem for the fundamental natural frequency:

$$\omega = \frac{m\pi}{L} \sqrt{\frac{G + \frac{\xi^2}{\epsilon}}{\rho} \left(1 - \frac{\xi^4}{(G\epsilon + \xi^2)^2} \frac{1}{(1+r)^2}\right)},$$

where m is the number of interdigitated transducer (IDT) fingers along the propagation direction, r is the ratio of the permittivity of the material-under-test ϵ^u to that of the piezoelectric material ϵ, and L is the length of the sensor. Since all of these parameters are known except for ϵ^u, the natural frequency can be used as a measure of the sample electric permittivity.

The sensitivity of the sensor can be expressed as the shift in natural frequency for a unit change in permittivity:

$$S_\omega = \frac{\partial \omega}{\partial \epsilon^u}. \tag{14}$$

The black colored curve in Fig. 2 shows the sensitivity of 400 μm long permittivity sensor made of shear-poled Lead-Zirconium-Titanate Navy Type-I (PZT4) with $m = 10$ actuation fingers along the propagation direction as a function of the relative permittivity of the material-under-test $\epsilon^u/\epsilon_\circ$. Sensitivity reaches an optimal value for materials with relative permittivity similar to ∼400. On the other hand, aqueous media have a relative permittivity similar to ∼80. It is impractical to address this mismatch by changing the sensor material, rather we explore the possibility of addressing it by exploiting the electrostrictive properties of piezoelectrics.

For piezoelectrics where the electrostrictive constant is not negligible ($M \neq 0$), we can estimate the effective permittivity $\tilde{\epsilon}$ by linearizing Eq. (12) around a given

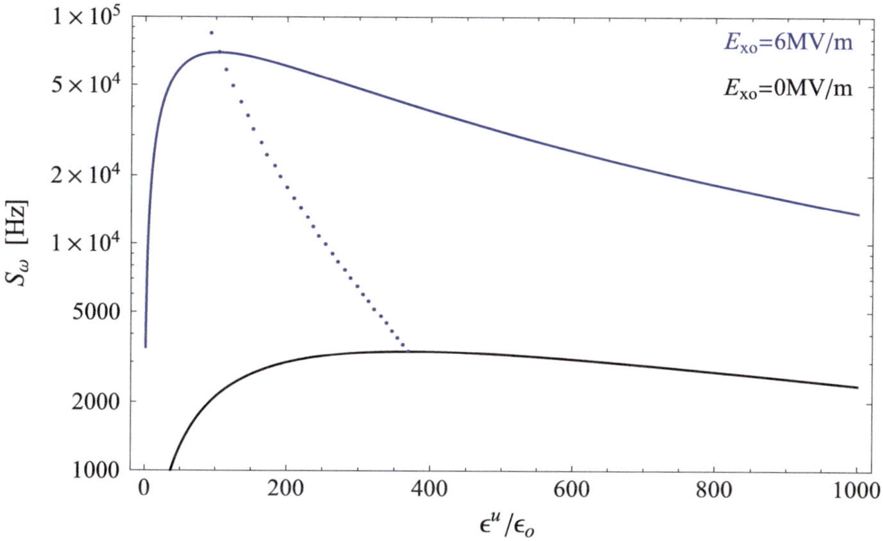

Fig. 2 Sensitivity of 400 μm long permittivity sensor made of PZT4 with $m = 10$ actuation fingers under no bias (black) and a DC electric field of $E_{x\circ} = 6$ MV/m and $E_{y\circ} = 0$ (blue). The dotted line tracks the maximum position

DC electric field $(E_{x\circ}, E_{y\circ})$ and zero pre-strain $(S_\circ = 0)$, by taking the derivative of electric displacement D_x with respect E_x:

$$\tilde{\epsilon} = \frac{\partial D_x}{\partial E_x}\bigg|_{(E_{x\circ}, E_{y\circ})} = \epsilon - 6\xi M E_{y_\circ} - 6GM^2(E_{y_\circ}^2 + E_{x_\circ}^2).$$

Therefore, it is possible to tune the permittivity of a piezoelectric material with non-negligible electrostriction by applying a DC electric field to it. The blue colored curve in Fig. 2 shows the sensitivity of the sensor described above as a function of the relative permittivity of the material-under-test where an electrostrictive constant of $M = 9.2 \times 10^{-18}$ m^2/V^2 was introduced under a biasing electric field of $E_{x\circ} =$ 6 MV/m and $E_{y\circ} = 0$. We note that the applied field has resulted in tuning the optimal sensitivity to a similar range of that for aqueous media and increased the sensor sensitivity, thereby opening the door to potential biomedical applications.

4 Nonlinear System Analysis

The right-hand side of Eqs. (3) is of order $O(\hat{\epsilon}^4)$ and, therefore, negligible. As a result, we adopt a quasi-static representation of the magnetic component of the field:

$$H_z(x, y, t) = H_\circ(x, y) \tag{15}$$

and reduce Eq. (3) to:

$$\frac{\partial E_x}{\partial y} - \frac{\partial E_y}{\partial x} = 0.$$

The electric field can, thus, be expressed purely in terms of the scalar potential $\psi(x, y)$ representing the voltage:

$$E_x = \frac{\partial \psi}{\partial x}, \quad E_y = \frac{\partial \psi}{\partial y}. \tag{16}$$

Using Eqs. (10), (11), (15), and (16) to substitute for the field variables in Eq. (2), we obtain the governing equation of motion as:

$$\ddot{u} + \beta_1 \nabla^2 u + \beta_2 \nabla^2 \psi + \beta_3 \psi_{,y}(\psi_{,xx} - \psi_{,yy}) - 2\beta_3 \psi_{,x} \psi_{,xy} = 0, \tag{17}$$

where the i subscript indicates partial derivative with respect to the i coordinate. The corresponding magnetic field is found from Eq. (4) as the time derivative of the electric displacement vector $\mathbf{D}$:

$$\begin{aligned} H_{\circ,x} = &\frac{\partial}{\partial t}\Big(\beta_4 u_{,y} + \beta_5 \psi_{,y} + \beta_6 u_{,x} \psi_{,x} + \beta_7 u_{,y} \psi_{,y} + \beta_8 \psi_{,x}^2 \psi_{,y} \\ &+ \beta_9 \psi_{,x}^2 + \beta_{10} \psi_{,y}^2 + \beta_8 \psi_{,y}^3\Big) \end{aligned} \tag{18}$$

$$\begin{aligned} H_{\circ,y} = &-\frac{\partial}{\partial t}\Big(\beta_4 u_{,x} + \beta_5 \psi_{,x} + \beta_6 u_{,x} \psi_{,y} - \beta_7 u_{,y} \psi_{,x} + \beta_8 \psi_{,x} \psi_{,y}^2 \\ &+ \beta_{11} \psi_{,y} \psi_{,x} + \beta_8 \psi_{,x}^3\Big). \end{aligned} \tag{19}$$

Differentiating equation (18) with respect to y and Eq. (19) with respect to x and subtracting the result, we eliminate the magnetic field to obtain:

$$\begin{aligned} \frac{\partial}{\partial t}(\nabla \cdot \mathbf{D}) = &\frac{\partial}{\partial t}\Big(\beta_4 \nabla^2 u + \beta_5 \nabla^2 \psi + 2\beta_6 u_{,x} \psi_{,xy} + \beta_7 u_{,y}(\psi_{,yy} - \psi_{,xx}) \\ &+ (\beta_6 - \beta_7) u_{,xy} \psi_{,x} + (\beta_6 u_{,xx} + \beta_7 u_{,yy}) \psi_{,y} + \beta_{11} \psi_{,xx} \psi_{,y} \\ &+ 2\beta_{10} \psi_{,y} \psi_{,yy} + (2\beta_9 + \beta_{11}) \psi_{,xy} \psi_{,x} + \beta_8 (3\psi_{,xx} + \psi_{,yy}) \psi_{,x}^2 \\ &+ \beta_8 (\psi_{,xx} + 3\psi_{,yy}) \psi_{,y}^2 + 4\beta_8 \psi_{,y} \psi_{,xy} \psi_{,x}\Big) = 0. \end{aligned} \tag{20}$$

We rewrite the displacement and electric fields as the summation of static and dynamic components:

$$u(x, y, t) = u^s(x, y) + u^d(x, y, t) \tag{21}$$

$$\psi(x, y, t) = \psi^s(x, y) + \psi^d(x, y, t). \tag{22}$$

The static displacement field $u^s(x, y)$ can be found by setting the time derivative in Eq. (17) equal to zero and solving the equilibrium equation:

$$\beta_1 \nabla^2 u^s + \beta_2 \nabla^2 \psi^s + \beta_3 \psi^s_{,y}(\psi^s_{,xx} - \psi^s_{,yy}) - 2\beta_3 \psi^s_{,x}\psi^s_{,xy} = 0 \tag{23}$$

subject to $\psi^s(x, y)$ the spatially distributed potential field imposed by the DC voltage applied to the metallic electrodes patterned along the sensor surface. As a first approximation, it is represented by an empirically fitted function in the propagation x-direction and an exponentially decaying function in the y-direction:

$$\psi^s(x, y) = e^{-\kappa_v^2 y}\Big(1.115 \cos\left(\frac{\pi m x}{L}\right) - 0.103 \cos\left(\frac{3\pi m x}{L}\right) - 0.019 \cos\left(\frac{5\pi m x}{L}\right) + 0.005 \cos\left(\frac{7\pi m x}{L}\right)\Big).$$

We substitute with Eqs. (21) and (22) in Eqs. (17) and (20) and use Eq. (23) in the result to obtain the equations of motion around the static equilibrium (u^s, ψ^s). In order to solve this dynamic system, we use a Galerkin expansion to express the system variables in terms of its mode shapes $\phi_j(x)$ as:

$$u^d(x, y, t) = \sum_{j=1}^{N} q_j(t)\kappa_{u_j} e^{-\kappa_{u_j}^2 y} \phi_j(x) \tag{24}$$

$$\psi^d(x, y, t) = \sum_{j=1}^{N} p_j(t)\kappa_{\psi_j} e^{-\kappa_{\psi_j}^2 y} \phi_j(x) + \frac{V(t)}{2} e^{-\kappa_v^2 y} \phi_1(x) \tag{25}$$

where $V(t)$ is the AC component of the applied voltage. We found the mode shapes by solving the eigenvalue problem of the system (Eqs. (17) and (20)) as:

$$\phi_j(x) = e^{ijm\pi \frac{x}{L}} + e^{-ijm\pi \frac{x}{L}}, \tag{26}$$

where j is an integer mode number.

In this work, the number of modes was taken to be $N = 2$. Multiplying the resulting Galerkin residuals with the corresponding left eigenfunctions and setting the integral over the domain equal to zero, we obtained a reduced-order model of the system in terms of the modal amplitudes as follows:

$$\ddot{q}_1 + \gamma_4 q_1 + \gamma_1 p_1 + \gamma_2 p_2 + \gamma_3 p_1 p_2 + \gamma_6 p_2 V = \gamma_5 V \tag{27}$$

$$\ddot{q}_2 + \gamma_{10} q_2 + \gamma_7 p_1 + \gamma_9 p_2 + \gamma_8 p_1^2 + \gamma_{12} p_1 V = \gamma_{11} V + \gamma_{13} V^2 \tag{28}$$

$$\gamma_{14}\dot{p}_1 + \gamma_{17}\dot{p}_2 + \gamma_{21}\dot{q}_1 + \gamma_{23}\dot{q}_2 + 2\gamma_{15}\dot{p}_1 p_1 + 3\gamma_{16}\dot{p}_1 p_1^2 + \gamma_{18}\dot{p}_2 p_1$$
$$+ \gamma_{18}\dot{p}_1 p_2 + 2\gamma_{19}\dot{p}_2 p_2 + \gamma_{20}\dot{p}_1 p_2^2 + 2\gamma_{20}\dot{p}_2 p_1 p_2 + \gamma_{22} p_2 \dot{q}_1 + \gamma_{22}\dot{p}_2 q_1$$

$$
\begin{aligned}
&+ \gamma_{24} p_1 \dot{q}_2 + \gamma_{24} \dot{p}_1 q_2 + \gamma_{32} \dot{p}_1 V^2 + \gamma_{26} \dot{p}_1 V + 2\gamma_{27} \dot{p}_1 p_1 V + \gamma_{28} \dot{p}_2 V \\
&+ 2\gamma_{29} \dot{p}_2 p_2 V + 2\gamma_{32} p_1 \dot{V} V + \gamma_{26} p_1 \dot{V} + \gamma_{27} p_1^2 \dot{V} + \gamma_{28} p_2 \dot{V} + \gamma_{29} p_2^2 \dot{V} \\
&+ \gamma_{30} \dot{q}_2 V + \gamma_{30} q_2 \dot{V} = 3\gamma_{33} \dot{V} V^2 + 2\gamma_{31} \dot{V} V + \gamma_{25} \dot{V} \qquad (29)
\end{aligned}
$$

$$
\begin{aligned}
&\gamma_{34} \dot{p}_1 + \gamma_{36} \dot{p}_2 + \gamma_{40} \dot{q}_1 + \gamma_{42} \dot{q}_2 + 2\gamma_{35} \dot{p}_1 p_1 + \gamma_{37} \dot{p}_2 p_1 + \gamma_{37} \dot{p}_1 p_2 \\
&+ \gamma_{38} \dot{p}_2 p_1^2 + 2\gamma_{38} \dot{p}_1 p_1 p_2 + 3\gamma_{39} \dot{p}_2 p_2^2 + \gamma_{41} p_1 \dot{q}_1 + \gamma_{41} \dot{p}_1 q_1 + \gamma_{49} \dot{p}_2 V^2 \\
&+ \gamma_{44} \dot{p}_1 V + \gamma_{45} \dot{p}_2 V + \gamma_{46} \dot{p}_2 p_1 V + \gamma_{46} \dot{p}_1 p_2 V + 2\gamma_{49} p_2 \dot{V} V + \gamma_{44} p_1 \dot{V} \\
&+ \gamma_{45} p_2 \dot{V} + \gamma_{46} p_1 p_2 \dot{V} + \gamma_{47} \dot{q}_1 V + \gamma_{47} q_1 \dot{V} = 2\gamma_{48} \dot{V} V + \gamma_{43} \dot{V} \qquad (30)
\end{aligned}
$$

5 Conclusion

We propose a permittivity sensor based on the propagation of Bleustein–Gulyaev waves along the surface of a piezoelectric layer. It is designed to sense the permittivity of an aqueous medium via a wave propagating at a speed close to sound speed, thereby allowing us to reduce the sensor size, and operating frequencies by several orders of magnitude. We found that it is not feasible to realize piezoelectric materials with optimal sensitivity in a range corresponding to that of aqueous media. Instead, we propose to tap into the electrostriction material nonlinearities of piezoelectrics to tune the sensor in order to realize optimal sensitivity in aqueous media range.

References

1. Folgero, K., Friiso, T., Hilland, J., Tjomsland, T.: A broad-band and high-sensitivity dielectric spectroscopy measurement system for quality determination of low-permittivity fluids. Meas. Sci. Technol. **6**(7), 995 (1995)
2. Birchak, J.R., Gardner, C.G., Hipp, J.E., Victor, J.M.: High dielectric constant microwave probes for sensing soil moisture. Proc. IEEE **62**(1), 93–98 (1974)
3. Meissner, T., Wentz, F.J.: The complex dielectric constant of pure and sea water from microwave satellite observations. IEEE Trans. Geosci. Remote Sens. **42**(9), 1836–1849 (2004)
4. Chen, T., Dubuc, D., Poupot, M., Fournie, J.J., Grenier, K.: Accurate nanoliter liquid characterization up to 40 GHz for biomedical applications: toward noninvasive living cells monitoring. IEEE Trans. Microwave Theory Tech. **60**(12), 4171–4177 (2012)
5. Taeb, A., Basha, M.A., Gigoyan, S., Marsden, M., Safavi-Naeini, S.: Label-free DNA sensing using millimeter-wave silicon WGM resonator. Opt. Express **21**(17), 19467–19472 (2013)
6. Qiao, G., Wang, W., Duan, W., Zheng, F., Sinclair, A.J., Chatwin, C.R.: Bioimpedance analysis for the characterization of breast cancer cells in suspension. IEEE Trans. Biomed. Eng. **59**(8), 2321–2329 (2012)
7. Grant, E., McClean, V., Nightingale, N., Sheppard, R., Chapman, M.: Dielectric behavior of water in biological solutions: studies on myoglobin, human low-density lipoprotein, and polyvinylpyrrolidone. Bioelectromagnetics **7**(2), 151–162 (1986)

8. Jamal, F.I., Guha, S., Eissa, M.H., Borngräber, J., Meliani, C., Ng, H.J., Kissinger, D., Wessel, J.: Low-power miniature k-band sensors for dielectric characterization of biomaterials. IEEE Trans. Microwave Theory Tech. **65**(3), 1012–1023 (2017)
9. Karami, M., Rezaei, P., Kiani, S., Sadeghzadeh, R.A.: Modified planar sensor for measuring dielectric constant of liquid materials. Electron. Lett. **53**(19), 1300–1302 (2017)
10. Bleustein, J.L.: A new surface wave in piezoelectric materials. Appl. Phys. Lett. **13**(12), 412–413 (1968)
11. Gulyaev, Y.V.: Electroacoustic surface waves in solids. ZhETF Pisma Redaktsiiu **9**, 63 (1969)
12. Guo, F., Wang, G., Rogerson, G.: Inverse determination of liquid viscosity by means of the Bleustein–Gulyaev wave. Int. J. Solids Struct. **49**(15), 2115–2120 (2012)
13. Kiełczyński, P., Szalewski, M., Balcerzak, A., Rostocki, A., Tefelski, D.: Application of SH surface acoustic waves for measuring the viscosity of liquids in function of pressure and temperature. Ultrasonics **51**(8), 921–924 (2011)
14. Kadota, M., Ago, J., Horiuchi, H.: A Bleustein–Gulyaev–Shimizu wave resonator having resonances for TV and VCR traps. IEEE Trans. Microwave Theory Tech. **44**(12), 2758–2762 (1996)

Nonlinear Metamaterials with Multiple Local Mechanical Resonators: Analytical and Numerical Analyses

Mohammad Bukhari and Oumar Barry

Abstract This paper examines the role of stiffness nonlinearity on a periodic one-dimensional chain with multiple local resonators. The cells of the chain consist of lumped masses connected through nonlinear springs. Each cell is embedded with multiple local resonators having different parameters. In one case the local resonators are assumed to be linear and in another case they are nonlinear. The dispersion equation for the system is derived analytically by the method of multiple scales (MMS). The results are validated via comparison with those in the literature and numerically via Matlab. The nonlinearity shows enhancement in the bandgap regions, especially with increasing number of local resonators.

Keywords Acoustics metamaterial · Perturbation techniques · Wave propagation

1 Introduction

The study of metamaterials has gained lots of attention in recent years due to their exceptional material properties and characteristics and their wider engineering applications. Metamaterials are a new class of artificial composites that derive their unique dynamic properties from both engineered local configurations and material constituents [1]. They were originally developed for electromagnetic and optical wave propagation and later the technology was extended to acoustic and mechanical waves. Most if not all metamaterials required the presence of periodic features with the potential of exhibiting interesting dynamic phenomena such as resonance or instability within the host structure. These interesting dynamic features can be judiciously employed for suppressing noise and vibration, harvesting energy,

M. Bukhari (✉) · O. Barry
Department of Mechanical Engineering, Virginia Tech, Blacksburg, VA, USA
e-mail: bukhari@vt.edu; obarry@vt.edu

W. Lacarbonara et al. (eds.), *New Trends in Nonlinear Dynamics*,
https://doi.org/10.1007/978-3-030-34724-6_2

non-destructive testing structures for defects, improving image resolution, and ameliorating the performance of antennas and many other engineering structures and devices [3].

Numerous investigators have examined the linear behavior of metamaterials focusing on acoustic-induced vibration suppression. Some metamaterials can be represented as discrete or continuous systems with embedded local mechanical resonators consisting of mass-spring-damper systems. These mechanical locally resonant metamaterials exhibit bandgap formation at wavelengths much larger than the lattice size [2]. The bandgaps can be made wider by embedding multiple resonators inside the cells [4, 5]. Beyond their linear interesting properties, nonlinear metamaterials may show superior performance in terms of wave propagation properties [6]. The weakly nonlinear discrete periodic structures shift the dispersion curves. This shift may result in wider bandgaps associated with softening or hardening nonlinearity [7]. The derivation of nonlinear dispersion equations can be carried out using different perturbation techniques [8] such as the Lindstedt–Poincare or multiple scales techniques [9].

In this work, we study the effect of stiffness nonlinearity on one-dimensional wave propagation in a periodic structure (i.e., spring-mass chain) with multiple local resonators. The parameters of the local resonators are different in order to realize multiple bandgap formations. We investigate two different cases of nonlinearity: cubic spring nonlinearity connecting the chains and cubic spring nonlinearity in the local resonators. We also study both hardening and softening nonlinearities and investigate their roles on bandgap formations. Closed-form expressions are presented for the nonlinear dispersion equation for any number of local resonators. This is achieved using the method of multiple scales (MMS) to solve the weakly nonlinear system of governing equations of motion. The obtained expressions can serve as a benchmark for predicting the bandgap formations of weakly nonlinear acoustic metamaterials. This work also provides general guidelines for exploiting nonlinearity in order to achieve better vibration mitigation.

2 Derivation of the Dispersion Equation

Figure 1 shows a schematic diagram of the proposed nonlinear acoustics metamaterial that is represented by a chain of mass-spring systems with embedded local resonators. Each unit cell consists of a rigid mass, m, connected to other cells through linear and nonlinear spring coefficients, k, and $\epsilon \Gamma$, respectively. Inside each cell, there are multiple local resonators with a mass, m_i, and linear or nonlinear spring with linear coefficient, k_i, and nonlinear coefficient, $\epsilon \Gamma_i$. The nondimensional free oscillation equations for each cell with s number of local resonators can be expressed as

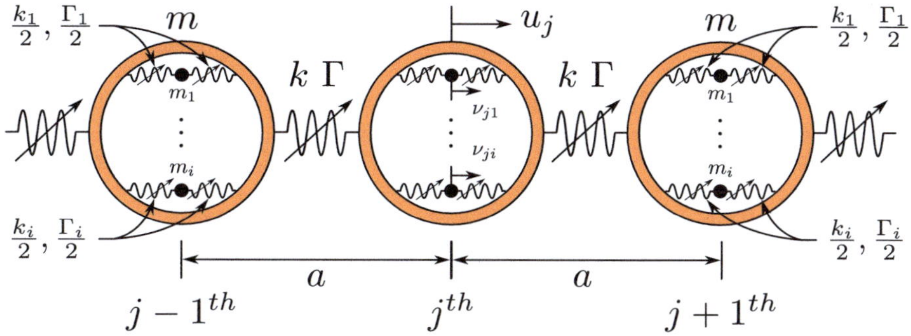

Fig. 1 The nonlinear acoustics metamaterial

$$\ddot{u_n} + 2u_n - u_{n-1} - u_{n+1} + \epsilon \bar{\Gamma}((u_n - u_{n-1})^3 + (u_n - u_{n+1})^3)$$

$$+ \sum_{i=1}^{s} \bar{k_i}(u_n - v_{ni}) + \sum_{i=1}^{s} \epsilon \bar{\Gamma_i}(u_n - v_{ni}) = 0 \tag{1}$$

$$\frac{\omega_n^2}{\omega_{di}^2}\ddot{v}_{ni} + (v_{ni} - u_n) + \epsilon \bar{\Gamma_i}(v_{ni} - u_n)^3 = 0, \tag{2}$$

where the dimensionless parameters are

$$\tau = \omega_n t;\ \bar{\Gamma} = \frac{\Gamma}{\omega_n^2 m};\ \bar{k_i} = \frac{k_i}{\omega_n^2 m} \tag{3}$$

and ω_n and ω_{di} are defined as $\omega_n = \sqrt{k/m}$ and $\omega_{di} = \sqrt{k_i/m_i}$.

Using MMS, we can assume expansions for the displacements in the form of

$$u_n(t, \epsilon) = u_{n0}(T_0, T_1) + \epsilon\ u_{n1}(T_0, T_1) \tag{4}$$

$$v_{ni}(t, \epsilon) = v_{ni0}(T_0, T_1) + \epsilon\ v_{ni1}(T_0, T_1), \tag{5}$$

where $T_0 = \tau$ is the fast time scale and $T_1 = \epsilon\tau$ is the slow time scale. Since the time is expressed in two independent variables, the time derivative can be presented by using the chain rule as

$$(\ddot{}) = D_0^2 + 2\epsilon D_0 D_1 + \cdots, \tag{6}$$

where $D_n = \frac{\partial}{\partial T_n}$.

2.1 Nonlinear Chain

In this case, we are only interested to study the effect of spring nonlinearity connecting the cells (i.e., the case of nonlinear resonator is discussed in the next section since its dispersion relation is different from that of the nonlinear chain case). Consequently, we set $\bar{\Gamma}_i = 0$ (i.e., nonlinear springs in the local resonators are zero). Substituting Eqs. (4)–(6) into Eqs. (1)–(2) and collecting the similar coefficients of ϵ, one can get

Order ϵ^0

$$D_0^2 u_{n0} + 2u_{n0} - u_{(n-1)0} - u_{(n+1)0} + \sum_{i=1}^{s} \bar{k}_i (u_{n0} - v_{ni0}) = 0 \tag{7}$$

$$\frac{\omega_n^2}{\omega_{di}^2} D_0^2 v_{ni0} - (u_{n0} - v_{ni0}) = 0 \tag{8}$$

Order ϵ

$$D_0^2 u_{n1} + 2u_{n1} - u_{(n-1)1} - u_{(n+1)1} + \sum_{i=1}^{s} \bar{k}_i (u_{n1} - v_{ni1})$$

$$= -2D_0 D_1 u_{n0} - \bar{\Gamma}((u_{n0} - u_{(n-1)0})^3 + (u_{n0} - u_{(n+1)0})^3) \tag{9}$$

$$\frac{\omega_n^2}{\omega_{di}^2} D_0^2 v_{ni1} - (u_{n1} - v_{ni1}) = -2\frac{\omega_n^2}{\omega_{di}^2} D_0 D_1 v_{ni0}. \tag{10}$$

At order ϵ^0 the problem is linear; therefore, the solution can be expressed as

$$u_n = Ae^{i(nk - \omega T_0)} + c \cdot c \tag{11}$$

$$v_{ni} = B_i e^{i(nk - \omega T_0)} + c \cdot c, \tag{12}$$

where $c \cdot c$ denotes complex conjugate, $k = qa$ is the dimensionless wave number, and q represents the wave number. A and B_i stand for the wave amplitude of the outer and inner masses, respectively. By substituting Eqs. (11)–(12) into Eqs. (7)–(8), the linear dispersion equation can be expressed as

$$-\omega^2 + (2 - 2\cos k) + \sum_{i=1}^{s} \bar{k}_i (1 - K_{\omega i}) = 0, \tag{13}$$

where $K_{\omega i} = \frac{1}{1 - \omega_n^2 \omega^2 / \omega_{di}^2}$. This linear dispersion equation represents both cases; however, the nonlinear dispersion relations are different. For nonlinear problem, we need to study the equations at order ϵ.

By rearranging the equations at order ϵ, we obtain

$$X(D_0^2 u_{n1} + 2u_{n1} - u_{(n-1)1} - u_{(n+1)1}) + \sum_{i=1}^{s} \frac{X\bar{k}_i \omega_n^2/\omega_{di}^2}{1 - \omega_n^2\omega^2/\omega_{di}^2} D_0^2 u_{n1}$$

$$= -2\sum_{i=1}^{s} \frac{X\bar{k}_i \omega_n^2/\omega_{di}^2}{1 - \omega_n^2\omega^2/\omega_{di}^2} D_0 D_1 v_{ni0} + X(-2D_0 D_1 u_{n0}$$

$$- \bar{\Gamma}((u_{n0} - u_{(n-1)0})^3 + (u_{n0} - u_{(n+1)0})^3), \tag{14}$$

where $X = \prod_{i=1}^{s}(\omega_n^2/\omega_{di}^2 D_0^2 + 1)$. Introducing Eqs. (11)–(12) into Eq. (14) yields

$$X(D_0^2 u_{n1} + 2u_{n1} - u_{(n-1)1} - u_{(n+1)1}) + \sum_{i=1}^{s} \frac{X\bar{k}_i \omega_n^2/\omega_{di}^2}{1 - \omega_n^2\omega^2/\omega_{di}^2} D_0^2 u_{n1}$$

$$= \left[2i\omega \sum_{i=1}^{s} \frac{X\bar{k}_i \omega_n^2/\omega_{di}^2}{1 - \omega_n^2\omega^2/\omega_{di}^2} A' K_{\omega i} + X(2i\omega A'\right.$$

$$\left. -12\bar{\Gamma} A^2 \bar{A}(1 - \cos k)^2\right] e^{i(nk-\omega T_0)} + NST, \tag{15}$$

where NST denotes non-secular terms, $A' = \frac{dA}{dT_1}$, and $\bar{A}$ is the complex conjugate of A. We note here that X becomes $X = \prod_{i=1}^{s}(1 - \omega^2\omega_n^2/\omega_{di}^2)$ after applying the operator D_0. The left-hand side of Eq. (15) has a nontrivial solution; therefore, the secular terms on the right-hand side must be eliminated by solving the solvability conditions [8] defined as the coefficients of $e^{i(nk-\omega T_0)}$.

Substituting the polar form $A = \frac{1}{2}\alpha e^{i\beta}$ into the solvability condition and separating the real and imaginary part, the modulation equations for the amplitude and phase can be expressed as

$$\omega \sum_{i=1}^{s} \frac{\bar{k}_i X \omega_n^2/\omega_{di}^2}{1 - \omega_n^2\omega^2/\omega_{di}^2} \alpha' K_{\omega i} + X\omega\alpha' = 0 \tag{16}$$

$$- \omega \sum_{i=1}^{s} \frac{\bar{k}_i X \omega_n^2/\omega_{di}^2}{1 - \omega_n^2\omega^2/\omega_{di}^2} \alpha\beta' K_{\omega i} - X\omega\alpha\beta' - \frac{3}{2} X\bar{\Gamma}\alpha^3(1 - \cos k)^2 = 0. \tag{17}$$

From the amplitude equation above, one can find that α is constant (i.e., $\alpha = \alpha_0$). From the phase equation, we can obtain

$$\beta = -\frac{3\bar{\Gamma}\alpha^2(1 - \cos k)^2}{2\omega\left(1 + \sum_{i=0}^{s} \frac{\bar{k}_i\omega_n^2/\omega_{di}^2}{1-\omega_n^2\omega^2/\omega_{di}^2} K_{\omega i}\right)} T_1. \tag{18}$$

Since $T_1 = \epsilon\tau$, the nonlinear frequency, ω_{nl}, associated with k is

$$\omega_{nl} = \omega + \epsilon\beta'. \tag{19}$$

2.2 *Nonlinear Local Resonators*

For this case, we are only interested in the nonlinearity from the local resonators hence we ignore the nonlinear spring coefficient of the cells (i.e., $\bar{\Gamma} = 0$). Following the techniques described in the previous section, one can express the phase as

$$\beta = -\frac{\sum_{i=1}^{s}\left[\frac{3}{8}\alpha^2(1-K_{\omega_i})^3\bar{\Gamma}_i\left(\frac{\bar{k}_i}{1-\omega_n^2\omega^2/\omega_{di}^2}-1\right)\right]}{\omega\left(1+\sum_{i=1}^{s}\frac{\bar{k}_i\omega_n^2/\omega_{di}}{1-\omega_n^2\omega^2/\omega_{di}^2}K_{\omega i}\right)}T_1. \tag{20}$$

Unlike the case of nonlinear chain, one should note here that the correction β is not explicitly a function of wave number. Moreover, the expression is different from that obtained in [10]. This is because the contribution of the resonators on the left-hand side from the equations at order ϵ was taken into account [6, 7].

3 Results and Discussion

For the numerical simulations, we select $\omega_0 = \omega_{d1} = 10^3$ for the case of single resonator. For the case of two local resonators we set $\omega_{d2} = 1.5\ \omega_0$ to obtain multiple bandgaps. The values of the nondimensional stiffness of the resonator are chosen to be $\bar{k}_i = \omega_{di}^2/\omega_0^2$. Also, the numerical simulation is based on $\epsilon\alpha^2\bar{\Gamma} = \epsilon\alpha^2\bar{\Gamma}_i = 0.06$. The band structure can be obtained by numerically integrating a chain (i.e., 100 cells were used in the simulation) excited at the middle (i.e., at $n = 50$) by a harmonic force. Then, we determine the wave number by picking the maximum value of the 2-D spectrum; such that the wave number is associated with the spatial frequency at the excitation frequency. Since the expression is derived for plane waves, the end condition is chosen to be a perfectly matched layer (PML); such that no reflected waves exist in the simulation. Following [9], this can be achieved by defining a linear viscous damping in the chain such as

$$c(n) = C_{\max}\left(\frac{n}{N}\right)^3, \tag{21}$$

where N is the number of simulated cells.

Figure 2a compares the results of the multiple scales method to those obtained using direct numerical integration, as well as, to those obtained in the literature using

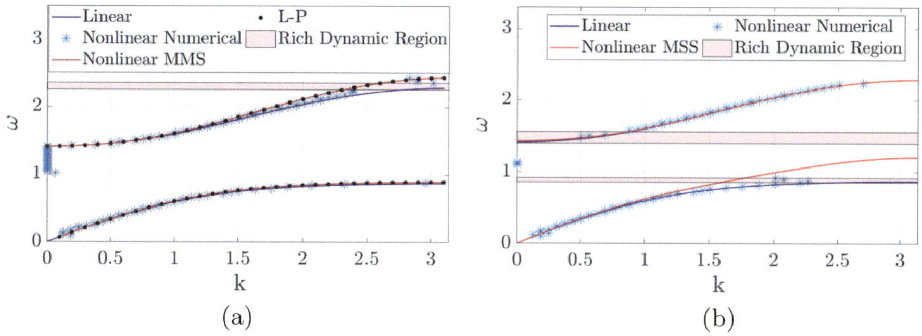

Fig. 2 Bandgap structure for single local resonator. (**a**) A chain connected by nonlinear spring. (**b**) Nonlinear resonators

Lindstedt–Poincare (L–P). The results show good agreement between the dispersion curves obtained by MMS and the results obtained by Lindstedt–Poincare (L–P) [6]. The comparison of the MMS results to the numerical results also shows a very good agreement in terms of detecting the boundaries of the bandgaps. But MMS fails to capture the rich dynamic region (i.e., inside the red patch), which is referred to pseudo-gap region [6]. It should be noted that pseudo-gap region here appears for different ranges as that observed in [6] because the input here is not wave packet. The wave indeed propagates through the structure, but the output appears at different multiple frequencies due to nonlinear interaction of the cells.

Figure 2b shows the analytical approximation (i.e., MMS) and numerical results of the band structure for the case of a single nonlinear local resonator. We can observe that the two methods slightly differ; particularly, when the frequency of the local resonator $\omega = \omega_{d1}$. Also, the comparison in terms of wider bandgap formation between nonlinear and linear local resonators is inconclusive in Fig. 2b. Higher order approximations or different analytical approximation methods such as the complexification averaging or the homotopy method may be required to obtain better accuracy and help provide better insight into the performance of the nonlinear local resonators.

Figure 3a shows the band structure of a chain connected by nonlinear springs with multiple local resonators. The results indicate that the MMS is a good predictor of the bandgap boundaries for multiple local linear resonators embedded in cells with nonlinear spring connections. The accuracy of the MMS approximation can even be further improved if higher order perturbation is considered. The results in this figure also show that the pseudo-gap almost vanishes; however, a narrow rich dynamic region still exists. For the case of multiple nonlinear local resonators, the same conclusion from Fig. 2b can be drawn. In that, Fig. 3b also reveals that the MMS is not very accurate in predicting the bandgap boundaries at frequencies confined between the resonators frequencies. The first order approximation of the MMS fails to predict the dynamic in the case of multiple nonlinear local resonators

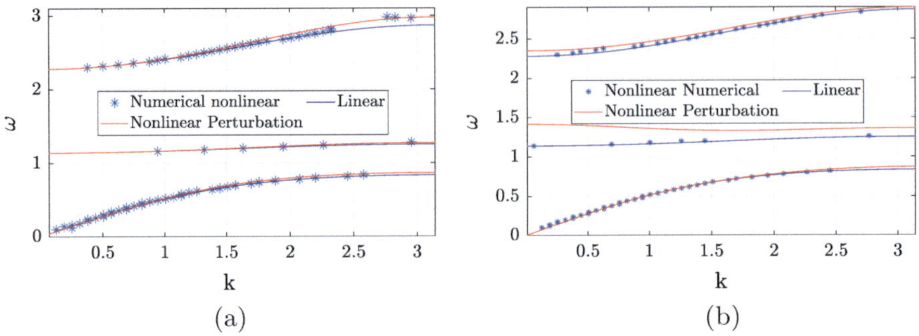

Fig. 3 Bandgap structure for two local resonators. (**a**) A chain connected by nonlinear spring. (**b**) Nonlinear local resonators

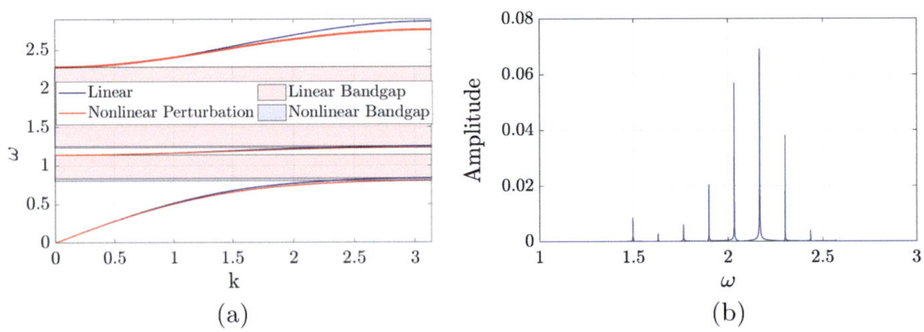

Fig. 4 Utilizing nonlinear chain toward vibration attenuation. (**a**) Softening nonlinearity. (**b**) Frequency shift

embedded in cells with linear spring connections. Hence, higher order perturbations or other analytical methods may be required to provide better approximations.

Finally, Fig. 4a shows that softening nonlinearity in the spring connecting the chain with multiple local linear resonators can be used to widening the bandgap. Therefore, the type of nonlinearity can be utilized in tuning the bandgap boundaries. Although softening nonlinearity is more desirable for vibration attenuation, solitary waves can only be realized with hardening nonlinearity [11]. In Fig. 4b, we can observe how the wave can appear at different secondary resonances due to softening or hardening nonlinearity. This frequency shift can be exploited in designing acoustics diode or acoustics rectifier [6, 12].

In the present study, we only handled the cubic type of nonlinearity. For other types of nonlinearities, one can approximate the nonlinearity by using Taylor expansion and rewrite the equation in terms of cubic polynomial. Similar procedures can then be used to derive the corresponding dispersion relations.

4 Conclusion

In this paper, we derived closed-form expressions of the dispersion equations describing the wave propagation in one-dimensional nonlinear acoustics metamaterial using the method of multiple scales. The obtained equations can be used in studying the dispersion curves for multiple local resonators unlike the one in the literature, which is limited to a single local resonator. The numerical simulations showed that these closed-form expressions accurately predict the band structure in the case of a chain with nonlinear spring connections and linear local resonators; whereas they fail to accurately predict the case of nonlinear local resonators embedded in cell with linear spring connection, particularly near the local resonator frequency. Numerical examples also demonstrate that wider forbidden regions can be achieved with multiple local resonators. This is an indication that superior vibration mitigation can be realized by the proposed softening or hardening nonlinearity.

Acknowledgement The authors would like to thank Mr. David Petrushenko for his help in generating figures and the start-up funding provided by Virginia Tech.

References

1. Hussein, M.I., Leamy, M.J., Ruzzene, M.: Dynamics of phononic materials and structures: historical origins, recent progress, and future outlook. Appl. Mech. Rev. **66**(4), 040802 (2014)
2. Liu, Z., Zhang, X., Mao, Y., Zhu, Y.Y., Yang, Z., Chan, C.T., Sheng, P.: Locally resonant sonic materials. Science **289**(5485), 1734–1736 (2000)
3. Bertoldi, K., Vitelli, V., Christensen, J., van Hecke, M.: Flexible mechanical metamaterials. Nat. Rev. Mater. **2**(11), 17066 (2017)
4. Huang, G.L., Sun, C.T.: Band gaps in a multiresonator acoustic metamaterial. J. Vib. Acoust. **132**(3), 031003 (2010)
5. Zhu, R., Liu, X.N., Hu, G.K., Sun, C.T., Huang, G.L.: A chiral elastic metamaterial beam for broadband vibration suppression. J. Sound Vib. **333**(10), 2759–2773 (2014)
6. Zhou, W.J., Li, X.P., Wang, Y.S., Chen, W.Q., Huang, G.L.: Spectro-spatial analysis of wave packet propagation in nonlinear acoustic metamaterial. J. Sound Vib. **413**, 250–269 (2018)
7. Narisetti, R.K., Leamy, M.J., Ruzzene, M.: A perturbation approach for predicting wave propagation in one-dimensional nonlinear periodic structures. J. Vib. Acoust. **132**, 031001 (2010)
8. Nayfeh, A.H.: Introduction to Perturbation Techniques. Wiley, New York (2011)
9. Manktelow, K., Leamy, M.J., Ruzzene, M.: Multiple scales analysis of wave-wave interactions in a cubically nonlinear monoatomic chain. Nonlinear Dyn. **63**, 193–203 (2011)
10. Kulkarni, P.P., Manimala, J.M.: Nonlinear and inertant acoustic metamaterials and their device implications. In: Dynamic Behavior of Materials, vol. 1, pp. 217–234. Springer, Cham (2018)
11. Ganesh, R., Gonella, S.: Spectro-spatial wave features as detectors and classifiers of nonlinearity in periodic chains. Wave Motion **50**(4), 821–835 (2013)
12. Ma, C., Parker, R.G., Yellen, B.B.: Optimization of an acoustic rectifier for uni-directional wave propagation in periodic mass-spring lattices. J. Sound Vib. **332**(20), 4876–4894 (2013)

Nonlinear Vibration Analysis of Metamaterial Honeycomb Sandwich Structures with Negative Poisson's Ratio

Shaotao Zhu, Jing Li, Ji Zhou, and Tingting Quan

Abstract The research on existence, bifurcation, and number of periodic solutions is closely related to Hilbert's 16th problem. The main goal of this chapter is to investigate the nonlinear dynamic response and periodic vibration characteristic of a simply supported concave hexagonal honeycomb sandwich plate with negative Poisson's ratio. The plate is subjected to its in-plane and transverse excitation. The curvilinear coordinate frame, Poincaré map, and improved Melnikov function are proposed to detect the existence and number of the periodic solutions. The theoretical analyses indicate the existence of periodic orbits and can guarantee at most four periodic orbits under certain conditions. Numerical simulations are performed to verify the theoretical results. The relative positons as well as the vibration characteristics can also be clearly found from the phase portraits. The periodic motion for the equation is closely related to the amplitude modulated periodic vibrations of the plate. The results will provide theoretical guidance to nonlinear vibration control for the metamaterial honeycomb sandwich structures.

Keywords Metamaterials · Honeycomb sandwich structures · Nonlinear vibration · Bifurcation of multiple periodic solutions · Negative Poisson's ratio

S. Zhu · J. Li (✉)
College of Applied Sciences, Beijing University of Technology, Beijing, P. R. China
e-mail: zhushaotao@emails.bjut.edu.cn; leejing@bjut.edu.cn

J. Zhou
State Key Laboratory of New Ceramics and Fine Processing, School of Materials Science and Engineering, Tsinghua University, Beijing, P. R. China

T. Quan
College of Applied Sciences, Beijing University of Technology, Beijing, P. R. China

School of Science, Tianjin Chengjian University, Tianjin, P. R. China

W. Lacarbonara et al. (eds.), *New Trends in Nonlinear Dynamics*,
https://doi.org/10.1007/978-3-030-34724-6_3

1 Introduction

The theory of nonlinear dynamics has developed rapidly in the past half century. The theory of periodic solutions for nonlinear dynamic systems deals with periodic vibrations of practical problems that appear in almost all disciplines of science and engineering. The research on existence, bifurcation, and number of periodic solutions is closely related to Hilbert's 16th problem which is not only significant for its theoretical development, but also plays an important role in solving practical problems.

In practical applications, the metamaterials with negative Poisson's ratio, known as "auxetic" [1], have attracted great interest nowadays [2, 3]. Chen and Feng [4] investigated the dynamic behaviors of a thin laminated plate embedded with auxetic layers subject to in-plane excitation. Duc and Cong [5] studied the nonlinear dynamic response of auxetic sandwich plates subjected to blast and mechanical loads. Li et al. [6] investigated the nonlinear bending behavior of sandwich beams with functionally graded negative Poisson's ratio honeycomb core in thermal environments.

In this chapter, we focus on the nonlinear dynamics of a simply supported concave hexagonal aluminum matrix honeycomb sandwich plate subjected to its in-plane and transverse excitation. The Melnikov function [7] is improved to detect the existence and number of the periodic solutions. The upper bound of the number of periodic solutions as well as their relative positions and vibration patterns can be clearly found from the numerical results.

2 Existence of Periodic Motions for Metamaterial Honeycomb Sandwich Plate

2.1 Dynamic Model, Averaged Equation, and Melnikov Function

A simply supported four-edge metamaterial honeycomb sandwich plate subjected to the transverse and in-plane excitations is shown in Fig. 1. A Cartesian coordinate is located in the middle surface of the plate as shown in Fig. 1a. The core of the plate is chosen as concave hexagonal honeycomb with negative Poisson's ratio and the unit cell of the core is shown in Fig. 1b. The length of the inclined and horizontal cell rib are l_1 and l_2, respectively, and the inclined angle is θ, assuming that the unit cell has uniform thickness t and t/l_2 is small.

We mainly focus on the transverse motion of the plate. A two degrees of freedom nonlinear differential equation for the dimensionless is obtained as follows [8].

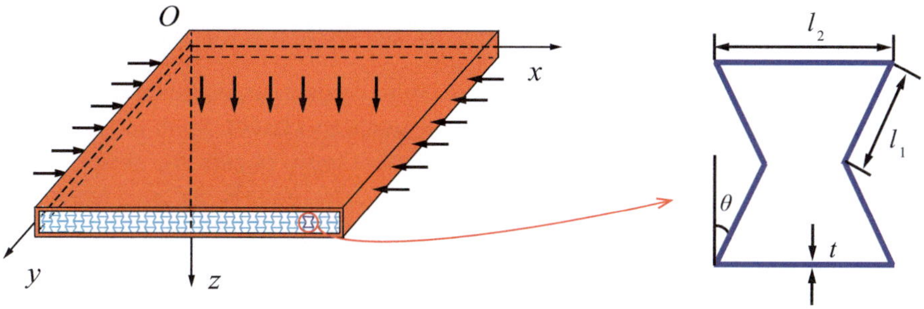

Fig. 1 The model of the metamaterial plate and the unit cell of the core: (**a**) mechanical model of the plate; (**b**) unit cell of the core with inclined rib length l_1, horizontal rib length l_2, inclined angle θ, and thickness t

$$\begin{aligned}&\ddot{w}_1+\omega_1^2 w_1+\alpha_1\mu\dot{w}_1+\alpha_2 w_1\cos(\Omega_2 t)+\alpha_3 w_1^3+\alpha_4 w_2^3\\&+\alpha_5 w_1 w_2^2+\alpha_6 w_1^2 w_2=f_1\cos(\Omega_1 t)\\&\ddot{w}_2+\omega_2^2 w_2+\beta_1\mu\dot{w}_2+\beta_2 w_2\cos(\Omega_2 t)+\beta_3 w_2^3+\beta_4 w_1^3\\&+\beta_5 w_1^2 w_2+\beta_6 w_1 w_2^2=f_2\cos(\Omega_1 t)\end{aligned}\tag{1}$$

where w_1 and w_2 represent the amplitudes of the first two modes, respectively, ω_1 and ω_2 are two different linear natural frequencies, μ is the damping coefficient, f_1, f_2 are forcing excitations, Ω_1 and Ω_2 are the frequencies of transverse and in-plane excitation, α_i, β_i $(i=1, \cdots, 6)$ are nondimensional coefficients denoted as $\boldsymbol{\alpha}=(\alpha_1, \cdots, \alpha_6)$, $\boldsymbol{\beta}=(\beta_1, \cdots, \beta_6)$ for convenience. We focused on the case of 1:1 internal resonance and primary parametric resonance. In this resonant case, $\omega_1^2=\Omega_1^2+\varepsilon\sigma_1$, $\omega_2^2=\Omega_2^2+\varepsilon\sigma_2$, where σ_1 and σ_2 are two detuning parameters and we assume that $\Omega_1=\Omega_2=1$. Introducing the scales transformations, $\alpha_i\to\varepsilon\alpha_i$, $\beta_i\to\varepsilon\beta_i$, $f_j\to\varepsilon f_j$, $i=1, 2, \cdots, 6$; $j=1, 2$ where ε is a small parameter. Therefore, by the methods of multiple scales, we will obtain the four dimensional averaged equation.

$$\dot{\boldsymbol{x}}=\boldsymbol{A}\boldsymbol{x}+\boldsymbol{F}(\boldsymbol{x},\boldsymbol{y}),\quad \dot{\boldsymbol{y}}=\boldsymbol{B}\boldsymbol{y}+\boldsymbol{G}(\boldsymbol{x},\boldsymbol{y})\tag{2}$$

where $\boldsymbol{x}=(x_1,x_2)^{\mathrm{T}}\in\mathbf{R}^2$, $\boldsymbol{y}=(y_1,y_2)^{\mathrm{T}}\in\mathbf{R}^2$, $\boldsymbol{F}=(F_1,F_2)^{\mathrm{T}}$ and $\boldsymbol{G}=(G_1,G_2)^{\mathrm{T}}$ are vector-valued polynomials in variables of (x_1,x_2,y_1,y_2).

$$\boldsymbol{A}=\begin{pmatrix}0 & m\\ -m & 0\end{pmatrix},\quad \boldsymbol{B}=\begin{pmatrix}0 & n\\ -n & 0\end{pmatrix},\quad m=\frac{1}{8}\left(\sigma_1^2+\mu^2\alpha_1^2\right),\quad n=\frac{1}{8}\left(\sigma_2^2+\mu^2\beta_1^2\right)$$

Introducing the rescaling transformations $\boldsymbol{F}\to\varepsilon\boldsymbol{F}$, $\boldsymbol{G}\to\varepsilon\boldsymbol{G}$, where $0<\varepsilon<<1$. Then system (2) can be rewritten as

$$\dot{\boldsymbol{x}}=\boldsymbol{J}DH_1(\boldsymbol{x})+\varepsilon\boldsymbol{F}(\boldsymbol{x},\boldsymbol{y}),\quad \dot{\boldsymbol{y}}=\boldsymbol{J}DH_2(\boldsymbol{y})+\varepsilon\boldsymbol{G}(\boldsymbol{x},\boldsymbol{y})\tag{3}$$

where $\boldsymbol{J} = \begin{pmatrix} 0 & 1 \\ -1 & 0 \end{pmatrix}$, $H_i : \mathbf{R}^2 \to \mathbf{R}$ $(i = 1, 2)$, and $DH_1(\boldsymbol{x}) = (\partial H_1/\partial x_1, \partial H_1/\partial x_2)^{\mathrm{T}}$, $DH_2(\boldsymbol{y}) = (\partial H_2/\partial y_1, \partial H_2/\partial y_2)^{\mathrm{T}}$. When $\varepsilon = 0$, system (3) degenerates to two uncoupled Hamilton systems with Hamilton function $H_1(\boldsymbol{x}) = m\left(x_1^2 + x_2^2\right)/2$, $H_2(\boldsymbol{y}) = n\left(y_1^2 + y_2^2\right)/2$. Then each system has a family of periodic orbits, $\Gamma_{h_1} = \left\{\boldsymbol{x}_{h_1} | H_1(\boldsymbol{x}) = h_1\right\}$ and $\Gamma_{h_2} = \left\{\boldsymbol{y}_{h_2} | H_2(\boldsymbol{y}) = h_2\right\}$, $(h_1, h_2 \in \mathbf{K}$, an open interval in $\mathbf{R})$. Suppose that the family of periodic orbits Γ_{h_1} and Γ_{h_2} can be expressed as

$$\begin{aligned} &x_1 = \sqrt{2h_1/m}\cos(mt), \quad x_2 = \sqrt{2h_1/m}\sin(mt) \\ &y_1 = \sqrt{2h_2/n}\cos\left(n\left(t + t_0\right)\right), \quad y_2 = \sqrt{2h_2/n}\sin\left(n\left(t + t_0\right)\right) \end{aligned} \tag{4}$$

The corresponding periods of the orbits are $T_1(h_1) = 2\pi/m$, $T_2(h_2) = 2\pi/n$. Introducing the curvilinear coordinates in the neighborhood of the invariant torus $\Gamma_{h_1} \times \Gamma_{h_2}$, define a global cross section Σ in the phase space, and construct the kth iteration of Poincaré map $P^k : \Sigma \to \Sigma$. Then the Melnikov function $\boldsymbol{M} = (M_1, M_2, M_3)^{\mathrm{T}}$ is

$$\begin{aligned} &M_1 = m\int_0^{2\pi}\left(x_1F_1 + x_2F_2\right)\mathrm{d}t, \quad M_2 = n\int_0^{2\pi}\left(y_1G_1 + y_2G_2\right)\mathrm{d}t \\ &M_3 = \int_0^{2\pi}\frac{(y_2G_1 - y_1G_2)h_1 - (x_2F_1 - x_1F_2)h_2}{2h_1h_2}\mathrm{d}t \end{aligned} \tag{5}$$

2.2 Existence of Periodic Motions

Given that $n = 2m = 2$, thus $T_1(h_1) = 2T_2(h_2) = 2\pi$. Let $\boldsymbol{M}$ be zero, yields

$$\begin{aligned} &2a_{13}h_1^2 + 2a_9h_1h_2 - a_5h_1\sqrt{h_2}\sin\left(2t_0\right) = 0 \\ &4b_{11}h_1h_2 + b_7h_2^2 - 2b_4h_1\sqrt{h_2}\sin\left(2t_0\right) = 0 \\ &kh_1 + lh_2 + 2\left(b_4h_1 + a_5h_2\right)h_2^{-1/2}\cos\left(2t_0\right) + C = 0 \end{aligned} \tag{6}$$

where $k = b_{15} + 2b_{10} - 4a_{12}$, $l = b_6 - a_{14} - 2a_8$, $C = 2(b_2 - 2a_2)$, a_i, b_i are respectively the coefficients of the terms of F_j, G_j $(i = 2, \cdots, 17;\ \ j = 1, 2)$. Suppose that $b_4 = 0$ and $a_{13}b_7 - 4a_9b_{11} \neq 0$, thus Eq. (6) is equivalent to

$$\begin{aligned} &\sqrt{h_1} = 2a\sqrt{b}\sin\left(2t_0\right), \quad \sqrt{h_2} = 2a\sin\left(2t_0\right) \\ &4\left(A^2 + B^2\right)\sin^4\left(2t_0\right) - 4\left(B^2 - AC\right)\sin^2\left(2t_0\right) + C^2 = 0 \end{aligned} \tag{7}$$

where $a = \frac{a_5b_{11}}{4a_9b_{11} - a_{13}b_7}$, $b = -\frac{b_7}{4b_{11}}$, $A = 2a^2(l + kb)$, $B = 2a_5a$. Denote

$$f(x) = 4\left(A^2 + B^2\right)x^2 - 4\left(B^2 - AC\right)x + C^2 = 0, \quad x = \sin^2\left(2t_0\right) \tag{8}$$

the number of positive solutions of Eq. (8) may be 0, 1, or 2. Correspondingly, the number of solutions for (t_0, h_1, h_2) of Eq. (7) may be 0, 2, or 4. So, if we choose some appropriate parameters, the number of periodic solutions of system (3) will reach the upper bound. To obtain the maximum number of solutions, $A^2 + B^2$ and $B^2 - AC$ should not be zero, and $\Delta = 16B^2(B^2 - C^2 - 2AC) > 0$. Furthermore, according to the relations between roots and coefficients, the two solutions of the quadratic Eq. (8) are both positive if and only if $B^2 - AC > 0$ and $C \neq 0$.

From the above analysis, the condition for Eq. (8) to have two positive solutions is obtained as follows.

$$\begin{gathered} \alpha_1\beta_1\beta_3\beta_5 < 0, \quad \beta_2^2 - 2\alpha_2^2 \neq 0, \quad 3\beta_4 f_1 + \beta_5 f_2 = 0 \\ \frac{\alpha_6 f_1 + \alpha_5 f_2}{4\alpha_5\beta_5 - 9\alpha_3\beta_3} > 0, \quad f_1^2 > \frac{64\alpha_5^2\beta_5^2\left(AC + \sqrt{(A^2+B^2)C^2}\right)}{B^2(\alpha_6\beta_5 - 3\alpha_5\beta_4)^2} \end{gathered} \tag{9}$$

It is just the condition that system (3) has four periodic orbits (under this condition, the number of periodic solutions reach the upper bound).

3 Numerical Simulations

In this subsection, we will choose a group of parameters to perform the numerical simulations to verify the theoretical result and present the distributions of the periodic orbits. The group of parameters is chosen as $\varepsilon = 0.001$, $\mu = 0.2$, $\sigma_1 = 2$, $\sigma_2 = 0$, $\boldsymbol{\alpha} = (10, 2, -2, -1, 3, 1)$, $\boldsymbol{\beta} = (20, 1, -5, 0.5, 3, 1)$.

From (9), we will calculate that $f_2 = -f_1/2$ and $f_1 > 8\sqrt{7}$. For a general case, we choose $f_1 = 24 > 8\sqrt{7}$. The phase portraits of the periodic motions as well as the orbits in their neighborhoods are shown in Fig. 2, where (a) and (b) represent the phase portraits projected on $\boldsymbol{x}$ and $\boldsymbol{y}$ plane, respectively. There are four periodic orbits which are distinguished clearly in Fig. 3. Figure 3a represents the phase portraits of the periodic orbits projected on plane (x_2, y_2), while (b)–(d) represent the phase portraits projected into various spaces. The moving patterns as well as the relative positions of the periodic orbits can also be recognized. Figure 4 shows the portraits of the wave propagation in different directions. Apparently, the moving period of x_1 and x_2 is twice that of y_1 and y_2.

4 Conclusions

In this chapter, nonlinear dynamic behaviors of a simply supported concave honeycomb sandwich plate subjected to its in-plane and transverse excitation are investigated. The novelty of this chapter is that the honeycomb core is chosen as metamaterial with negative Poisson's ratio which is widely applied in aircraft and aerospace engineering. The multiple periodic motions can be clearly found from both the theoretical and numerical results under certain conditions. The upper bound

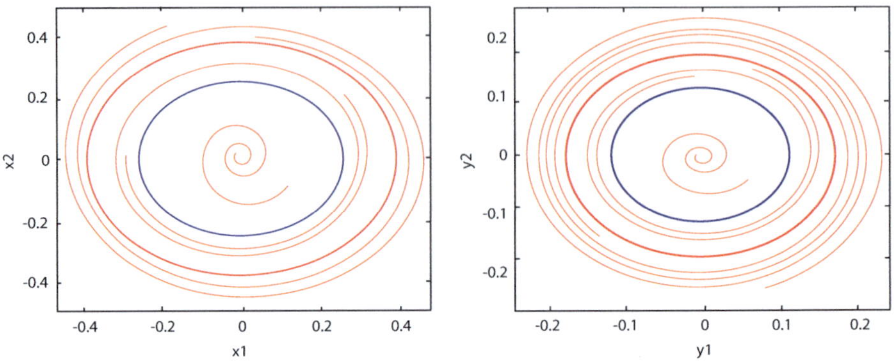

(a) (b)

Fig. 2 Phase portraits of distributions of orbits projected on different planes: (**a**) projections on (x_1, x_2) plane; (**b**) projections on (y_1, y_2) plane

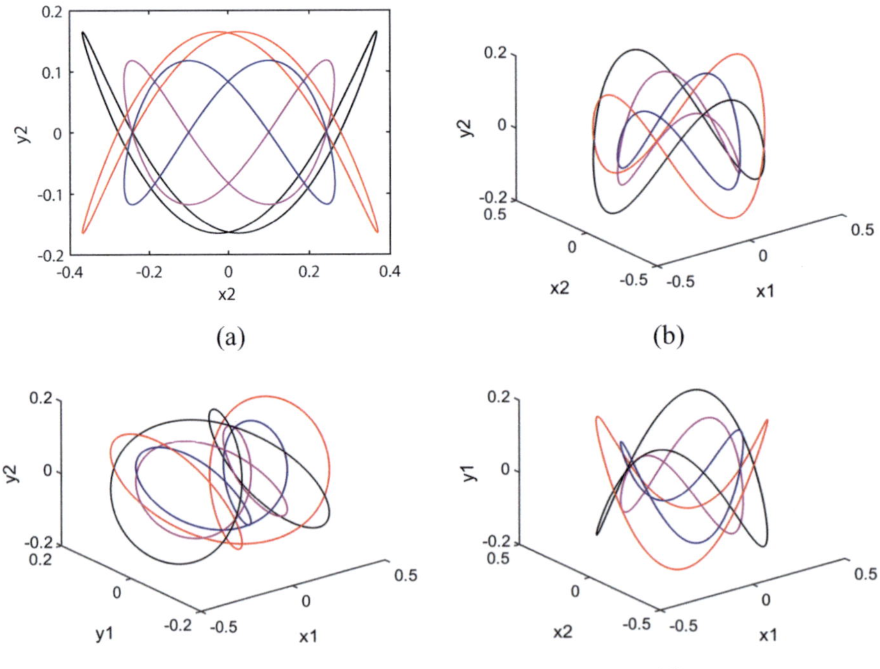

(a) (b)

(c) (d)

Fig. 3 Phase portraits of the four periodic orbits: (**a**) projections on (x_2, y_2) plane; (**b**) projections in (x_1, x_2y_2) space; (**c**) projections in (x_1, y_1y_2) space; (**d**) projections in (x_1, x_2y_1) space

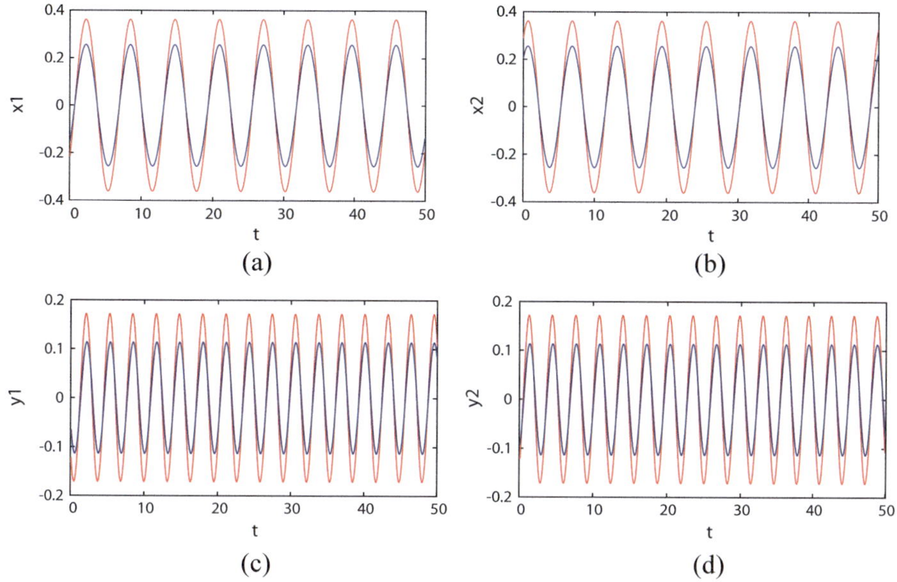

Fig. 4 Portraits of the wave propagation: (**a**) in x_1 direction; (**b**) in x_2 direction; (**c**) in y_1 direction; (**d**) in y_2 direction

of the number of periodic orbits is 4. The periodic motion of averaged equation may lead to amplitude modulated periodic oscillations for the model under certain conditions. Results obtained will provide theoretical guidance to nonlinear vibration control for metamaterials honeycomb sandwich structures. In further study, we should pay attention to the stabilities of periodic solutions for the system.

Acknowledgements The research project is supported by National Natural Science Foundation of China (11772007, 11372014, 11802200, 11571026) and also supported by Beijing Natural Science Foundation (1172002, Z180005), the International Science and Technology Cooperation Program of China (2014DFR61080).

Appendix

Assuming that $H_{1,j} = F_j$, $H_{2,j} = G_j$, $c_{1,k} = a_k$, $c_{2,k} = b_k$ ($j = 1, 2;\ k = 1, 2, \cdots, 17$). Then $\boldsymbol{F} = (F_1, F_2)^{\mathrm{T}}$ and $\boldsymbol{G} = (G_1, G_2)^{\mathrm{T}}$ of Eq. (2) can be expressed as

$$
\begin{aligned}
H_{i,1} = c_{i,16} - c_{i,2}\left((2-i)\,x_2 + (i-1)\,y_2\right) + 2c_{i,3}y_1y_2 + c_{i,5}\left(x_2y_1 + x_1y_2\right) \\
+ 2c_{i,4}x_1x_2 + \left(c_{i,6}y_2 + c_{i,7}y_1 + c_{i,8}x_2 + c_{i,9}x_1\right)\left(y_1^2 + y_2^2\right) \\
+ \left(c_{i,10}y_2 + c_{i,11}y_1 + c_{i,12}x_2 + c_{i,13}x_1\right)\left(x_1^2 + x_2^2\right) \\
+ \left(c_{i,14}y_2 + 2c_{i,9}y_1 + c_{i,15}x_2 + 2c_{i,11}x_1\right)\left(x_1y_1 + x_2y_2\right)
\end{aligned}
$$

$$
\begin{aligned}
H_{i,2} = c_{i,17} - 5c_{i,2}\left((2-i)\,x_1 + (i-1)\,y_1\right) + c_{i,3}\left(y_1^2 + 3y_2^2\right) + c_{i,4}\left(x_1^2 + 3x_2^2\right) \\
+ c_{i,5}\left(x_1 y_1 + 3x_2 y_2\right) + \left(c_{i,7} y_2 - c_{i,6} y_1 + c_{i,9} x_2 - c_{i,8} x_1\right)\left(y_1^2 + y_2^2\right) \\
+ \left(c_{i,11} y_2 - c_{i,10} y_1 + c_{i,13} x_2 - c_{i,12} x_1\right)\left(x_1^2 + x_2^2\right) \\
+ \left(2c_{i,9} y_2 - c_{i,14} y_1 + 2c_{i,11} x_2 - c_{i,15} x_1\right)\left(x_1 y_1 + x_2 y_2\right)
\end{aligned}
$$

where the coefficients are respectively

$a_2 = \alpha_2^2/24$	$a_3 = (\alpha_5 f_1 + 3\alpha_4 f_2)/16$	$a_4 = (3\alpha_3 f_1 + \alpha_6 f_2)/16$
$a_5 = (\alpha_6 f_1 + \alpha_5 f_2)/8$	$a_6 = 3\alpha_4(\sigma_1 + \sigma_2)/8$	$a_7 = 3\mu\alpha_4(3\beta_1 - \alpha_1)/8$
$a_8 = \alpha_5(2\sigma_1 - \sigma_2)/4$	$a_9 = \mu\alpha_5\beta_1/4$	$a_{10} = \alpha_6(3\sigma_2 - \sigma_1)/8$
$a_{11} = \mu\alpha_6(\alpha_1 + \beta_1)/8$	$a_{12} = 3\alpha_3\sigma_1/4$	$a_{13} = 3\mu\alpha_1\alpha_3/4$
$a_{14} = \alpha_5\sigma_2/2$	$a_{15} = \alpha_6(3\sigma_1 - \sigma_2)/4$	$a_{16} = \mu\alpha_1 f_1/16$
$a_{17} = \sigma_1 f_1/16$	$b_2 = \beta_2^2/24$	$b_3 = (\beta_6 f_1 + 3\beta_3 f_2)/16$
$b_4 = (3\beta_4 f_1 + \beta_5 f_2)/16$	$b_5 = (\beta_6 f_2 + \beta_5 f_1)/8$	$b_6 = 3\beta_3\sigma_2/4$
$b_7 = 3\mu\beta_1\beta_3/4$	$b_8 = \beta_6(3\sigma_1 - \sigma_2)/8$	$b_9 = \mu\beta_6(\alpha_1 + \beta_1)/8$
$b_{10} = \beta_5(-\sigma_1 + 2\sigma_2)/4$	$b_{11} = \mu\alpha_1\beta_5/4$	$b_{12} = 3\beta_4(\sigma_1 + \sigma_2)/8$
$b_{13} = 3\mu\beta_4(3\alpha_1 - \beta_1)/8$	$b_{14} = \beta_6(-\sigma_1 + 3\sigma_2)/4$	$b_{15} = \beta_5\sigma_1/2$
$b_{16} = \mu\beta_1 f_2/16$	$b_{17} = \sigma_2 f_2/16$	

References

1. Evans, K.E.: Auxetic polymers: a new range of materials. Endeavour. **15**, 170–174 (1991)
2. Yu, X.L., Zhou, J., Liang, H.Y., Jiang, Z.Y., Wu, L.L.: Mechanical metamaterials associated with stiffness, rigidity and compressibility: a brief review. Prog. Mater. Sci. **94**, 114–173 (2018)
3. Greaves, G.N., Greer, A.L., Lakes, R.S., Rouxel, T.: Poisson's ratio and modern materials. Nat. Mater. **10**, 823–837 (2011)
4. Chen, X., Feng, Z.H.: Dynamic behaviour of a thin laminated plate embedded with auxetic layers subject to in-plane excitation. Mech. Res. Commun. **85**, 45–52 (2017)
5. Duc, N., Cong, P.H.: Nonlinear dynamic response and vibration of sandwich composite plates with negative Poisson's ratio in auxetic honeycombs. J. Sandw. Struct. Mater. **20**, 692–717 (2018)
6. Li, C., Shen, H.S., Wang, H.: Nonlinear bending of sandwich beams with functionally graded negative Poisson's ratio honeycomb core. Compos. Struct. **212**, 317–325 (2019)
7. Li, J., Quan, T.T., Zhang, W.: Bifurcation and number of subharmonic solutions of a 4d non-autonomous slow-fast system and its application. Nonlinear Dyn. **92**, 721–739 (2018)
8. Chen, J.E., Zhang, W., Guo, X.Y., Sun, M.: Theoretical and experimental studies on nonlinear oscillations of symmetric cross-ply composite laminated plates. Nonlinear Dyn. **73**, 1697–1714 (2013)

Wave Propagation Phenomena in Nonlinear Elastic Metamaterials

Federica Mezzani, Amir Sajjad Rezaei, and Antonio Carcaterra

Abstract The present chapter provides a general method to deal with nonlinear integro-differential equations, based on statistical linearization and Fredholm's approach. In this context, the elastic metamaterial is characterized by long-range nonlocal interactions besides a nonlinear short-range constitutive relationship. Results are analytically obtained and unveil the birth of unconventional propagation.

Keywords Elastic metamaterials · Nonlinearity · Long-range interaction · Wave propagation · Nonlocality

1 Introduction

This work is aimed at presenting a general approach to nonlinear and long-range constitutive elastic relationships, based on statistical linearization and Fredholm's equation that enable the investigation of the dispersion relationship. Eventually, a simple example is introduced to show a first digression about the potential application of the method.

Metamaterials enjoy widespread attention due to the unexpected results they produce in many applications. In electromagnetics, metamaterials are frequently related to anomalous dissipation and diffraction properties of electromagnetic media that lead to negative group velocity, light stopping, and fast light [1]. Superluminal propagation has been observed through an acoustic experimental setup [2, 3]. In mechanics, metamaterials change the connectivity scheme of a structure and induce micropolar, higher-gradient, and nonlocal elasticity. In this context, metamaterials are thought as conventional elastic materials equipped with long-range interactions, the source of integral contributions, and nonlinear constitutive relationships. The effect of these long-range interactions has been investigated

F. Mezzani (✉) · A. S. Rezaei · A. Carcaterra
Department of Mechanical and Aerospace Engineering, Sapienza University of Rome, Rome, Italy
e-mail: federica.mezzani@uniroma1.it

W. Lacarbonara et al. (eds.), *New Trends in Nonlinear Dynamics*,
https://doi.org/10.1007/978-3-030-34724-6_4

in [4–7], but never tackling the nonlinearity. Long-range interactions change the topology of the connections and represent a breakthrough in the conventional concept of particle–particle interaction between closest neighbors, which leads to classical wave propagation. When the connection is extended either to one-to-all particles or to all-to-all particles, the introduced modification becomes source of a more sophisticated propagation behavior. Similar effects have been noticed only in quantum physics [8, 9], in Vlasov's theory [10]. Another remarkable example is in acoustics, where long-range electrical interactions can control the acoustic fields and well-known examples are plasmas and charged gases. In any case, the relation between nonlocality and wave propagation has not been intensively developed, and unconventional propagating phenomena are not as common as those appearing in electromagnetic applications.

Statistical linearization is a known procedure [11–13], often applied to random and complex systems [14]. An example is the analysis of a catenary anchor leg mooring (CALM) systems [15]. The chance to obtain analytical solutions is achieved by the definition of a specific kernel for the long-range interactions: indeed, they are modeled as elastic connections between far particles; however, the range of interaction is limited by a rectangular window, as in [7], but here analyzed for its one-dimensional counterpart. Eventually, the obtained parametric problem is treated as a Fredholm's equation [16] and the dispersion relationship is calculated accordingly. Remarkable insights can be derived: (1) the applied random force affects the propagation characteristics, namely the background noise modifies the phase and group velocity of the waveguide, (2) after stochastic linearization, the system shows periodic variation of the elastic parameters along the waveguide axis, and (3) the nonlinear properties generate the wavenumbers coupling.

2 Prototype Equation and Statistical Linearization

A mathematical model, based on the nonlocal elasticity theory of Eringen, is considered. For a three-dimensional, continuous, unbounded medium, equipped with long-range interactions, the Navier-Cauchy equation of motion becomes:

$$\rho \boldsymbol{u}_{tt}(\boldsymbol{x},t) + \frac{E}{2(1+\nu)}\left[\nabla^2 \boldsymbol{u}(\boldsymbol{x},t) - \frac{1}{1-2\nu}\nabla\left(\nabla\cdot\boldsymbol{u}(\boldsymbol{x},t)\right)\right] + \boldsymbol{g}(\boldsymbol{u}) + \int_{\boldsymbol{\xi}\in \boldsymbol{R}^3} \boldsymbol{F}(\boldsymbol{x},\boldsymbol{\xi})\, d\boldsymbol{V} = \boldsymbol{f}(\boldsymbol{x},t) \quad (1)$$

The long-range interaction appears as an integral term. It represents the summation of the long-range interaction forces, exerted on the particle originally at $\boldsymbol{x}$, due to all the particles at $\boldsymbol{\xi}$. $\boldsymbol{g}(x)$ resembles the possible sources of local nonlinearities and has differential nature.

Nonlinear equations can be attacked by the statistical linearization-SL approach [11, 12]. Equation (1) can be written as:

$$\boldsymbol{L}(\boldsymbol{u}) + \boldsymbol{g}(\boldsymbol{u}) = \boldsymbol{f}(t) \tag{2}$$

where $\boldsymbol{L}$ and $\boldsymbol{g}$ are a linear integro-differential and a nonlinear differential operator, respectively, $\boldsymbol{u}(\boldsymbol{x}, t)$ is the displacement of the elastic system, $\boldsymbol{f}(t)$ a random external force. The SL approach replaces the nonlinear term $\boldsymbol{g}(\boldsymbol{u})$ term with an equivalent linear operator $\boldsymbol{L}_{\mathbf{eq}}$, leading to:

$$\boldsymbol{L}(\boldsymbol{u}) + \boldsymbol{L}_{\mathbf{eq}}(\boldsymbol{u}, \boldsymbol{p}) = \boldsymbol{f}(t). \tag{3}$$

The vector $\boldsymbol{p}$ is a set of parameters that can be suitably selected to make Eq. (3) as close as possible to Eq. (2). A direct comparison between Eqs. (2) and (3) produces the *error equation*:

$$\boldsymbol{e}(\boldsymbol{u}, \boldsymbol{p}) = \boldsymbol{L}_{\mathbf{eq}}(\boldsymbol{u}, \boldsymbol{p}) - \boldsymbol{g}(\boldsymbol{u}) \tag{4}$$

The SL requires the minimization of the mean-square $E\{\bullet\}$ of the *error equation* in terms of the parameters $\boldsymbol{p}$:

$$\frac{\partial E\left\{\boldsymbol{e}^{\mathrm{T}}(\boldsymbol{p})\,\boldsymbol{e}(\boldsymbol{p})\right\}}{\partial \boldsymbol{p}} = 0. \tag{5}$$

The solution of this last equation determines the optimal vector $\boldsymbol{p}^*$ and one can analyze the equivalent linear integro-differential equation $\boldsymbol{L}(\boldsymbol{u}) + \boldsymbol{L}_{\mathbf{eq}}(\boldsymbol{u}, \boldsymbol{p}^*) = \boldsymbol{f}(t)$, instead of the nonlinear integro-differential Eq. (2).

3 Statistical Linearization of the Navier-Cauchy Equation with Long-Range Forces

The analysis in the previous section leads to Eq. (5) the form of which is specifically investigated here under some simplified assumptions.

The equation of motion, for a one-dimensional system, with kernel $H(x - \xi)$ satisfying the action-reaction principle, i.e., $H(x - \xi) = -H(\xi - x)$, and decaying with the distance, namely $\lim_{|x-\xi| \to +\infty} H(x - \xi) = 0$, is of the form:

$$\begin{aligned} &\rho u_{tt}(x,t) + E_0 u_{xx}(x,t) - 3E_1 u_x^2(x,t)\, u_{xx}(x,t) \\ &\quad - \int \tilde{k}\,[u(x,t) - u(\xi,t)]\, H(x-\xi)\, d\xi = f(t)\delta(x) \end{aligned} \tag{6}$$

meaning:

$$\boldsymbol{L}\left(\boldsymbol{u}\right) \equiv \rho u_{tt}\left(x,t\right) + E_0 u_{xx}\left(x,t\right) - \int \tilde{k}\left[u\left(x,t\right) - u\left(\xi,t\right)\right] H\left(x-\xi\right) d\xi \quad (7)$$

$$\boldsymbol{g}\left(\boldsymbol{u}\right) \equiv -3E_1 u_x^2\left(x,t\right) u_{xx}\left(x,t\right) \quad (8)$$

where * indicates the convolution operation, and we assume:

$$\boldsymbol{L_{eq}}\left(\boldsymbol{p}\right) \equiv p u_{xx}\left(x,t\right) \quad (9)$$

$\tilde{k}$ resembles the stiffness modulation of the elastic connections. About the form of the kernel $H(x)$, it is not necessary at the moment to introduce any further specification.

The nonlinear term $3E_1 u_x^2\left(x,t\right) u_{xx}\left(x,t\right)$ is derived considering a nonlinear stress-strain relationship:

$$\sigma\left(\varepsilon\right) = E_0\varepsilon + E_1\varepsilon^3 \quad (10)$$

It is apparent then that the error equation is:

$$\boldsymbol{e}\left(\boldsymbol{p}\right) \equiv p u_{xx}\left(x,t\right) - 3E_1 u_x^2\left(x,t\right) u_{xx}\left(x,t\right) \quad (11)$$

and it is immediate to evaluate the parameter p through Eq. (5), which leads to:

$$p = 3E_1 \frac{E\left\{u_x^2 u_{xx}^2\right\}}{E\left\{u_{xx}^2\right\}} \quad (12)$$

The complete characterization of the parameter p requires the definition of the coefficients $E\left\{u_x^2 u_{xx}^2\right\}$ and $E\left\{u_{xx}^2\right\}$, which becomes simpler when it is reasonable to assume those variables equipped with a Gauss-like type of probability density function. For this reason, we limit our attention to the case f is a random force with flat power spectral density S_F and a Gauss-like distribution. Additionally, it is assumed to be a point force applied at the origin of the x axis.

A modal expansion of the solution helps in finding simpler results. Without loss of generality we assume $u(x,t) = \phi(x)q(t)$, including only one single mode. Under these conditions, Eq. (12) provides $p = 3E_1\phi'^2(x)\sigma_q^2$.

Therefore, Eq. (3) assumes the form:

$$\rho\phi(x)\ddot{q}(t) - \left[E_0 + 3E_1\phi'^2(x)\sigma_q^2\right]\phi''(x)q(t)$$
$$- q(t)\int \tilde{k}\left[\phi(x) - \phi\left(\xi\right)\right] H\left(x-\xi\right) d\xi = f(t)\delta(x). \quad (13)$$

To evaluate the coefficient σ_q^2, the following expression can be considered [17]:

$$\sigma_u^2(x)=\phi^2(x)\sigma_q^2=\frac{1}{2\pi}\int_{-\infty}^{+\infty} S_u\left(x,\omega,\sigma_q^2\right)d\omega=\frac{S_F}{2\pi}\int_{-\infty}^{+\infty}\left|FRF\left(x,\omega,\sigma_q^2\right)\right|^2 d\omega \tag{14}$$

with σ_q^2 the variance of $q(t)$, S_u the power spectral density of u, $FRF(x, \omega)$ the complex frequency response of the system at x when the input force is at $x = 0$. The frequency response can be easily obtained by Eq. (13). Eventually, Eq. (14) appears to be an equation in terms of the variance σ_q^2.

4 Parametric Problem and Fredholm's Equation

Once Eq. (14) is solved, the value σ_q^2 can be substituted into Eq. (13). Replacing in it $u(x, t) = \phi(x)q(t)$, and for a rectangular window on length a, around the origin, we obtain:

$$\rho\ddot{u}(x,t)-\left[E_0+3E_1\phi'^2(x)\sigma_q^2\right]u''(x,t)-2\tilde{k}au(x,t)+H(x)*u(x,t)=f(t)\delta(x) \tag{15}$$

This is a linear integro-differential equation. However, the term $3E_1\phi'^2(x)\sigma_q^2u''(x,t)$ exhibits a parametric dependence on x. This can be regarded as an extra-stiffness $E_p = 3E_1\phi'^2(x)\sigma_q^2$ that superimposes to E_0. For example, if $\phi(x)$ is the first mode, then $\phi'^2(x)$ is a square of a sinusoid, and it perturbates the constant stiffness E_0, as in Fig. 1.

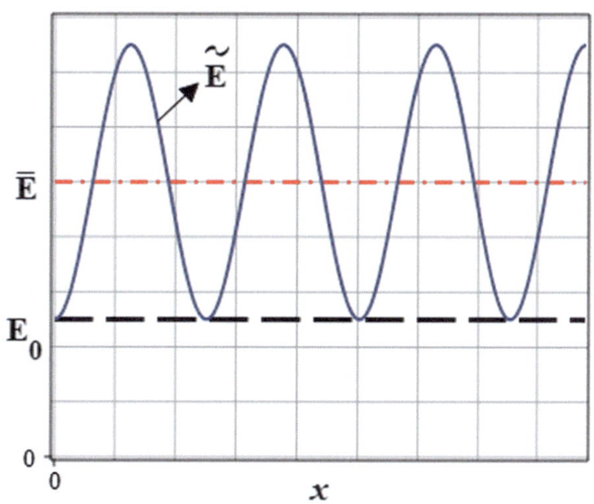

Fig. 1 Effective stiffness $\tilde{E}$

We present here a method based on the Fredholm's integral equation to approach Eq. (15). Applying the Fourier transform both in space and time, we obtain:

$$-\rho\omega^2\hat{\hat{u}} + k^2 E_0 \hat{\hat{u}} + 3E_1 \hat{\phi}'^2 \sigma_q^2 * \left(k^2 \hat{\hat{u}}\right) - 2\tilde{k}a\hat{\hat{u}} + \tilde{k}\hat{\hat{u}}\hat{H} = \hat{F} \tag{16}$$

where $\hat{H} = 2\tilde{k}\sin(ka)/k$ for the chosen window. For any $x \neq 0$ we have the homogeneous integral equation:

$$-\rho\omega^2\hat{\hat{u}} + k^2 E_0 \hat{\hat{u}} + 3E_1 \hat{\phi}'^2 \sigma_q^2 * \left(k^2 \hat{\hat{u}}\right) - 2\tilde{k}a\hat{\hat{u}} + \tilde{k}\hat{\hat{u}}\hat{H} = 0 \tag{17}$$

whose explicit form is:

$$3E_1\sigma_q^2 \int_{-\infty}^{+\infty} \hat{\phi}'^2 \left(k - k'\right) k'^2 \hat{\hat{u}} \left(k'\right) dk' = \left[\rho\omega^2 - k^2 E_0 + 2\tilde{k}a - \tilde{k}\hat{H}\right] \hat{\hat{u}}. \tag{18}$$

The discretization of the previous equation leads to:

$$\sum_j 3E_1\sigma_q^2 \hat{\phi}'^2 \left(k_i - k_j\right) k_j^2 \hat{\hat{u}} \left(k_j\right) \Delta k - \left[\rho\omega^2 - k_i^2 E_0 + 2\tilde{k}a - \tilde{k}\hat{H}\right] \hat{\hat{u}} \left(k_i\right) = 0 \tag{19}$$

namely,

$$\boldsymbol{A}\boldsymbol{u} = \boldsymbol{0}, \text{ where } \boldsymbol{u} = \begin{Bmatrix} u\left(k_1\right) \\ u\left(k_2\right) \\ \vdots \\ u\left(k_n\right) \end{Bmatrix}, \text{ with}$$

$$[\boldsymbol{A}]_{ij} = 3E_1\sigma_q^2 \hat{\phi}'^2 \left(k_i - k_j\right) k_j^2 \Delta k - \left(\rho\omega^2 - k_j^2 E_0 + 2\tilde{k}a - \tilde{k}\hat{H}\right) \delta_{ij} \tag{20}$$

We are interested in nontrivial and not identically to zero solutions. This technically happens if the *Fredholm determinant* associated to $\boldsymbol{A}$ vanishes, from which we can derive the dispersion relationship.

The condition for the determinant to vanish does not lead to equations independent for each k_i and the wavenumbers mix up. Given a value for the frequency, the vanishing determinant condition does not imply each k_i is determined. The determinant expression is:

$$\det\boldsymbol{A} = \boldsymbol{f}\left(k_1, k_2, .., k_n, \omega, \sigma_q^2, \hat{H}\right) = 0 \tag{21}$$

This means that, assigned a value for ω, σ_q^2 and $\hat{H}$, the values for the wavenumbers are not uniquely determined. Rather, one can select arbitrarily $N - 1$ wavenumber values and determine consequently the value for the N-th remaining wavenumber. Since, at any frequency, the values for the $N - 1$ wavenumbers can be arbitrarily assigned, the N-th can range in a set, describing a segment at each frequency.

The dispersion relationship depends on: (1) the presence of $\hat{H}$ due to long-range interactions, (2) the presence of σ_q^2 as an effect of the background noise in the structure.

From Eq. (21) we can learn what follows: firstly it is not possible to excite individual and independent wavenumbers with a given frequency, and this is reminiscent of nonlinear systems (e.g., super-harmonics and sub-harmonics); secondly, the level of the random excitation $\boldsymbol{f}$ affects the propagation characteristics since it alters the value of the coefficient by σ_q^2. Accordingly, Fig. 2 shows the effect of this coefficient on the parametric stiffness, E_p.

A last consideration can be extrapolated by both Figs. 1 and 2. The effective stiffness $\tilde{E}$, taking into account both the standard elastic modulus E_0 and its modulation E_p due to the nonlinearities, is a periodic function. This suggests a localization phenomenon, namely the Anderson localization, related to stop band, low pass and high pass behaviors.

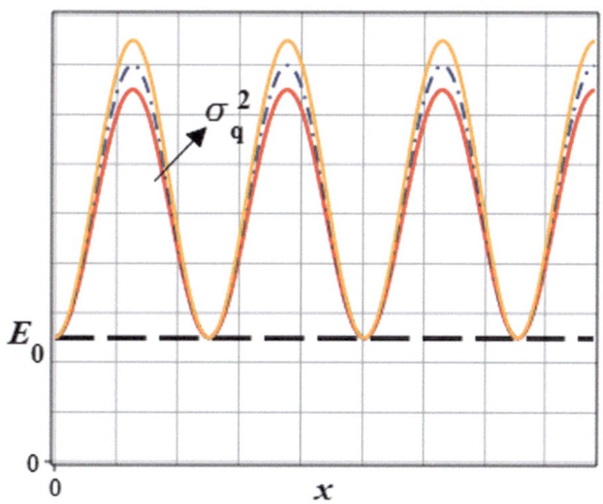

Fig. 2 Effect of σ_q^2 on $\tilde{E}$

5 Homogenization of the Stiffness and Parametric Dispersion Relationship

To complete the investigation, we report here the analysis of the dispersion relationship associated to Eq. (15). The aim is to highlight the effect of the random external force on the propagation characteristics. It is then reasonable to consider the homogenized value of the effective stiffness $\tilde{E}$, i.e., its average value $\overline{E}$, deduced by Fig. 1. Using the same type of long-range interactions, the nondimensional dispersion relationship is:

$$\Omega^2 = K^2 - 2\chi\,(1 - \mathrm{sinc}\,K) \tag{22}$$

where $\Omega = \omega a\sqrt{\rho/\overline{E}}$ is the nondimensional frequency, $K = ka$ the nondimensional wavenumber, and $\chi = \tilde{k}a^3/\overline{E}$ is the nondimensional parameter comparing the effect of the long-range interactions and of the nonlinearities. The advantage of such type of formulation stands in the chance to discuss the dynamic behavior in terms of the parameter χ only. Figure 3 shows the trend of the dispersion relationship varying with χ, compared with the standard one. All curves start with a very high slope to converge, for high values of the nondimensional wavenumber, to the conventional D'Alembert propagation. Since the derivative of the dispersion relationship is the group velocity, this implies that the steeper the slope, the higher the group velocity, disclosing superfast propagation of the envelope, till the limit of superluminal propagation when the curve has a vertical tangent, i.e., the case of σ_{q4}^2. Moreover, since $\overline{E}$ is a function of the random force, through the coefficient σ_q^2, it can be

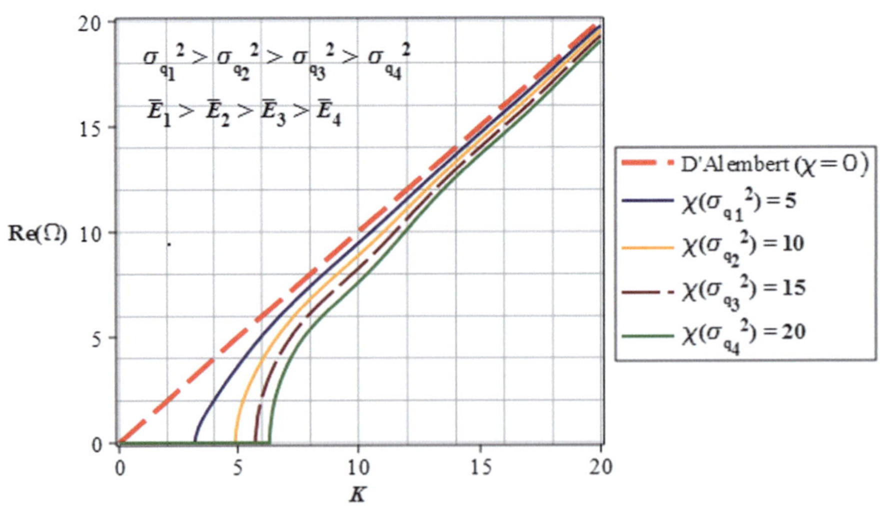

Fig. 3 Nondimensional dispersion relationship

discussed also in terms of σ_q^2: the higher the value of σ_q^2, the higher the homogenized stiffness $\overline{E}$ and the higher the value of σ_q^2, the closer is the behavior to the standard propagation. Eventually, the smaller is σ_q^2, the higher is the group velocity.

6 Conclusions

In this chapter, the case of a one-dimensional waveguide, in which the presence of long-range interactions together with the nonlinear nature of the constitutive law, is investigated. The long-range interactions are modeled as elastic connections occurring between distant particles but only within a delimited region of length a. The nonlinear constitutive law accounts not only for the first order of the strain but it includes also the third. The resulting equation of motion, approached with the statistical linearization process, has space-dependent coefficients. To analyze the obtained parametric problem, the Fredholm's approach is applied and by solving the associated determinant the dispersion relationship is found. It depends on the nature of the long-range interaction, through the term $\hat{H}$, and of the nonlinearity, related to the coefficient σ_q^2. This brings to light important insights: (1) the level of the random excitation directly affects the propagation characteristics, (2) as recurring of nonlinear systems, it is not possible with a single frequency to excite independent and individual wavenumbers. Moreover, the shape of the equivalent stiffness unveils Anderson localization and thus, possible wave-stopping phenomena.

References

1. Laude, V., Korotyaeva, M. E.: Stochastic band structure for waves propagating in periodic media or along waveguides. arXiv preprint arXiv:1801.09914 (2018)
2. Robertson, W.M., Pappafotis, J., Flannigan, P., Cathey, J., Cathey, B., Klaus, C.: Sound beyond the speed of light: measurement of negative group velocity in an acoustic loop filter. Appl. Phys. Lett. **90**(1), 014102 (2007)
3. Mugnai, D., Ranfagni, A., Ruggeri, R.: Observation of superluminal behaviors in wave propagation. Phys. Rev. Lett. **84**(21), 4830 (2000)
4. Carcaterra, A., Coppo, F., Mezzani, F., Pensalfini, S.: Long-range retarded elastic metamaterials: wave-stopping, negative and hypersonic group velocity. Phys. Rev. Appl. **11**(1), 014041 (2019)
5. Mezzani, F., Coppo, F., Carcaterra, A.: Long-range coupling of waveguides. In: Proceedings of ISMA 2018—International Conference on Noise and Vibration Engineering and USD 2018—International Conference on Uncertainty in Structural Dynamics (2018)
6. Mezzani, F., Coppo, F., Pensalfini, S., Roveri, N., Carcaterra, A.: Twin-waves propagation phenomena in magnetically-coupled structures. Proc. Eng. **199**, 711–716 (2017)
7. Coppo, F., Rezaei, A.S., Mezzani, F., Pensalfini, S., Carcaterra, A.: Wave path in an elastic membrane with selective nonlocality. In: Proceedings of ISMA 2018—International Conference on Noise and Vibration Engineering and USD 2018—International Conference on Uncertainty in Structural Dynamics (2018)

8. Vaidya, V.D., Guo, Y., Kroeze, R.M., Ballantine, K.E., Kollár, A.J., Keeling, J., Lev, B.L.: Tunable-range, photon-mediated atomic interactions in multimode cavity QED. Phys. Rev. X. **8**(1), 011002 (2018)
9. Puri, S., Andersen, C.K., Grimsmo, A.L., Blais, A.: Quantum annealing with a network of all-to-all connected, two-photon driven Kerr nonlinear oscillators. arXiv preprint arXiv:1609.07117 (2016)
10. Vlasov, A.A.: On high-frequency properties of electron gas. J. Exp. Theor. Phys. **8**(3), 291–318 (1938)
11. Kozin, F.: The method of statistical linearization for non-linear stochastic vibrations. In: Ziegler, F., Schuëller, G.I. (eds.) Nonlinear Stochastic Dynamic Engineering Systems. IUTAM Symposium (International Union of Theoretical and Applied Mechanics). Springer, Berlin (1988)
12. Bernard, P., Wu, L.: Stochastic linearization: the theory. J. Appl. Probab. **35**(03), 718–730 (1998)
13. Bernard, P.: Stochastic linearization: what is available and what is not. Comput. Struct. **67**(1–3), 9–18 (1998)
14. Roberts, J.B., Spanos, P.D.: Random Vibration and Statistical Linearization. Courier Corporation, New York (2003)
15. Culla, A., Carcaterra, A.: Statistical moments predictions for a moored floating body oscillating in random waves. J. Sound Vib. **308**, 44–66 (2007)
16. Fredholm, I.: Solution d'un problème fondamental de la théorie de l'élasticité. Arkiv. Mat. Astro. Fysik 2. **28**, 3–8 (1906)
17. Meirovitch, L.: Elements of Vibration Analysis. McGraw-Hill Science, Engineering & Mathematics Publisher, New York (1975)

Numerical Simulations in Nonlinear Elastic Metamaterials with Nonlocal Interaction

Francesco Coppo, Federica Mezzani, Sara Pensalfini, and Antonio Carcaterra

Abstract The paper presents a numerical investigation of an elastic metamaterial, i.e., a one-dimensional elastic waveguide, equipped with nonlocal (long-range) and nonlinear interactions. The dynamic behavior of the newly defined structure is described by nonlinear integro-differential equation of motion. Numerical simulations, comparing the linearized model and the nonlinear one, unveil the arising of wave-stopping and backward propagation phenomena.

Keywords Elastic metamaterials · Nonlinear · Wave stopping · Negative group velocity · Long-range interactions

1 Introduction

Metamaterials are finely designed materials capable of achieving unusual behaviors thanks to the particular geometries and connection topologies in which they are arranged. Besides, thanks to the fast development, in the last decades, of micro- and nano-manufacturing technologies [1, 2], the fields of possible applications have rapidly increased. Under the name metamaterials, several kinds of materials can be classified and, among many, electromagnetic [3], acoustic [4], and elastic [5] are the ones enjoying a wider attention from the scientific community. The singular emerging phenomena involve fast light propagation [6], wave-stopping [7–9], unusual dissipation [10], and backward propagation [11–13]. The complexity of the models describing such systems, usually involving convolution terms and integral-differential equations, prevented thorough analytical dissertations on the propagation behavior, in a general panorama in which even literature provides only few examples of works dealing with nonlocal interactions [14, 15].

F. Coppo · F. Mezzani (✉) · S. Pensalfini · A. Carcaterra
Department of Mechanical and Aerospace Engineering, Sapienza University of Rome, Rome, Italy
e-mail: francesco.coppo@uniroma1.it; federica.mezzani@uniroma1.it

W. Lacarbonara et al. (eds.), *New Trends in Nonlinear Dynamics*,
https://doi.org/10.1007/978-3-030-34724-6_5

This work completes the investigation presented in [16], confined within the hypothesis of the linearization. Without the chance of achieving analytical solutions, this paper presents numerical simulations aimed at comparing the linear model and the nonlinear one, expanding the concept of a linear, wider connectivity [17] to a nonlinear one. The idea is not only to confirm the remarkable results obtained in [16] and in [18, 19], but also to establish them as general description of the dynamic behavior of an elastic metamaterial, since no linearization procedure is here applied.

2 Model for a Long-Range Waveguide

In a classical D'Alembert waveguide, modelled according to the principles of the local elasticity theory, each elementary particle is connected with its closest neighbors only. However, as the interaction length increases, nonlocal interactions emerge and the assumptions of the elastic theory are not effective anymore.

This work, following the guidelines presented in [16], wants to interpret the effects of nonlocal connections, beyond the linearized model, through the mean of numerical simulations.

Nonlocal connections, i.e., distance-dependent interactions, which can appear in the shape of magnetostatic, electrostatic, as well as gravitational forces, alter the intrinsic pattern of a system, with direct consequences on its dynamic response. The equation of motion for a one-dimensional, continuous, infinite waveguide, modified because of the introduction of long-range forces, already presented in [16], assumes the form:

$$\rho \frac{\partial^2 w}{\partial t^2} - E \frac{\partial^2 w}{\partial x^2} - \int_{-\infty}^{+\infty} F(r) d\xi = 0, \tag{1}$$

where w is the displacement, ρ and E are the mass density and the Young's modulus, respectively, $F(r)$ is the force exerted on the particle at x, due to the interaction with the particle at ξ, and r is the actual distance between the particles and specifically:

$$r(x, \xi, t) = x|_{t=0} + w(x,t) - \left[\xi|_{t=0} + w(\xi, t)\right] = x - \xi + w(x,t) - w(\xi, t). \tag{2}$$

The analysis of the dynamic behavior demands the definition of a model for the interaction forces, which, even for a three-dimensional system, are chosen to be parallel to the distance vector and so that the action–reaction principle holds:

$$\boldsymbol{F}(\boldsymbol{r}) = f(|\boldsymbol{r}|)\boldsymbol{r}. \tag{3}$$

The sign of $f(|\boldsymbol{r}|)$ rules the nature of the interaction: attractive for negative $f(|\boldsymbol{r}|)$, repulsive if positive. Moreover, among many possibilities, a Gauss-like trend is selected for $f(|\boldsymbol{r}|)$, which, for a one-dimensional domain, becomes:

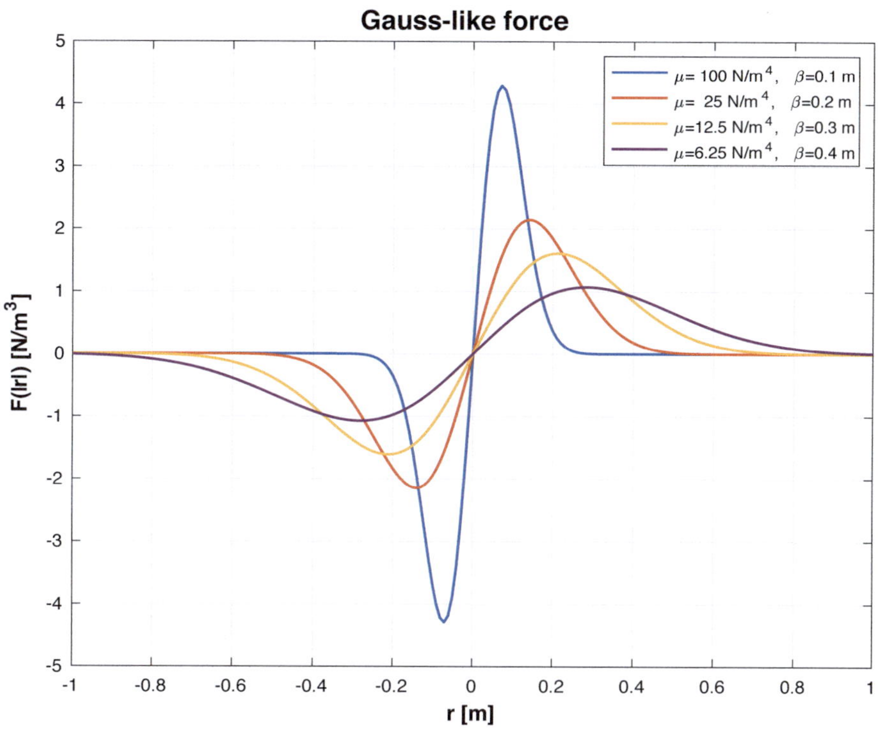

Fig. 1 Gauss-like force as function of the distance

$$f(|r|) = \mu e^{-\left(\frac{r}{\beta}\right)^2}, \tag{4}$$

where μ and β represent the magnitude and the characteristic length of the long-range forces, respectively. This shape, more than guaranteeing the antisymmetric configuration of $F(r)$, implies a force vanishing with the distance, property common to many physical long-range forces, as the previously mentioned gravitational, Coulomb, etc.

Figure 1 shows the trend of $f(|r|)$ varying with μ and β.

For this model of nonlocal interactions and assuming $\varepsilon(x, \xi, t) = w(x, t) - w(\xi, t)$, Eq. (1) becomes:

$$\rho \frac{\partial^2 w}{\partial t^2} - E \frac{\partial^2 w}{\partial x^2} - \int_{-\infty}^{+\infty} \mu(x - \xi + \varepsilon) e^{-\left(\frac{x-\xi+\varepsilon}{\beta}\right)^2} d\xi = 0. \tag{5}$$

Since Eq. (1) is strongly nonlinear, it cannot be solved analytically and no closed form solutions can be provided. In fact, the investigation presented in [16], aimed at finding analytical solutions, is bounded to the linearized problem discussion.

Next sections examine the numerical approach adopted to extract the propagation characteristics of both the linearized and the nonlinear models.

3 Numerical Simulations

To perform numerical simulations, the finite counterpart of the waveguide is considered. The rod is assumed clamped at the ends and subjected to a repulsive Gauss-like long-range interactions, as shown in Fig. 2.

Table 1 reports the physical properties of the system.

The simulations look at the free problem under initial conditions capable of exciting several wavelength of the system.

The results of the simulations in terms of displacement $w(x, t)$ are collected in a space-time map. A double discrete Fourier transform is applied [20] transforming the space-time variable $w(x, t)$ into the wavenumber-frequency dependent variable $W(k, \omega)$.

The contour lines of $W(k, \omega)$ are plotted and from the exam of its crest, a curve emerges interpreted as the equivalent of a dispersion relationship.

3.1 Linearized Model

Under the hypothesis of small ε, namely for $w \ll \beta$ (see [16]), the system represented in Eq. (5) can be simplified as:

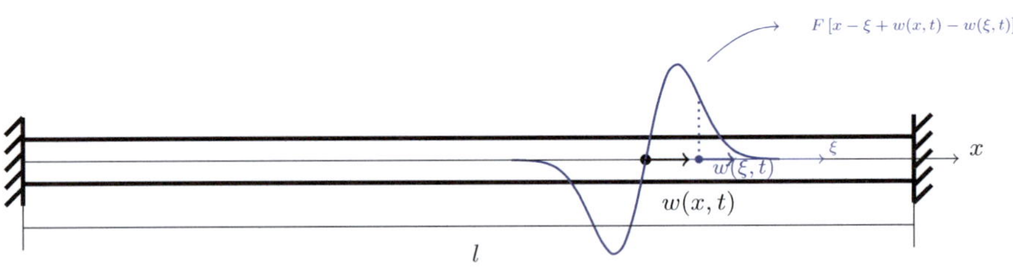

Fig. 2 Longitudinal finite waveguide model

Table 1 Physical parameters

Mass density ρ	1320 kg/m^3
Young's modulus E	1.6 MPa
Rod length l	100 m
Long-range interaction magnitude μ	$1.05 \cdot 10^6$ N/m^4
Long-range interaction length β	6.36 m

Table 2 Simulations parameters

Δx	0.5 m
Δt	2.9 ms
T_{sim}	22.98 s

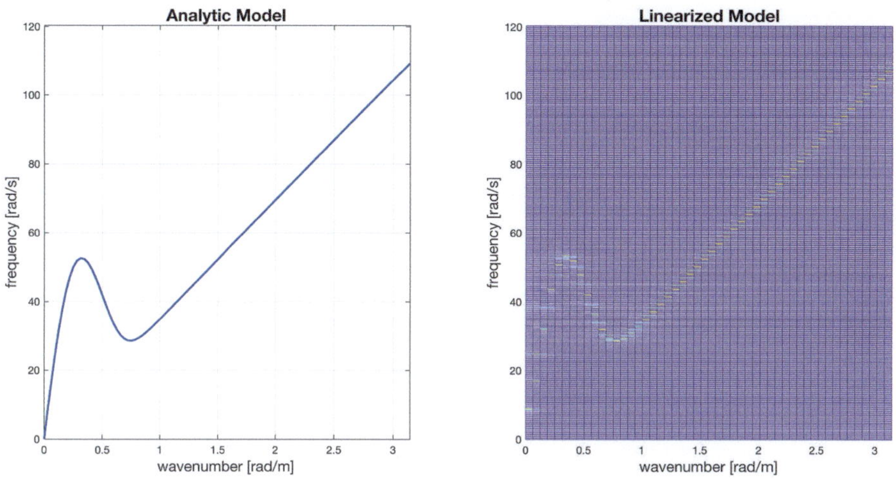

Fig. 3 Dispersion relationship of the simplified model: analytic result (left) and simulations (right)

$$\rho \frac{\partial^2 w}{\partial t^2} - E \frac{\partial^2 w}{\partial x^2} + h * w = 0, \tag{6}$$

where $h = \mu \left[1 - 2\left(\frac{x}{\beta}\right)^2\right] e^{-\left(\frac{x}{\beta}\right)^2}$, and $*$ is the convolution operator (space domain), where for the simulation we use (Table 2).

The convolution term leads to the presence of an additional stiffness, and the system is simulated with finite differences approach for the space discretization, and with Runge–Kutta algorithms for the study of the time evolution.

Figure 3 shows a comparison between analytical and numerical solutions for the dispersion relationship in the linear case, where the numerical analysis uses also the double Fourier transform method described above.

3.2 Nonlinear Simulations

The system response is simulated without any small displacement hypothesis.

Figure 4 on the right, with ripples traveling apart of the envelope, shows a highly dispersive propagation.

The dispersion relationship, depicted in Fig. 5, is obtained by applying the double FFT to the equation of motion. It describes the propagation behavior and being

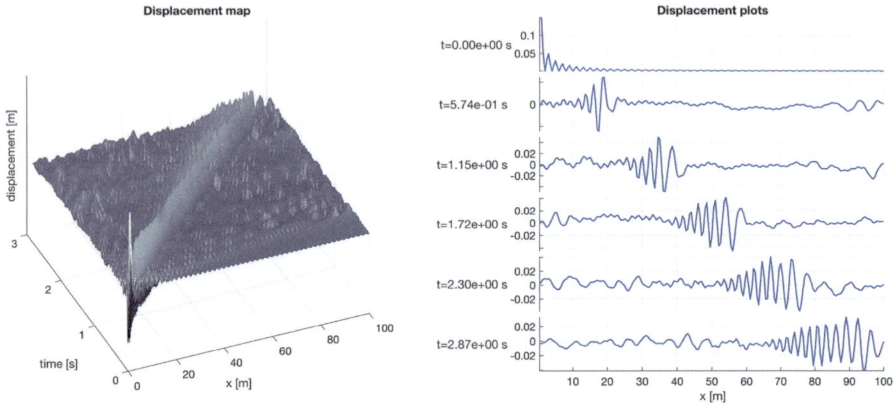

Fig. 4 Displacement map of nonlinear model (left), and some frame-shots (right)

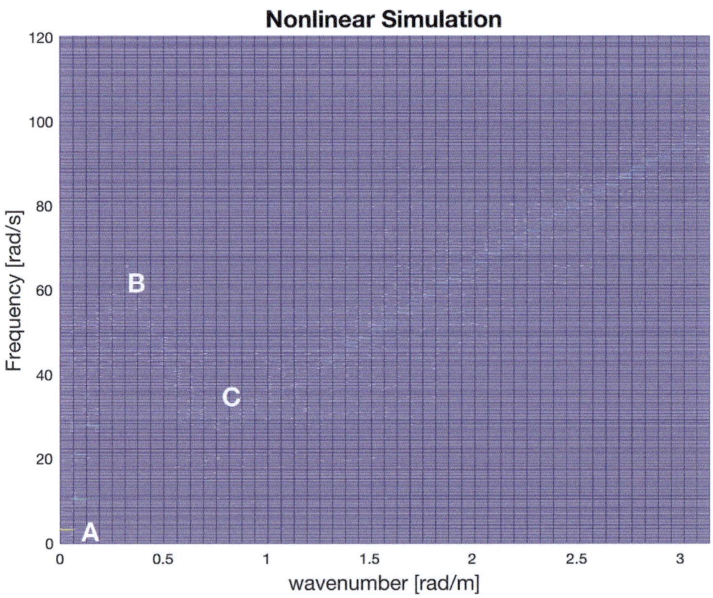

Fig. 5 Dispersion relationship (surface plot) of the nonlinear model

related to the group velocity c_g, it is straightforward to identify the different regimes. In fact, the dispersion curve shows the presence of three main branches, indicated with the letters A, B, and C in Fig. 5.

Since $c_g = \frac{\partial \Omega}{\partial K}$, where $\Omega = \Omega(K)$ is the general expression of the non-dimensional dispersion relationship, the AB segment, characterized by a positive slope, is associated to a positive group velocity; in B, the curve exhibits a maximum, related to wave-stopping. From B to C, since the slope is negative, so is the group velocity. In C, corresponding to the minimum of the dispersion relationship, $c_g = 0$ and a new wave-stopping condition emerge. Beyond C, the curve converge to the standard propagation regime, typical of the D'Alembert waveguide.

From the comparison between these two curves, it is apparent how these phenomena do not occur in a standard D'Alembert waveguide, whose propagation regime is well depicted by the straight line in Fig. 5. As the system consists of a simple longitudinal waveguide, but further equipped with nonlinear long-range forces, it is possible to conclude that the variations in the propagating behavior, with respect to the conventional one, can only be addressed to the introduction of such forces.

4 Conclusions

The paper describes the onset of unconventional propagation phenomena in long-range nonlinear elastic metamaterials, continuing the investigation presented by the authors in [16]. The physics of such media leads to strongly nonlinear and nonlocal integral-differential models. In general, these equations do not admit analytical solutions and only numerical simulations can provide insights into the dynamic response.

Numerical simulations are performed for the linearized model, corroborating the achievements analytically obtained for the simplified model in [16], and for the nonlinear problem, verifying the assumption that nonlocal interactions lead to unusual propagation regimes. Forward propagation occurs at a speed higher than the one proper of conventional D'Alembert waveguide, wave-stopping phenomenon appears for two specific combinations of (Ω_i, K_i), eventually backward propagation occurs when a regime of negative group velocity is established.

References

1. Bückmann, T., Kadic, M., Schittny, R., Wegener, M.: Phys. Status Solidi (b) **252**(7), 1671 (2015). https://doi.org/10.1002/pssb.201451698
2. Greer, J.R., Park, J.: Additive manufacturing of nano-and microarchitected materials. Nano Lett. **18**(4), 2187–2188 (2018)
3. Nagerl, A., Watkins, J.: Electromagnetic meta-materials. US Patent Application 14/322,453, 2014
4. Brunet, T., Poncelet, O., Aristégui, C., Leng, J., Mondain-Monval, O.: J. Acoust. Soc. Am. **141**(5), 3811 (2017)
5. Carcaterra, A., Akay, A., Bernardini, C.: Mech. Syst. Signal Process. **26**, 1 (2012)
6. Laude, V., Korotyaeva, M.E.: (2018). arXiv:1801.09914
7. Yanik, M.F., Fan, S.: Phys. Rev. Lett. **92**(8), 083901 (2004)
8. Tsakmakidis, K.L., Pickering, T.W., Hamm, J.M., Page, A.F., Hess, O.: Phys. Rev. Lett. **112**(16), 167401 (2014)
9. Chang, D., Safavi-Naeini, A.H., Hafezi, M., Painter, O.: New J. Phys. **13**(2), 023003 (2011)

10. Carcaterra, A., Akay, A.: Phys. Rev. E **84**(1), 011121 (2011)
11. McDonald, K.T.: Am. J. Phys. **69**(5), 607 (2001)
12. Martinez, O., Gordon, J., Fork, R.: JOSA A **1**(10), 1003 (1984)
13. Gehring, G.M., Schweinsberg, A., Barsi, C., Kostinski, N., Boyd, R.W.: Science **312**(5775), 895 (2006)
14. Eringen, A.C.: Int. J. Eng. Sci. **10**(5), 425 (1972)
15. Tarasov, V.E.: J. Phys. A Math. Gen. **39**(48), 14895 (2006)
16. Carcaterra, A., Coppo, F., Mezzani, F., Pensalfini, S.: Phys. Rev. Appl. **11**(1) (2019). https://doi.org/10.1103/physrevapplied.11.014041
17. Carcaterra, A., Roveri, N., Akay, A.: In Proceedings of ISMA2018 International Conference on Noise and Vibration Engineering, USD2016 International Conference on Uncertainty in Structural Dynamics (2018)
18. Coppo, F., Rezaei, A.S., Mezzani, A., Pensalfini, F., Carcaterra, S.: In Proceedings of ISMA2018 International Conference on Noise and Vibration Engineering, USD2016 International Conference on Uncertainty in Structural Dynamics (2018)
19. Mezzani, F., Coppo, F., Carcaterra, A.: In Proceedings of ISMA2018 International Conference on Noise and Vibration Engineering, USD2016 International Conference on Uncertainty in Structural Dynamics (2018)
20. Alleyne, D., Cawley, P.: J. Acoust. Soc. Am. **89**(3), 1159 (1991)

Nonlinear Dynamic Response of Nanocomposite Cantilever Beams

Michela Talò, Biagio Carboni, Giovanni Formica, Giulia Lanzara, Matthew Snyder, and Walter Lacarbonara

Abstract The nonlinear dynamic response of carbon nanotube (CNT) nanocomposite cantilevers is experimentally and theoretically investigated. Nanocomposite cantilevers made of a thermoplastic polymer and high aspect ratio CNTs are subject to a primary resonance base excitation. The experimentally obtained frequency-response curves highlight the effects of the CNT/polymer stick-slip energy dissipation on the nonlinear macroscopic dynamic response of the nanocomposite beams. The hysteresis arising from the nanostructural stick-slip gives rise to a change of nonlinearity dominated by the flexural curvature hardening effect towards a softening behavior at low amplitudes. The CNT/polymer frictional sliding hysteresis is here described by a hysteretic restoring force in the context of the nonlinear Euler–Bernoulli beam theory. An initial parametric analysis shows the capability of the model to capture qualitatively the softening–hardening frequency response trend.

Keywords Frequency response · Softening/hardening · Carbon nanotube nanocomposite · Bouc–Wen model of hysteresis · Path following

M. Talò (✉) · B. Carboni · W. Lacarbonara
Department of Structural and Geotechnical Engineering, Sapienza University of Rome, Rome, Italy
e-mail: michela.talo@gmail.com

G. Formica
Department of Architecture, University of Roma Tre, Rome, Italy

G. Lanzara
Department of Engineering, University of Roma Tre, Rome, Italy

M. Snyder
Department of Engineering Mechanics, United States Air Force Academy, CO, USA

W. Lacarbonara et al. (eds.), *New Trends in Nonlinear Dynamics*,
https://doi.org/10.1007/978-3-030-34724-6_6

1 Introduction

Due to their multifunctionality, CNT nanocomposites are drawing a great deal of attention as high performance structural materials for a variety of applications in dynamic environments. Ultra-lightweight nanocomposites, made of engineering polymers integrated with carbon nanofillers, are already employed to realize dynamic devices, such as micro-resonators, micro-accelerometer, chemical and biosensors [1–4], to mention but a few.

In the perspective of a wider exploitation of CNT-reinforced polymer materials in aerospace, automotive, and civil engineering, this work aims to highlight some interesting nonlinear properties of these materials.

This paper is primarily focused on the nonlinear dynamic response of nanocomposite cantilever beams excited near the primary resonance of their lowest bending modes. The study investigates the effects of hysteresis arising from the CNT/polymer interfacial sliding on the macroscopic nonlinear dynamic response. The primary energy dissipation source in nanocomposites is represented by the stick-slip frictional sliding motion of the polymer chains surrounding the CNTs [5, 6]. The hysteretic behavior of CNT/polymer nanocomposites becomes increasingly more significant as more interfacial surface area becomes available for the frictional slippage. The interfacial surface area increases with the CNTs content and the CNTs aspect ratio as observed in [7, 8]. Such hysteretic phenomenon combined with the linear viscous damping of the polymer system may yield a softening nonlinearity in the material dynamic response, as also shown for a variety of physical/mechanical systems [9] or structures with initial curvature [10]. To this end, the nonlinear dynamic response of the nanocomposite cantilevers is here experimentally investigated and compared with the response of a neat polymer beam. The nanocomposite beams are made of the thermoplastic polybutylene terephthalate polymer (PBT) with the addition of single-walled CNTs. The experimental frequency-response curves are compared with those numerically obtained by implementing a path following algorithm. The nonlinear Euler–Bernoulli beam theory and the Galerkin discretization method are employed to derive the equation of motion of the cantilevers exhibiting a hardening nonlinearity. In addition, the nanostructural stick-slip energy dissipation shows its effects through a softening-type nonlinearity at low oscillation amplitudes followed by a hardening response at higher amplitudes. To model such response, observed here for the first time, a modification of the reduced-order equation of motion is sought by adding a modal restoring force according to a modified Bouc–Wen model of hysteresis.

2 Experimental Dynamic Response

To investigate the dynamic response of polymer and nanocomposite beams, an experimental study is carried out subjecting the beams to a transverse harmonic base excitation. The frequency-response curves for the primary resonance of the lowest bending mode are experimentally acquired.

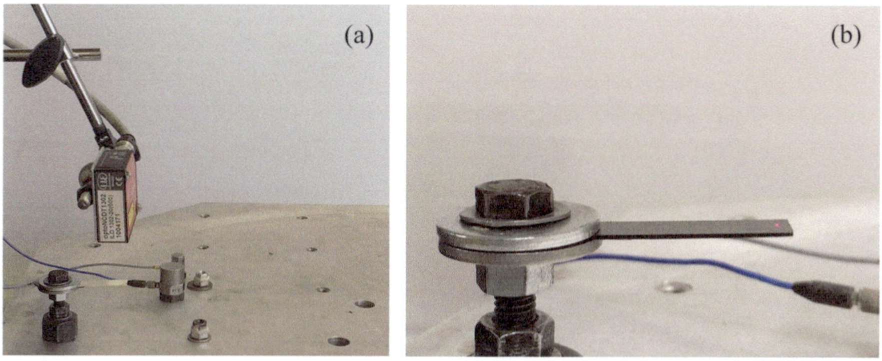

Fig. 1 (**a**) Experimental setup with the mounted polymer cantilever beam and (**b**) the nanocomposite cantilever beam

2.1 *Experimental Setup and Testing*

The adopted test sequences involve a frequency sweep around the first natural frequency of the cantilever beams at different base excitation amplitudes. The pure polymer and nanocomposite beams are manufactured according to the procedure described in [11] considering a 2% nanofiller weight fraction. Pictures of the samples and of the experimental setup are shown in Fig. 1. Both polymer and nanocomposite beams exhibit a uniform rectangular cross section with area $A = bh = 9.78 \times 0.79\,\text{mm}^2$ and length $L = 34.5\,\text{mm}$. The PBT polymer mass density is $\rho_m = 1{,}310\,\text{kg/m}^3$. The addition of 2 wt% CNTs with a mass density of $\rho_c = 1{,}750\,\text{kg/m}^2$ does not significantly affect the mass of the samples, therefore $\rho_m A$ is assumed as the mass per unit length for all beams. The beams are clamped to an electrodynamic shaker equipped with a head expander which can apply a harmonic excitation with a maximum acceleration amplitude of 10.3 g and a maximum frequency of 2 kHz.

The shaker excitation is controlled and monitored by means of two piezoelectric accelerometers (maximum acceleration of 50 g) placed on the shaker head expander. The response of the cantilevers is measured using a laser displacement sensor pointing at the tip of the beams and acquiring displacements within the range of ± 10 mm with a resolution of 0.02%. The accelerometer and laser signals are monitored in the time and frequency domains by the signal analyzer LabView.

The frequencies of the vibration modes of the cantilever beams are estimated through an experimental modal analysis in order to identify the frequency ranges of interest. The lowest natural frequency is found to be $f_m = 135.7\,\text{Hz}$ for the pure polymer beam, and $f_{nc} = 153.2\,\text{Hz}$ for the 2 wt% CNT/PBT nanocomposite beam, respectively. This step yields the identification of the materials elastic moduli. Backward and forward frequency sweeps are carried out in the frequency bandwidth $[120-150]$ Hz for the pure polymer, and $[140-170]$ Hz for the nanocomposite beams, respectively. The frequency-response curves are acquired for acceleration amplitudes set to $[1, 2, 3]g$ with a linear sweep rate of 0.025 Hz/s.

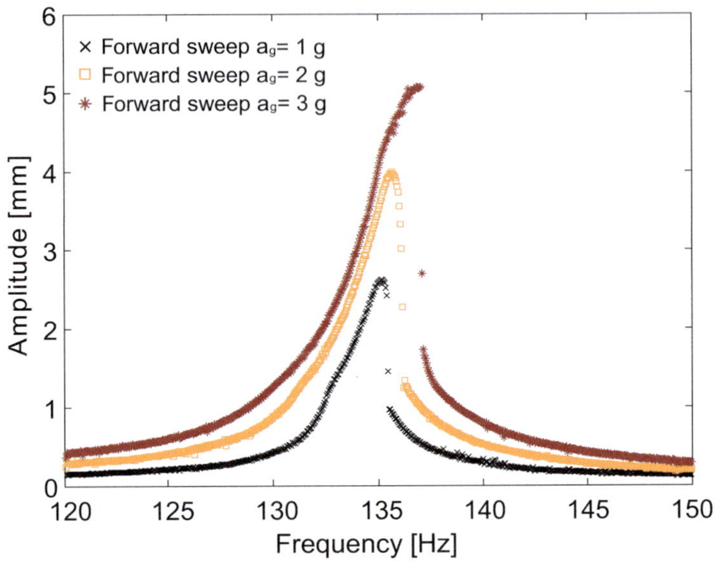

Fig. 2 Frequency-response curves for the PBT polymer cantilever beams experimentally obtained for different base excitation amplitudes

2.2 Experimental Results

The experimentally obtained frequency-response curves for the neat polymer and 2 wt% CNT/PBT nanocomposite beams are reported in Figs. 2 and 3, respectively. The nonlinear dynamic responses around the lowest natural frequency feature a positive stiffness variation when increasing the base excitation amplitude, thus giving rise to the expected hardening-type nonlinearity. Such emerging nonlinearity is exhibited by both the polymer and nanocomposite cantilevers. However, while the response of the neat polymer beam is of the hardening-type throughout the whole investigated range of excitation amplitudes, the observed flexural behavior of the nanocomposite beam suggests that when the beam oscillates in bending at low amplitudes, the loss of stiffness caused by the nanostructural interfacial sliding yields a softening feature while, at larger amplitudes, the nonlinear bending curvature effects prevail making the beam behave as a hardening-type oscillator. Such response demonstrates the onset of hysteresis in the material due to the stick-slip energy dissipation phenomenon at the CNT/polymer interfaces. The interfacial frictional sliding between the CNTs and the surrounding polymer chains collectively causes a macroscopic loss of stiffness.

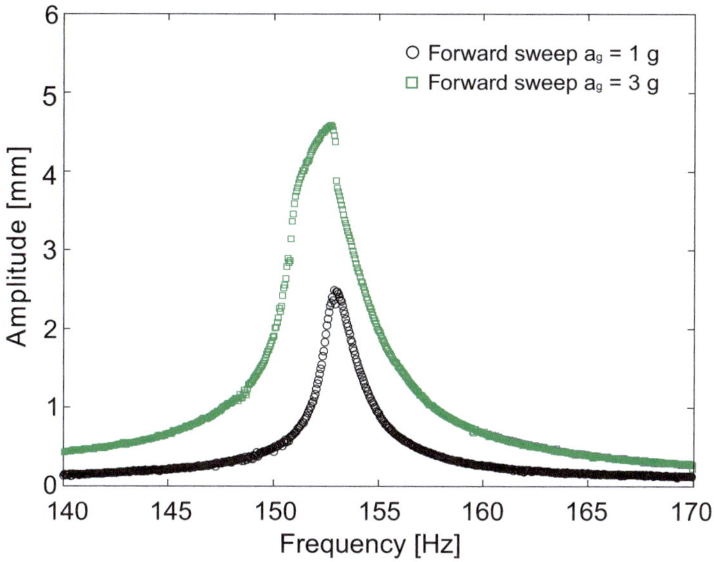

Fig. 3 Frequency-response curves for the 2 wt% CNT/PBT nanocomposite cantilever beams experimentally obtained for different base excitation amplitudes

3 Modeling the Nonlinear Dynamic Response

To describe the flexural motion of the neat polymer as well as the nanocomposite cantilever beams, the inextensible Euler–Bernoulli beam theory [10] is employed. By letting L denote the beam span, s the arclength along the beam centerline, ρ_m the mass density, J the moment of inertia, A the cross sectional area, E_m and E_{nc} the elastic moduli of the polymer and the nanocomposite materials, respectively, and ω_m the frequency of the polymer beam lowest bending mode, the nonlinear equation of motion expressed in terms of the transverse deflection $v(s, t)$ is cast in a suitable nondimensional form [10] by rescaling variables and parameters as follows:

$$s^* = \frac{s}{L}, \quad t^* = \omega_m t, \quad \omega_m = \omega_1^2 \sqrt{\frac{E_m J}{\rho_m A L^4}}, \quad v^* = \frac{v}{L},$$

$$c^* = \frac{cL^2}{\omega_1^2 \sqrt{E_m J \rho_m A}}, \quad k_E = \frac{E_{nc}}{E_m}, \quad f^* = f \frac{L^3}{E_m J \omega_1^4}$$

with $\omega_1 = 1.875$ being the first root of the characteristic equation associated with the boundary-value problem for free vibrations of a cantilever beam with unitary mass and bending stiffness, and $c = 2\xi \omega_m A \rho_m$ representing the viscous damping coefficient of the beam. For ease of notation, the stars will be dropped in the subsequent equations.

The cantilever is subject to a harmonic base motion y whose acceleration is $y_{tt} = Y\Omega^2 \cos \Omega t$ with frequency Ω near the frequency of the lowest bending mode. Such base excitation yields an equivalent force per unit reference length $f = -\rho_m A y_{tt}$ which, in nondimensional form, reads $f = Y\Omega^2 \cos \Omega t$ where Y and Ω are rescaled by the beam length and the frequency ω_m, respectively.

The resulting nondimensional equation of motion for the pure polymer cantilever is given by the following integral-partial differential equation truncated to third-order terms [10]:

$$
\begin{aligned}
& v_{tt} + cv_t + v_s \int_0^s (v_{\xi t}^2 + v_\xi v_{\xi tt})\, d\xi - \tfrac{1}{2} v_s^2 v_{tt} \\
& - v_{ss} \int_s^1 \Big(- v_{tt} v_\xi + \int_0^\xi (v_{\zeta t}^2 + v_\zeta v_{\zeta tt})\, d\zeta \Big)\, d\xi \\
& + k_E \big(v_{ss} + \tfrac{1}{2} v_s^2 v_{ss} \big)_{ss} - k_E v_{ss} \int_s^1 v_{\xi\xi} v_{\xi\xi\xi}\, d\xi = -y_{tt}
\end{aligned}
\tag{1}
$$

with boundary conditions given by

$$
v(0,t) = 0, \;\; v_s(0,t) = 0, \;\; v_{ss}(1,t) = 0, \;\; v_{sss}(1,t) = 0. \tag{2}
$$

By employing the Galerkin discretization method, a reduced-order model of the beam is derived. The deflection is expressed in the form $v(s,t) = \sum_{j=1}^{n} q_j(t) \Phi_j(s)$ where $\Phi_j(s)$ are the mode shapes and $q_j(t)$ are the modal coordinates.

Retaining one mode only, for example, the jth mode, and multiplying Eq. (1) by $\Phi_j(s)$ and integrating over the span of the beam yield the following reduced-order equation of motion:

$$
\ddot{q}_j + m_{2,j} \dot{q}_j^2 q_j + m_{3,j} \ddot{q}_j q_j^2 + c_j \dot{q}_j + k_j q_j + k_{3,j} q_j^3 = p_j \cos \Omega t, \tag{3}
$$

where the parameters k_j and $k_{3,j}$ denote the nondimensional jth linear and nonlinear modal stiffness, respectively; $m_{2,j}$ and $m_{3,j}$ are the coefficients of the nonlinear inertia terms; c_j is the modal damping coefficient, and p_j is the equivalent modal force magnitude.

When the beam is made of nanocomposite material, the reduced-order model looks the same as Eq. (3) except for the replacement of the linear modal restoring force $k_j q_j$ with the hysteretic restoring force f_j given in rate form [12] by

$$
\dot{f}_j = k_j \left\{ 1 - \left(1 - \frac{k_{m,j}}{k_j} \right) [\beta + \gamma \, \mathrm{sign}((h_j - k_{m,j} q_j) \dot{q}_j)] \left(\frac{h_j - k_{m,j} q_j}{f_{o,j}} \right)^n \right\} \dot{q}_j, \tag{4}
$$

where k_j is the linear modal stiffness, $k_{m,j}$ is the post-elastic stiffness, $f_{o,j}$ controls the onset of the CNT/polymer stick-slip in the jth mode, highlighting the transition from the elastic to the post-elastic behavior. The nondimensional parameters γ, β, and n regulate the shape and smoothness of the loading and unloading branches as in the Bouc–Wen model of hysteresis [7, 12]. The nondimensional coefficients are found by defining the geometric and constitutive parameters of the cantilever beams.

3.1 Frequency-Response Curves via Path Following

A pseudo-arclength path following scheme is implemented to construct the frequency-response curves given by Eqs. (3) and (4). From the experimental tests, the resulting elastic moduli estimated according to $E = (\rho_m A L^4 \omega^2)/(\omega_1^4 J_m)$ for the polymer and the 2 wt% CNT/PBT nanocomposites beams are found to be $E_m = 2{,}100$ MPa, and $E_{nc} = 2{,}675$ MPa, respectively. Accordingly, the nondimensional coefficients of the Galerkin-reduced model are evaluated. The coefficients $m_{2,1}$, $m_{3,1}$, and p_1 do not vary with the material of the cantilevers since they are independent of the beams constitutive features and are estimated to be $m_{2,1} = 4.60$, $m_{3,1} = 2.44$, $p_1 = 12.36$. The nondimensional coefficients c_1, $k_{3,1}$, and $k_{1,1}$ are found to vary for the investigated materials, being k_E=1 for the neat polymer beam, and $k_E = 1.2738$ for the 2 wt% CNT/PBT beam. Therefore, $c_1 = 2\xi$, $k_{3,1} = 13.81$, and $k_1 = 12.36$ for the polymer beam; $c_1 = 2.26\xi$, $k_{3,1} = 17.59$, and $k_1 = 15.75$ for the 2 wt% CNT/PBT beam. The equivalent damping ratio ξ is experimentally found to range 1.1–2.6%. Note that the nonlinear stiffness coefficient is positive and larger than the linear stiffness.

The ad hoc implemented path following algorithm [2] allows to trace the branch of stable and unstable periodic solutions for the investigated nanocomposite cantilevers. By restricting the analysis to the lowest mode (i.e., $j = 1$), and setting $k_{3,1}/k_1 = 0.8$, $\gamma = 0.5$, $f_{o,1} = 0.1$, the obtained frequency-response curves in Fig. 4 are found to be in qualitative agreement with the experimentally obtained curves showing the peculiar softening behavior at low amplitudes and the hardening behavior at larger amplitudes. As shown in Fig. 4 (right), a significant increase of the cubic coefficient with respect to the linear stiffness turns the nanocomposite beam behavior into hardening also at low amplitudes. However, $k_{3,1}$ is associated with the nonlinear bending effects and cannot be directly controlled while the degree of hysteresis is controlled by the CNTs content.

4 Conclusions

The nonlinear dynamic response of nanocomposite cantilevers to a base harmonic excitation was studied experimentally. The obtained frequency-response curves clearly show an interesting nonlinear behavior, namely, softening at low amplitudes and hardening at larger amplitudes. These nonlinear trends are the result of two concurrent nonlinearities of the opposite sign, the stiffness-degrading CNT/polymer

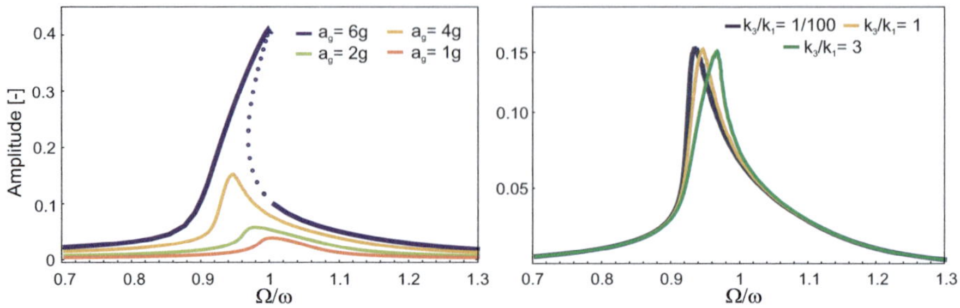

Fig. 4 Theoretically obtained nondimensional frequency-response curves for 2 wt% CNT/PBT nanocomposite cantilever beam for different base acceleration amplitudes a_g (left), and nonlinear stiffness coefficients k_3 (right). The damping ratio is set to $\xi = 2\%$

stick-slip and the stiffness-augmenting nonlinear bending curvature effects. The influence of the stick-slip energy dissipation on the nonlinear dynamic nanocomposites behavior is highlighted and clearly identified in the experimental results. Therefore, a model of the restoring force mimicking the Bouc–Wen model of hysteresis was introduced as an additional internal force in the equation of motion. The modification shows to be a promising strategy to describe the dynamic response of nanocomposite materials. The nonlinear flexural behavior was described by the nonlinear, inextensible Euler–Bernoulli beam theory treated by the Galerkin method. The predicted frequency-response curves are in good agreement with the experimental results. Ongoing efforts are targeted towards the identification of the nanocomposite material properties using the generated dynamic data and a genetic-type system identification method.

Acknowledgements The research leading to these results was supported by the European Research Council under the EU Seventh Framework Program (FP/2007-2013)/ERC Grant Agreement n. 308261 and by the European Office of Aerospace Research and Development/Air Force Office of Scientific Research Grant (Grant N. FA9550-141-0082 DEF).

References

1. Li, X., Song, M., Yang, J., Kitipornchai, S.: Primary and secondary resonances of functionally graded graphene platelet-reinforced nanocomposite beams. Nonlinear Dyn. **95**(3), 1807–1826 (2019)
2. Cetraro, M., Lacarbonara, W., Formica, G.: Nonlinear dynamic response of carbon nanotube nanocomposite microbeams. J. Comput. Nonlinear Dyn. **12**(3), 031007 (2017)
3. Rokni, H., Milani, A.S., Seethaler, R.J.: Size-dependent vibration behavior of functionally graded CNT-reinforced polymer microcantilevers: modeling and optimization. Eur. J. Mech. A. Solids **49**, 26–34 (2015)
4. Yas, M.H., Heshmati, M.: Dynamic analysis of functionally graded nanocomposite beams reinforced by randomly oriented carbon nanotube under the action of moving load. Appl. Math. Model. **36**(4), 1371–1394 (2012)

5. Koratkar, N.A., Suhr, J., Joshi, A., Kane, R.S., Schadler, L.S., Ajayan, P.M., Bartolucci, S.: Characterizing energy dissipation in single-walled carbon nanotube polycarbonate composites. Appl. Phys. Lett. **87**(6), 063102 (2005)
6. Suhr, J., Koratkar, N.A.: Energy dissipation in carbon nanotube composites: a review. J. Mater. Sci. **43**(13), 4370–4382 (2008)
7. Taló, M., Lacarbonara, W., Formica, G., Lanzara, G.: Hysteresis identification of carbon nanotube composite beams. In: ASME 2018 International Design Engineering Technical Conferences and Computers and Information in Engineering Conference, p. V004T08A021. American Society of Mechanical Engineers, New York (2018)
8. Lacarbonara, W., Taló, M., Carboni, B., Lanzara, G.: Tailoring of hysteresis across different material scales. In: Recent Trends in Applied Nonlinear Mechanics and Physics, pp. 227–250. Springer, Cham (2018)
9. Carboni, B., Lacarbonara, W., Brewick, P.T., Masri, S.F.: Dynamical response identification of a class of nonlinear hysteretic systems. J. Intell. Mater. Syst. Struct. **29**(13), 2795–2810 (2018). https://doi.org/10.1177/1045389X18778792
10. Lacarbonara, W.: Nonlinear Structural Mechanics: Theory, Dynamical Phenomena and Modeling. Springer, New York (2013)
11. Taló, M., Krause, B., Pionteck, J., Lanzara, G., Lacarbonara, W.: An updated micromechanical model based on morphological characterization of carbon nanotube nanocomposites. Compos. Part B **115**(Supplement C), 70–78 (2017)
12. Formica, G., Taló, M., Lanzara, G., Lacarbonara, W.: Parametric identification of carbon nanotube nanocomposites constitutive response. J. Appl. Mech. **86**(4), 041007 (2019). https://doi.org/10.1115/1.4042137

A Numerical Strategy for Multistable Nanocomposite Shells

Leonardo Leonetti, Giovanni Formica, Domenico Magisano, Michela Talò, Giovanni Garcea, and Walter Lacarbonara

Abstract This work presents an efficient continuation strategy based on the Riks method to describe the stable and of unstable branches of the response of carbon nanotubes (CNT)/polymer nanocomposite shells. Exploiting the superior properties of this class of nanostructured materials in the context of elastic instabilities and multistability has the potential to pave the way towards a variety of novel smart engineering applications. The equilibrium paths and the static bifurcations of CNT nanocomposites are numerically investigated highlighting the effects of material parameters such as the orientation and weight fraction of high aspect ratio CNTs integrated in a thermoplastic polymer.

Keywords Carbon nanotube nanocomposite · Multistability · Shells · Finite element method · Path following · Riks incremental method

1 Introduction

Over the last decades, elastic instabilities have no longer been regarded necessarily as a failure but as an additional functionality for a variety of smart engineering applications. Multistable composite shell structures made of a polymer matrix and reinforcing fibers, such as carbon, or glass fibers to mention but a few, have recently drawn attention as lightweight adaptive structures ensuring multiple stable

L. Leonetti · D. Magisano · G. Garcea
Department of Informatics, Electronics and System Engineering, University of Calabria, Rende, Italy

G. Formica
Department of Architecture, University of Roma Tre, Rome, Italy
e-mail: giovanni.formica@uniroma3.it

M. Talò (✉) · W. Lacarbonara
Department of Structural and Geotechnical Engineering, Sapienza University of Rome, Rome, Italy
e-mail: michela.talo@gmail.com

W. Lacarbonara et al. (eds.), *New Trends in Nonlinear Dynamics*,
https://doi.org/10.1007/978-3-030-34724-6_7

configurations when purposefully manufactured [1]. Composite laminates can be made to be bistable or even multistable by stacking plies of fiber-reinforced polymer in a designed nonsymmetric sequence.

Thanks to thermal effects induced during the manufacturing processes, multistable composites are able to snap from one stable configuration to another [2]. As a step forward in the field of morphing, multistable shell structures can be made of polymer matrices reinforced with a strengthening nanofiller, i.e., carbon nanotubes (CNT). CNT/polymer nanocomposites belong to the class of high-performance multifunctional materials. Nanocomposites have already been proved to be lightweight, resistant to corrosion and to high temperatures thanks to the use of engineering thermoplastic polymer matrices integrated with CNTs which makes them well reinforced, but also highly damped and electrically conductive [3–5]. Therefore, adding multistability to the desirable features of nanocomposite structures may represent a step forward towards the development of new smart structures. Thermoplastic polymer nanocomposites in the shape of shallow shells can be obtained through a manufacturing process that involves cooling from a high temperature. Thermally induced stresses occur during cooling and, as a result of these residual stresses combined with material imperfections, the initially flat laminates turn out to become curved upon final cooling [6]. The cooled equilibrium shape represents the shape which minimizes the potential energy of the nanocomposite shell. In addition, the shell may also have other stable configurations, thus showing bi-stability as in the case of the nanocomposite shell here investigated. As a result, if conveniently designed and manufactured, nanocomposite shells may exhibit nonlinear precritical load paths and undergo a snap-through or snap-back instability at their limit points [7].

Large efforts have been made to devise continuation strategies with the capability of finding limit (and more general bifurcation) points as well as tracing the equilibrium paths connecting these points. A relevant contribution was provided by the work of Riks in 1979 [8]. Based on the Newton method, the Riks incremental method introduces the arclength along the equilibrium path as a path following parameter. The present work aims to describe the equilibrium paths of CNT/polymer nanocomposite shallow shells. The nanocomposites are made of the high-performance thermoplastic polybutylene terephthalate (PBT) polymer and high aspect ratio, single walled carbon nanotubes.

The influence of the CNT orientation and CNT weight fraction on the shell bi-stability is investigated. In order to obtain the equilibrium paths describing both stable and unstable equilibrium states, the adopted numerical strategy implements a method based on the Riks scheme [8] thus allowing to trace the pre- and post-critical equilibrium paths. Such a method has been validated in several previous works and efficiently employed in nonlinear shell problems (see, e. g., [9, 10]). In addition, for the finite element implementation, a non standard formulation is here adopted, namely, the mixed solid-shell finite element, first proposed by Sze et al. in 1993 [11], and later developed and utilized in composite shells in [12, 13].

2 Constitutive Nanocomposites Behavior

The constitutive elastic response of the CNT/PBT nanocomposites is described through an updated micromechanical model based on the Eshleby–Mori–Tanaka approach and presented in [14]. By denoting the constitutive elastic tensors of the polymer matrix and the CNT nanofiller as $\mathbf{L}_m$ and $\mathbf{L}_c$, respectively, the nanocomposite equivalent elastic tensor is given by the expression

$$\mathbf{L} = \mathbf{L}_m + \phi_c \langle (\mathbf{L}_c - \mathbf{L}_m) : \mathbf{A}_c \rangle (\phi_m \mathbf{I} + \phi_c \langle \mathbf{A}_c \rangle)^{-1}, \tag{1}$$

where ϕ_m and ϕ_c are the matrix and CNT volume fractions, respectively, $\mathbf{I}$ is the identity tensor, and $\mathbf{A}_c$ is the dilute mechanical strain concentration tensor defined as $\mathbf{A}_c = \left[\mathbf{I} + \mathbf{S} : (\mathbf{L}_m)^{-1} : (\mathbf{L}_c - \mathbf{L}_m)\right]^{-1}$, in which $\mathbf{S}$ denotes the Eshelby tensor, and accounts for the shape, geometry, and variability of the CNTs length within the same nanocomposite [14]. To introduce the CNT orientation, which may be random or aligned along a preferential direction, the terms enclosed by angle brackets indicate the tensor transformation from the local nanofiller coordinates to the global coordinates system ($\langle \mathbf{A}_c \rangle = \mathbf{T}^\top \mathbf{A}_c \mathbf{T}$ accounting for the relative rotation about a prescribed coordinate axis in case of CNTs aligned along a given direction) as well as the additional averaging operation over all orientations in the case of randomly dispersed CNTs. The rotation tensor $\mathbf{T} = \mathbf{T}[\varphi, \beta, \gamma]$ represents the rotation tensor provided as a function of the Euler angles $\{\varphi, \beta, \gamma\}$. The resulting equivalent elastic tensor of the nanocomposite material proves to have transversely isotropic symmetry (with the isotropy plane being orthogonal to the longitudinal axis of the nanotubes), as also validated by molecular models [15].

3 Numerical Strategy

We consider a solid finite element and denote with $\boldsymbol{\zeta} = \{\xi, \eta, \zeta\}$ the convective coordinates as the natural coordinates for the finite element interpolation. The initial configuration is assumed as the reference configuration, described by $\mathbf{X}[\boldsymbol{\zeta}] \equiv \{X[\boldsymbol{\zeta}], Y[\boldsymbol{\zeta}], Z[\boldsymbol{\zeta}]\}$, while $\mathbf{x}[\boldsymbol{\zeta}]$ denotes the position in the current configuration. Let $\mathbf{d}[\zeta]$ denote the displacement field so that $\mathbf{x}[\boldsymbol{\zeta}] = \mathbf{X}[\boldsymbol{\zeta}] + \mathbf{d}[\boldsymbol{\zeta}]$. The position vector of a point inside the element and its displacement are interpolated within a trilinear 8-node hexahedron as follows:

$$\mathbf{X}[\boldsymbol{\zeta}] = \mathbf{N}_d[\boldsymbol{\zeta}]\mathbf{X}_e, \quad \mathbf{d}[\boldsymbol{\zeta}] = \mathbf{N}_d[\boldsymbol{\zeta}]\mathbf{d}_e \tag{2}$$

where $\mathbf{d}_e$ and $\mathbf{X}_e$ are vectors collecting the element nodal displacements and coordinates, respectively, and $\mathbf{N}_d[\boldsymbol{\zeta}]$ represents the matrix collecting the trilinear interpolation functions. Adopting Voigt notation, the Green-Lagrange covariant strain components are collected in the vector $\bar{\boldsymbol{\varepsilon}} \equiv [\bar{\varepsilon}_{\xi\xi}, \bar{\varepsilon}_{\eta\eta}, 2\bar{\varepsilon}_{\xi\eta}, \bar{\varepsilon}_{\zeta\zeta}, 2\bar{\varepsilon}_{\eta\zeta}, 2\bar{\varepsilon}_{\xi\zeta}]^\top$, which turns out to be

$$\bar{\boldsymbol{\varepsilon}} = \left(\mathcal{L}[\boldsymbol{\zeta}] + \frac{1}{2}\mathcal{Q}[\boldsymbol{\zeta}, \mathbf{d}_e]\right)\mathbf{d}_e, \tag{3}$$

where $\mathcal{L}[\boldsymbol{\zeta}] \equiv \mathcal{Q}[\boldsymbol{\zeta}, \mathbf{X}_e]$ and $\mathcal{Q}$ are the linear and quadratic part of the Green-Lagrange strain measure, respectively, expressed in terms of the finite element interpolation. In order to avoid shear and trapezoidal locking phenomena, both natural transverse normal strain and shear strain are actually defined by an assumed natural strain (ANS) formulation. By assuming here and henceforth that the Z-axis and the X-Y-plane are parallel to the ζ-axis and mid-surface of the shell, respectively, the covariant strains can be linearized with respect to ζ in the following form:

$$\bar{\boldsymbol{\varepsilon}} \approx \begin{bmatrix} \bar{\mathbf{e}}[\xi,\eta] + \zeta\ \bar{\boldsymbol{\chi}}[\xi,\eta] \\ \bar{\varepsilon}_{\zeta\zeta}[\xi,\eta] \\ \bar{\boldsymbol{\gamma}}[\xi,\eta] \end{bmatrix} \quad \text{with} \quad \begin{aligned} \bar{\mathbf{e}}[\xi,\eta] &= [\bar{\varepsilon}_{\xi\xi}\ \bar{\varepsilon}_{\eta\eta}\ 2\bar{\varepsilon}_{\xi\eta}] \\ \bar{\boldsymbol{\chi}}[\xi,\eta] &= [\bar{\varepsilon}_{\xi\xi},_{\zeta}\ \bar{\varepsilon}_{\eta\eta},_{\zeta}\ 2\bar{\varepsilon}_{\xi\eta},_{\zeta}] \\ \bar{\boldsymbol{\gamma}}[\xi,\eta] &= [2\bar{\varepsilon}_{\eta\zeta}\ 2\bar{\varepsilon}_{\xi\zeta}]. \end{aligned} \tag{4}$$

Denoting with $\bar{\boldsymbol{\sigma}} \equiv [\bar{\sigma}_{\xi\xi}, \bar{\sigma}_{\eta\eta}, 2\bar{\sigma}_{\xi\eta}, \bar{\sigma}_{\zeta\zeta}, 2\bar{\sigma}_{\eta\zeta}, 2\bar{\sigma}_{\xi\zeta}]^{\top}$ the vector collecting the contravariant stress components, from the stress-strain work we define the quantities conjugate with the above strain components, i.e.:

$$\mathcal{W} = \int_V \bar{\boldsymbol{\varepsilon}}^{\top}\bar{\boldsymbol{\sigma}}\,dV = \int_{\Omega}\left(\bar{\mathcal{N}}^{\top}\bar{\mathbf{e}} + \bar{\mathcal{M}}^{\top}\bar{\boldsymbol{\chi}} + \bar{s}_{\zeta\zeta}\bar{\varepsilon}_{\zeta\zeta} + \bar{\mathcal{T}}^{\top}\bar{\boldsymbol{\gamma}}\right),$$

where we used $\int_{\Omega}(\cdots) = 2\int_{-1}^{1}\int_{-1}^{1}(\cdots)\det(\mathbf{J})d\xi d\eta$, $\mathbf{J}$ being the Jacobian matrix $\mathbf{J}[\xi,\eta] = \left[\mathbf{X},_{\xi}\ \mathbf{X},_{\eta}\ \mathbf{X},_{\zeta}\right]$, that is assumed to be constant with respect to ζ. The evaluation of the generalized constitutive relations stems out from the integration of the stresses across the thickness direction to obtain the generalized stress resultants. The properties of the elastic nanocomposite material (either isotropic or transversely isotropic) are given in a local reference frame $\{\boldsymbol{i}_1, \boldsymbol{i}_2, \boldsymbol{i}_3\}$, with $\boldsymbol{i}_1$ collinear with the nanofiller orientation defined by a rotation angle θ around $\boldsymbol{i}_3$ parallel to the Z axis. The material stiffness matrix in the Cartesian element reference frame can be evaluated through a rotation tensor:

$$\boldsymbol{\sigma} = \mathbf{L}\boldsymbol{\varepsilon} \quad \text{with} \quad \mathbf{L} = \mathbf{R}[\theta]^{\top}\mathbf{L}'\mathbf{R}[\theta] = \begin{bmatrix} \mathbf{L}_p & \mathbf{L}_{pz} & \mathbf{L}_{pt} \\ \mathbf{L}_{pz}^{\top} & L_z, & \mathbf{L}_{zt} \\ \mathbf{L}_{pt}^{\top} & \mathbf{L}_{zt}^{\top} & \mathbf{L}_t \end{bmatrix}, \tag{5}$$

where

$$\mathbf{L}' = \begin{bmatrix} \mathbf{L}'_p & \mathbf{L}'_{pz} & \mathbf{L}'_{pt} \\ \mathbf{L}'^{T}_{pz} & L'_z & \mathbf{L}'_{zt} \\ \mathbf{L}'^{T}_{pt} & \mathbf{L}'^{T}_{zt} & \mathbf{L}'_t \end{bmatrix} \quad \mathbf{L}'_p \equiv \begin{bmatrix} L'_{1111} & L'_{1122} & L'_{1112} \\ L'_{1122} & L'_{2222} & L'_{2212} \\ L'_{1112} & L'_{2212} & L'_{1212} \end{bmatrix} \quad \mathbf{L}'_{pz} \equiv \begin{bmatrix} L'_{1133} \\ L'_{2233} \\ L'_{1233} \end{bmatrix}$$

$$\mathbf{L}'_{pt} \equiv \begin{bmatrix} L'_{1123} & L'_{1113} \\ L'_{2223} & L'_{2213} \\ L'_{1223} & L'_{1213} \end{bmatrix} \quad \mathbf{L}'_{zt} \equiv \left[L'_{3323}\ L'_{3313}\right] \quad \mathbf{L}'_t \equiv \begin{bmatrix} L'_{2323} & L'_{2313} \\ L'_{1323} & L'_{1313} \end{bmatrix} \quad L'_z = L'_{3333}.$$

The generalized stress resultants are obtained according to the following expression:

$$\begin{bmatrix} \mathcal{N} \\ s_{\zeta\zeta} \\ \mathcal{M} \\ \mathcal{T} \end{bmatrix} = \frac{1}{2}\int_{-1}^{1} \begin{bmatrix} \mathbf{L}_p & \mathbf{L}_{pz} & \zeta\mathbf{L}_p & \mathbf{L}_{pt} \\ \mathbf{L}_{pz}^{\top} & \mathbf{L}_{zz}^{\top} & \zeta\mathbf{L}_{zp} & \mathbf{L}_{zt} \\ \zeta\mathbf{L}_p^{\top} & \zeta\mathbf{L}_{zp}^{\top} & \zeta^2\mathbf{L}_p & \zeta\mathbf{L}_{pt} \\ \mathbf{L}_{pt}^{\top} & \mathbf{L}_{zt}^{\top} & \zeta\mathbf{L}_{pt}^{\top} & \mathbf{L}_t \end{bmatrix} \begin{bmatrix} \mathbf{e} \\ \varepsilon_{zz} \\ \chi \\ \gamma \end{bmatrix} d\zeta.$$

4 Numerical Results

As case study, stable and unstable solutions of a 2 wt% CNT/PBT nanocomposite disk that has the shape of a shallow dome are investigated. The assumed CNT weight fraction wt was experimentally proven to be sufficiently high to provide a reinforcement effect and at the same time sufficiently low to guarantee a good CNT dispersion degree [14]. Based on experimental data, the CNTs length is assumed to vary in the range [0.09–2.54] μm, while the Young moduli and Poisson ratios employed to define the elastic tensors for the PBT matrix and the CNTs are $E_m = 2.47$ GPa, $E_c = 850$ GPa, $\nu_m = 0.36$, and $\nu_c = 0.1$, respectively. The carbon nanotubes are assumed to be aligned in the shell plane (axes $\{x_1, x_2\}$) along direction $\varphi = 45°$, which is a counterclockwise angle with respect to x_1 (see Fig. 1). The nanocomposite shell equilibrium path is obtained according to a continuation strategy [7, 8]. The equilibrium path, depicted in Fig. 2, exhibits a loss of stability when reaching the first limit point denoted by (b) (Fig. 2 left). Before such limit point the shell is stable and occupies equilibrium states as those shown at point (a) of Fig. 2 (left). By increasing the adimensional load, the limit point is approached when the effective stiffness vanishes (one of the eigenvalues of the tangent stiffness coalesces to zero and, subsequently, becomes negative). At such limit point the structure suffers a snap-through instability associated with a sudden jump to a far-away equilibrium state. Hence, the bi-stable shell undergoes a large deformation and reaches another stable configuration indicated as point (c). Once the shell reaches the new stable configuration, if the adimensional load is decreased, the nanocomposite shell reaches the deformed state shown at point (d), which denotes the second limit point (Fig. 2 right). The structure suffers a reverse snap-through, thus jumping to the previous stable state. The intermediate path delimited by the two limit points represents the unstable branch denoted by the dashed line.

Thus, the adopted computational strategy allows to describe the unstable equilibrium states by tracing the branch where the negative geometric stiffness overcomes the elastic stiffness [7]. As a result, between the two red and blue dashed lines, defining the two limit loads, the nanocomposite shell may occupy one of the two stable configurations separated by the unstable equilibrium.

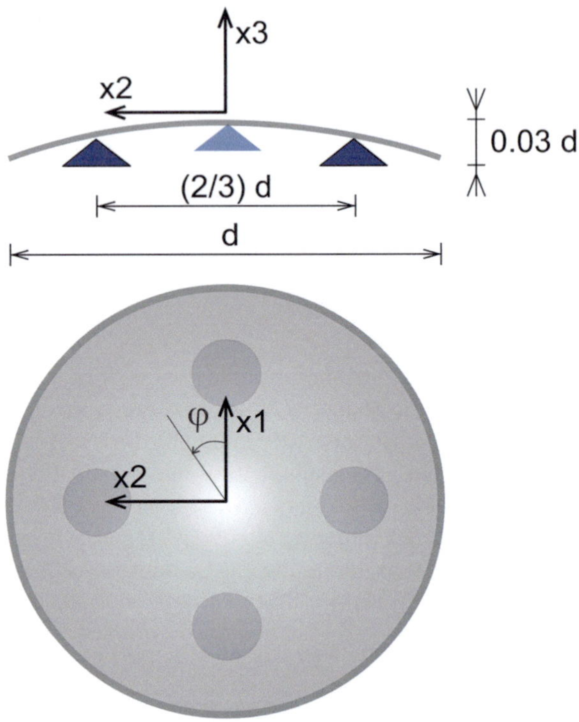

Fig. 1 Reference configuration of a shallow nanocomposite shell having a diameter $d = 60$ mm, and the carbon nanotubes aligned along the direction rotated by φ in the shell plane described by axes $\{x_1, x_2\}$. The shell is constrained by four supports described by the four circles

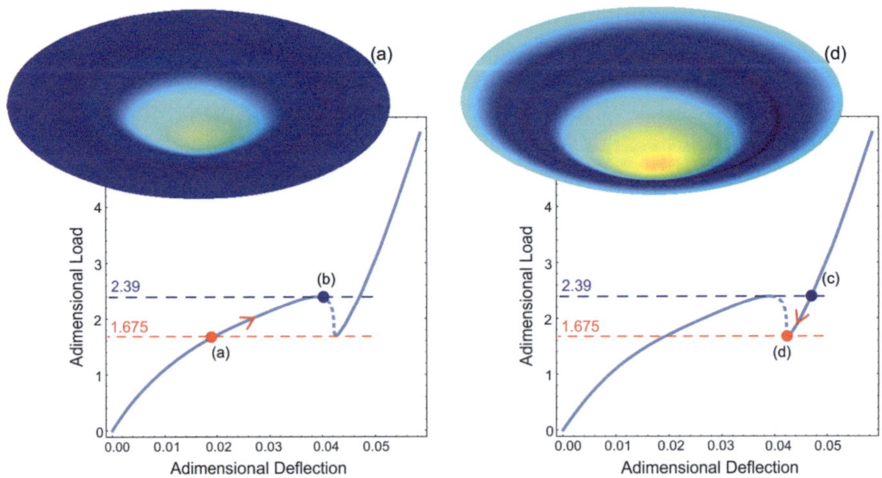

Fig. 2 Different deformed shapes of the 2% wt CNT/PBT shell for the same adimensional load of 1.675. Comparison between two configurations of the equilibrium path, corresponding to point (a) (left), and to point (d) (right), respectively

4.1 Sensitivity to CNT Weight Fraction and Orientation

The influence of the CNT weight fraction on the equilibrium path of the nanocomposite shell is investigated by varying the nanotubes content from 0.5 wt% to 5 wt%. Figure 3 shows the remarkable effect of the CNT weight fraction on the bi-stability of the nanocomposite. As expected, the limit load enabling the snap-through mechanism of the bi-stable structure grows with the CNT weight fraction and the snapping from one stable state to the other involves larger deflections for larger CNT contents. In addition, the effects of the CNTs alignment, denoted by the angle φ, on the equilibrium path of the nanocomposite shell are investigated together with the effect of random orientations of the CNTs in the shell plane (2D random) or in all directions (3D). Figure 4 shows the equilibrium path for the 2 wt% CNT/PBT nanocomposite in four significant cases. If the nanotubes are aligned along axis x_1 or for symmetry reasons, along axis x_2, i.e., $\varphi = 0°$ or $\varphi = 90°$, the multistability of the nanocomposite shell vanishes and the structure becomes mono-stable. On the other hand, a 2D and even more a 3D random orientation of the CNTs make the bi-stability more relevant. Such behavior may be due to the asymmetry (with respect to the mid-plane of the shell) in the spatial arrangement of the nanofiller randomly dispersed in the hosting polymer matrix.

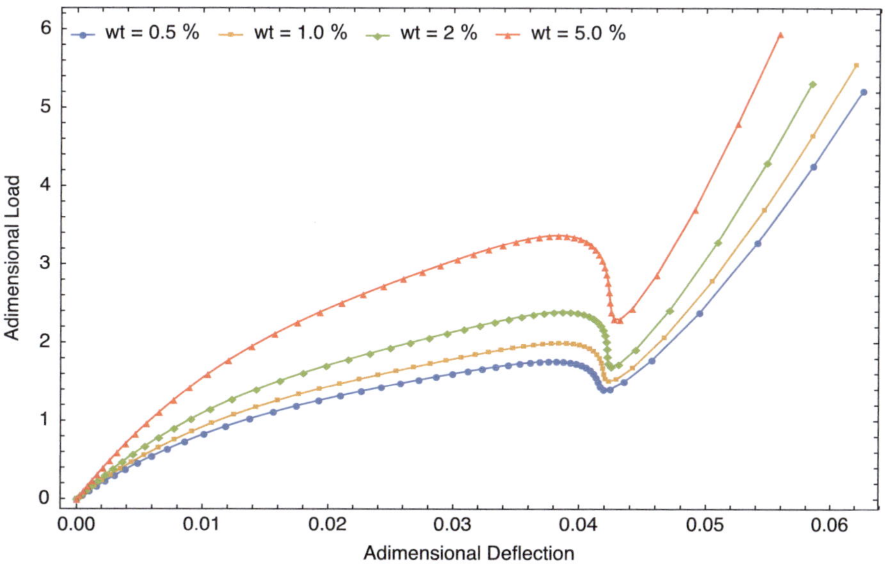

Fig. 3 Equilibrium paths for nanocomposite shells having CNTs aligned with $\varphi = 45°$ w.r.t. axis x_1 featuring various CNT weight fractions, wt = {0.5, 1.0, 2.0, 5.0}%

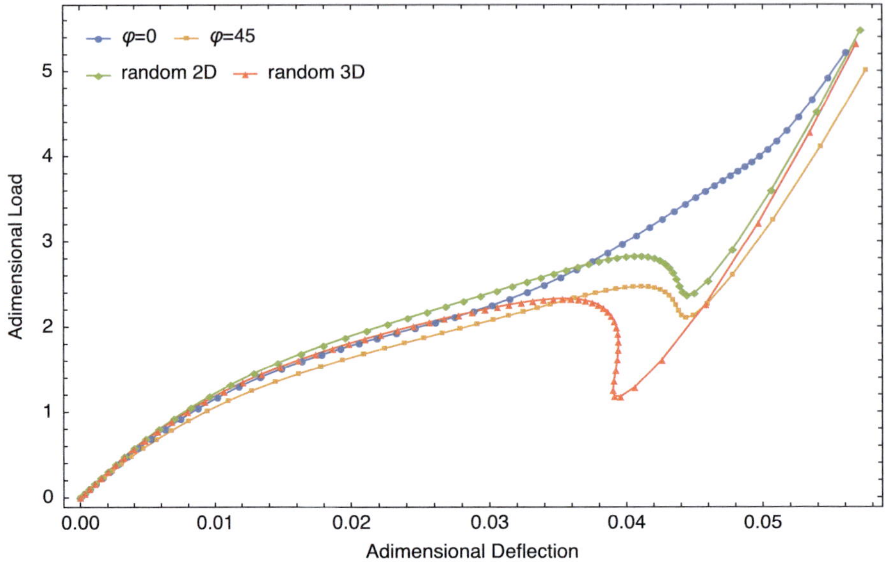

Fig. 4 Equilibrium paths for 2% wt CNT/PBT nanocomposite shells. Comparison between the most significant cases for the CNT orientation in the nanocomposite

5 Conclusions

Bistable nanocomposite shell structures are investigated and the material features found to be mostly influential on the onset of snap-through are identified. By tuning the CNT weight fraction and the CNT orientation in the hosting polymer matrix, significantly different equilibrium paths for the nanocomposite shells can be obtained. In particular, by increasing the CNT weight fraction and achieving a 3D random orientation of the CNTs, the deflection suffered by the nanocomposite shell to snap from one stable configuration to the other may be amplified. On the other hand, the alignment of the nanofiller along one of the in-plane main axes of the shell can cause the loss of bi-stability. The Riks method together with the use of mixed solid-shell finite elements enables the computation of the equilibrium paths. Changes of shape of the nanocomposite shell and both stable and unstable states are accurately described through numerical path following analysis. Therefore, multistability exhibited by such lightweight multifunctional nanocomposite structures—here efficiently predicted—turns out to be extremely promising in the perspective of new shape-adaptive applications, such as aerodynamic morphing applications.

Acknowledgements The research leading to these results was supported by the European Office of Aerospace Research and Development/Air Force Office of Scientific Research Grant (Grant N. FA9550-141-0082 DEF).

References

1. Groh, R.M.J., Pirrera, A.: Extreme mechanics in laminated shells: new insights. Extreme Mech. Lett. **23**, 17–23 (2018)
2. Hufenbach, W., Gude, M., Kroll, L.: Design of multistable composites for application in adaptive structures. Compos. Sci. Technol. **62**(16), 2201–2207 (2002)
3. Formica, G., Lacarbonara, W., Alessi, R.: Vibrations of carbon nanotube-reinforced composites. J. Sound Vib. **329**(10), 1875–1889 (2010)
4. Pandey, G., Thostenson, E.T.: Carbon nanotube-based multifunctional polymer nanocomposites. Polym. Rev. **52**(3), 355–416 (2012)
5. Formica, G., Taló, M., Lanzara, G., Lacarbonara, W.: Parametric identification of carbon nanotube nanocomposites constitutive response. J. Appl. Mech. **86**(4), 041007 (2019). https://doi.org/10.1115/1.4042137
6. Wagg, D., Bond, I., Weaver, P., Friswell, M. (eds.): Adaptive Structures: Engineering Applications. Wiley, Chichester (2007)
7. Lacarbonara, W.: Nonlinear Structural Mechanics: Theory, Dynamical Phenomena and Modeling. Springer, New York (2013)
8. Riks E.: An incremental approach to the solution of snapping and buckling problems. Int. J. Solids Struct. **15**(7), 529–551 (1979)
9. Garcea, G., Trunfio, G., Casciaro, R.: Mixed formulation and locking in path-following nonlinear analysis. Comput. Methods Appl. Mech. Eng. **165**(1–4), 247–272 (1998)
10. Magisano, D., Leonetti, L., Garcea, G.: Advantages of the mixed format in geometrically nonlinear analysis of beams and shells using solid finite elements. Int. J. Numer. Methods Eng. **109**(9), 1237–1262 (2017)
11. Sze, K., Ghali, A.: Hybrid hexahedral element for solids, plates, shells and beams by selective scaling. Int. J. Numer. Methods Eng. **36**(9), 1519–1540 (1993)
12. Magisano, D., Leonetti, L., Garcea, G.: Koiter asymptotic analysis of multilayered composite structures using mixed solid-shell finite elements. Compos. Struct. **154**, 296–308 (2016)
13. Magisano, D., Liang, K., Garcea, G., Leonetti, L., Ruess, M.: An efficient mixed variational reduced-order model formulation for nonlinear analyses of elastic shells. Int. J. Numer. Methods Eng. **113**(4), 634–655 (2018)
14. Taló, M., Krause, B., Pionteck, J., Lanzara, G., Lacarbonara, W.: An updated micromechanical model based on morphological characterization of carbon nanotube nanocomposites. Compos. B: Eng. **115**, 70–78 (2017)
15. Odegard, G.M., Gates, T.S., Wise, K.E., Park, C., Siochi, E.J.: Constitutive modeling of nanotube-reinforced polymer composites. Compos. Sci. Technol. **63**(11), 1671–1687 (2003)

Parametric Vibrations of Functionally Graded Sandwich Plates with Complex Forms

Kurpa Lidiya, Shmatko Tetyana, and Jan Awrejcewicz

Abstract Buckling behavior and parametric vibrations of sandwich plates with arbitrary forms and made of isotropic and functionally graded materials (FGM) are studied. Different types of lamination schemes were considered: a sandwich plate with FGM face sheets and isotropic (metal or ceramic) core and a plate with a FGM core and ceramics or metal on top and bottom face sheets. Effective material properties are computed according to Voigt's rule in thickness direction. To calculate mechanical characteristics for different types of lamination schemes, the analytical expressions were obtained. The formulation of the problem was carried out using the first-order shear deformation theory (FSDT) of the plate. A subcritical state of the plate was taken into account.

Keywords Parametric vibrations · Functionally graded sandwich plates · The R-functions theory

1 Introduction

Nowadays, functionally graded materials (FGM) are widely used in many fields of industry as heat-resistant thin-constructed elements. There are many published papers devoted to studying the dynamic and static behavior of FGM plates and shells as designed objects [1, 2] and others. The application of FGM can help reduce

K. Lidiya
Department of Applied Mathematics, National Technical University "KhPI", Kharkov, Ukraine
e-mail: kurpalidia@gmail.com

S. Tetyana (✉)
Department of Higher Mathematics, National Technical University "KhPI", Kharkov, Ukraine
e-mail: ktv_ua@yahoo.com

J. Awrejcewicz
Department of Automation, Biomechanics and Mechatronics, Lodz University of Technology, Lodz, Poland
e-mail: jan.awrejcewicz@p.lodz.pl

W. Lacarbonara et al. (eds.), *New Trends in Nonlinear Dynamics*,
https://doi.org/10.1007/978-3-030-34724-6_8

mechanically and thermally induced stresses caused by the material properties mismatch as well as improves the bonding strength in the case of coatings and laminated facings. This explains the growing interest in studying laminated composite and sandwich plates often used in modern engineering applications. Many theories and methods for the mathematic modeling and analysis of such elements have been proposed [3–5]. Among these methods, one of an important method is Ritz's method. It is known that one of the main difficulties related with the use of Ritz's method in case of the complex plate geometry is a choice of a basic system of functions that satisfy the boundary conditions. In this work, the R-functions theory [6] is used for solving this problem. Earlier, Ritz's variational method and the R-functions theory (RFM) have been effectively applied for investigation of vibration of layered plates and shallow shells [7, 8].

In this study, we first develop RFM to research of linear and parametric vibrations of FG sandwich plates using the first-order shear deformation theory (FSDT). The plate is composed of three layers, one or two of which are functionally graded in the thickness direction. Material properties of the FGM are calculated by power law. Pre-buckling state is taken into account. Proposed algorithm is applied to square plate with two cut outs. Initial nonlinear system with partial derivatives is reduced to Mathieu equation by method proposed authors earlier in [8].

2 Mathematical Problem

Let us consider a sandwich plate with a metal core and FGM facing (Type A) and a sandwich plate with a FGM core and the lower layer made of ceramic and upper layer made of metal (Type B). It is considered that the FGM layers are made of a mixture of ceramics and metals. The thickness of the layers from the bottom to the top $h^{(1)}$, $h^{(2)}$, $h^{(3)}$ can be varied. The general thickness h is a constant defined as a sum $h = h^{(1)} + h^{(2)} + h^{(3)}$. Let us assume that the plate is subjected to periodic in-plane load $p_N = p_{st} + p_{dyn} \cos \theta t$, where p_{st} is a static component, p_{dyn} is the amplitude of the periodic part, and θ is the frequency of the load. Note that all external forces are varied proportionally to the parameter λ. The material properties of the plate vary continuously and smoothly in the thickness direction. Effective material properties P_{eff}, like Young's modulus E, Poisson's ratio ν, and mass density ρ for FGM can be estimated by the following Voigt's law:

$$P_{\text{eff}}^{(r)} = (P_u - P_l)\, V_c^{(r)}(z) + P_l, \tag{1}$$

where P_u and P_l are corresponding properties of the upper and lower surfaces of the r-layer, respectively; $V_c^{(r)}(z)$ is the volume fraction of ceramics.

Parametric excitation of the plate subjected to periodic loads is investigated by the first-order shear deformation theory (FSDT) [9] taking into account a shear deformation. In this case, the governing differential equations of equilibrium for free vibration of a plate subjected to external in-plane loading can be expressed as

$$N_{11},_x + N_{12},_y = I_0 u,_{tt} + I_1 \psi_x,_{tt}, \quad N_{12},_x + N_{22},_y = I_0 v,_{tt} + I_1 \psi_y,_{tt}, \tag{2}$$

$$M_{11},_x + M_{12},_y - Q_x = I_2 \psi_x,_{tt} + I_1 u,_{tt}, \quad M_{12},_x + M_{22},_y - Q_y = I_2 \psi_y,_{tt} + I_1 v,_{tt}, \tag{3}$$

$$Q_x,_x + Q_y,_y + \left(N_{11}^0 w,_{xx} + 2N_{12}^0 w,_{xy} + N_{22}^0 w,_{yy} \right) = I_o w,_{tt}, \tag{4}$$

where I_0, I_1, I_2 are defined as follows

$$(I_0, I_1, I_2) = \sum_{r=1}^{3} \int_{z_r}^{z_{r+1}} \left(\rho^{(r)}\right) \left(1, z, z^2\right) dz,$$

where $\rho^{(r)}$ is the mass density of r-th layer, and the values $\{N\} = \left\{N_{11}^0, N_{22}^0, N_{12}^0\right\}^{\mathrm{T}}$ denote the force resultant in the pre-buckling state.

The in-plane force resultant vector $\{N\} = \{N_{11}, N_{22}, N_{12}\}^{\mathrm{T}}$, bending and twisting moments resultant vector $\{M\} = \{M_{11}, M_{22}, M_{12}\}^{\mathrm{T}}$, and transverse shear force resultant $\{Q\} = \{Q_x, Q_y\}^{\mathrm{T}}$ are calculated by integration along Oz-axes and can be recast to the following form:

$$\{N\} = [A]\{\varepsilon\} + [B]\{\chi\}, \quad \{M\} = [B]\{\varepsilon\} + [D]\{\chi\}, \tag{5}$$

where $[A]$, $[B]$, $[D]$ are square matrices of the third order with elements A_{ij}, B_{ij}, D_{ij}, $(i,j = 1, 2, 6)$ defined in the following way:

$$A_{ij} = \sum_{r=1}^{3} \int_{z_r}^{z_{r+1}} Q_{ij}^{(r)} dz, \quad B_{ij} = \sum_{r=1}^{3} \int_{z_r}^{z_{r+1}} Q_{ij}^{(r)} z dz, \quad D_{ij} = \sum_{r=1}^{3} \int_{z_r}^{z_{r+1}} Q_{ij}^{(r)} z^2 dz,$$

where $z_1 = -h/2, \quad z_2 = h_1, \quad z_3 = h_2, \quad z_4 = h/2$. Values $h_1, \ h_2$ are corresponding lower and upper borders of the middle layer. Expressions of integrand functions $Q_{ij}^{(r)}$ $(i, j = 1, 2, 6)$ have the following form

$$Q_{11}^{(r)} = Q_{22}^{(r)} = \frac{E^{(r)}}{1 - \left(\nu^{(r)}\right)^2}, \qquad Q_{12}^{(r)} = \frac{\nu^{(r)} E^{(r)}}{1 - \left(\nu^{(r)}\right)^2}, \qquad Q_{66}^{(r)} = \frac{E^{(r)}}{2\left(1 + \nu^{(r)}\right)},$$

where $E^{(r)}, \ \nu^{(r)}$ are defined by formulas (1).

Strain components $\{\varepsilon\} = \{\varepsilon_{11}, \varepsilon_{12}, \varepsilon_{22}\}^{\mathrm{T}}$, $\{\chi\} = \{\chi_{11}, \chi_{12}, \chi_{22}\}^{\mathrm{T}}$ at an arbitrary point of the plate are as follows:

$$\varepsilon_{11} = u,_x + \frac{1}{2} w,_x^2, \quad \varepsilon_{22} = v,_y + \frac{1}{2} w,_y^2, \quad \varepsilon_{12} = u,_y + v,_x + w,_x w,_y \tag{6}$$

$\varepsilon_{13}=w_{,x}+\psi_x,\ \ \varepsilon_{23}=w_{,y}+\psi_y,\ \ \chi_{11}=\psi_{x,x},\ \ \chi_{22}=\psi_{y,y},\ \ \chi_{12}=\psi_{x,y}+\psi_{y,x}.$

The transverse shear force resultants Q_x, Q_y are as follows:

$$Q_x = K_s^2 A_{33}\varepsilon_{13}, \quad Q_y = K_s^2 A_{33}\varepsilon_{23}, \tag{7}$$

where K_s^2 denotes the shear correction factor. In this chapter, we take $K_s^2 = 5/6$. Coefficients A_{ij}, B_{ij}, D_{ij} were obtained in an analytical form [8].

3 Method of Solution

Since the pre-buckling state can be inhomogeneous, first, we should determine the parameters $\{N\} = \left\{N_{11}^0, N_{22}^0, N_{12}^0\right\}^{\mathrm{T}}$. It is possible to prove that this problem can be reduced to a variational problem related to finding the minimum of the following functional:

$$\begin{aligned} I\left(u, v, w, \psi_x, \psi_y\right) = \tfrac{1}{2}\iint\limits_{\Omega} \Big(N_{11}^{(L)}\varepsilon_{11}^{(L)} + N_{22}^{(L)}\varepsilon_{22}^{(L)} + N_{12}^{(L)}\varepsilon_{12}^{(L)} + \\ + M_{11}^{(L)}\chi_{11} + M_{22}^{(L)}\chi_{22} + M_{12}^{(L)}\chi_{12} + Q_x\varepsilon_{13} + Q_y\varepsilon_{23}\Big) d\Omega + \int\limits_{\partial\Omega_1} N_n^{(L)}\left(ul + vm\right) d\Omega_1, \end{aligned} \tag{8}$$

where the terms with superscripts L correspond to the linear terms in formulas (5).

To find the buckling load the dynamical approach is applied. Then the problem is reduced to an equivalent variational problem of the minimization of the following functional

$$\begin{aligned} I\left(u, v, w, \psi_x, \psi_y\right) = \tfrac{1}{2}\iint\limits_{\Omega} \Big[N_{11}^{(L)}\varepsilon_{11}^{(L)} + N_{22}^{(L)}\varepsilon_{22}^{(L)} + N_{12}^{(L)}\varepsilon_{12}^{(L)} + \\ + M_{11}^{(L)}\chi_{11} + M_{22}^{(L)}\chi_{22} + M_{12}^{(L)}\chi_{12} + Q_x\varepsilon_{13} + Q_y\varepsilon_{23} + \\ + p_{\mathrm{st}}\left(N_{11}^0\left(w_{,x}\right)^2 + N_{22}^0\left(w_{,y}\right)^2 + N_{12}^0 w_{,x}w_{,y}\right)\Big] d\Omega - \\ - \tfrac{1}{2}\omega_L^2 \iint\limits_{\Omega}\left(I_0\left(u^2+v^2+w^2\right) + I_1\left(\psi_x^2+\psi_y^2\right)\right) d\Omega. \end{aligned} \tag{9}$$

The value of the parameter p_{st} increases when the natural frequency ω_{L} is a real number. The value of the buckling load N_{cr} is defined by the value of the parameter p_{st} corresponding to the smallest nonnegative value of the frequency. Minimization of the functionals (8 and 9) is performed using Ritz's method. The sequence of coordinate functions is constructed by the R-functions theory [6].

In order to solve the nonlinear vibration problem we develop the approach proposed in [7, 8]. As a result, we get ordinary differential equation, investigation of which is performed by Bolotin's approach. The equation has the following form:

$$\ddot{y}(t) + 2\varepsilon y(t) + \Omega_{\mathrm{L}}^2 \left(1 - 2k \cos \theta t\right) y(t) + y^2(t)\beta + y^3(t)\gamma = 0, \qquad (10)$$

where $2k = \frac{\alpha p_{\mathrm{dyn}}}{\Omega_{\mathrm{L}}^2}$, $\Omega_{\mathrm{L}}^2 = \omega_{\mathrm{L}}^2 - p_{\mathrm{st}}\alpha$ is the frequency of the plate compressed by the static load p_{st}, ω_{L} stands for the natural frequency of free vibration, ε is the damping ratio of the plate, and the expressions for the coefficients α, β, and γ are obtained in an analytical form, similarly as it has been done in the paper [8].

4 Numerical Results

In order to show the possibilities of the proposed method we investigate vibration and buckling analysis of a plate shown in Fig. 1. Consider the plate with fixed geometrical parameters: $b/a = 1$; $c/2a = 0.3$; $d/2a = 0.25$; $h/2a = 0.1$. The material properties of the FGM mixture Al/Al_2O_3 used in the study are as follows:

$$\begin{aligned}
&\mathrm{Al}: \quad E_{\mathrm{m}} = 70\ \mathrm{GPA}, \quad \nu_{\mathrm{m}} = 0.3, \quad \rho_{\mathrm{m}} = 2707\ \mathrm{kg/m};\\
&\mathrm{Al_2O_3}: \quad E_{\mathrm{c}} = 380\ \mathrm{GPA}, \quad \nu_{\mathrm{c}} = 0.3, \quad \rho_{\mathrm{c}} = 3800\ \mathrm{kg/m}.
\end{aligned}$$

Suppose that the plate is simply supported and uniformly compressed on the border of the region. Boundary conditions take the following form:

$$\begin{aligned}
&w = 0, \quad \forall (x, y) \in \partial\Omega, \quad u = 0, \quad \forall (x, y) \in \partial\Omega^{(u)},\\
&\partial\Omega^{(u)}: \ (y = -b, \ |x| \le a) \cup \ (y = d, \ c \le x \le a \ \cup \ -a \le x \le -c)\\
&\qquad\qquad \cup \ (y = b, \ |x| \le c),
\end{aligned}$$

$$v=0, \ \forall (x, y) \in \partial\Omega^{(v)}, \ \partial\Omega^{(v)}: \ (x = \pm a, \ -b \le y \le d) \cup (x = \pm c, \ d \le y \le b),$$

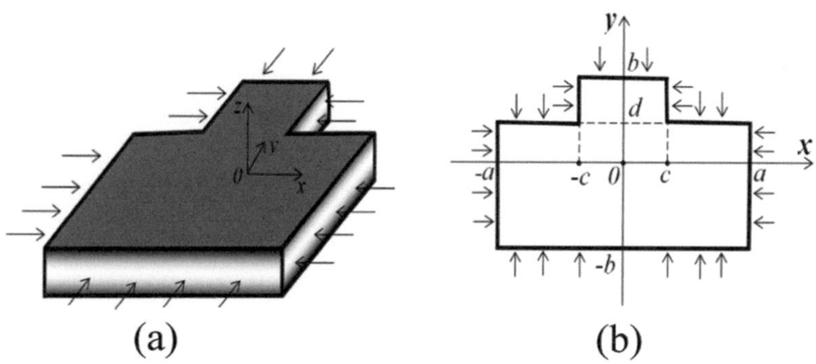

Fig. 1 Form of the plate (**a**) and its planform (**b**)

$$\psi_x = 0, \quad \forall (x, y) \in \partial\Omega^{(\psi_x)},$$
$$\partial\Omega^{(\psi_x)}: \ (y = -b, \ |x| \le a) \cup \ (y = d, \ c \le x \le a \ \cup \ -a \le x \le -c) \cup \ (y = b, \ |x| \le c),$$

$$\psi_y = 0, \quad \forall (x, y) \in \partial\Omega^{(\psi_y)}, \quad \partial\Omega^{(\psi_y)}: \ (x = \pm a, \ -b \le y \le d) \cup \ (x = \pm c, \ d \le y \le b).$$

In order to construct the admissible functions by the RFM [6], we should build the solution structure satisfying at least main boundary conditions. It can be chosen in the following way:

$$w = \omega^{(w)}\Phi_1, \quad u = \omega^{(u)}\Phi_2, \quad v = \omega^{(v)}\Phi_3, \quad \psi_x = \omega^{(\psi_x)}\Phi_4, \quad \psi_y = \omega^{\psi_y}\Phi_5, \tag{11}$$

where functions $\omega^{(u)}, \ \omega^{(v)}, \ \omega^{(w)}, \ \omega^{(\psi_x)}, \ \omega^{(\psi_y)}$ are constructed by the RFM. They have to vanish on those parts of the boundary, where the corresponding functions $u, \ v, \ w, \ \psi_x, \ \psi_y$ are equal to zero. Indefinite components $\Phi_i \in C^2(\Omega \cup \partial\Omega), \quad i = 1, 2, \ldots, 5$ of the structure are presented as an expansion in a series of a complete system (power polynomials). In order to check the reliability of the obtained results in case of a complex geometry, let us change the size of the domain in such a way that the domain will be close to a square. For example, let us put $c/2a = 0.48, \quad d/2a = 0.48, \quad h/2a = 0.1$. A comparison of the obtained values of buckling $\widehat{N}_{cr} = \frac{N_{cr}}{100E_0h^3}$ for the plate of the Type A and $p = 0.5$ with available results [10–12] for the square plate is shown in Table 1.

The values of results are almost the same. As can be seen from Table 1, the buckling load is slightly higher for the complex plate than for the square one, what is in agreement with the physical meaning.

The values of the buckling load and the natural frequency $\Lambda = \omega_L(2a)^2\sqrt{\rho_0/E_0}/h$ for a plate with geometrical parameters $c/2a = 0.3$, $d/2a = 0.25$ for Type A and B are presented in Table 2. The power law exponent p is varied.

Table 1 Buckling load $\widehat{N}_{cr}$ (Al/Al_2O_3, $E_0 = 1$ GPa, $\rho_0 = 1$ kg/m^3, Type A)

$c/a = d/a$	Method	Ratio of the thickness of layers $h^{(1)} - h^{(2)} - h^{(3)}$					
		1-0-1	2-1-2	2-1-1	1-1-1	2-2-1	1-2-1
0.96	RFM	3.673	3.961	4.104	4.210	4.398	4.603
1	RFM	3.667	3.958	4.101	4.205	4.392	4.597
	[10]	3.6828	3.9709	4.1127	4.2185	4.4052	4.6083
	[11]	3.6783	3.9676	1.1000	4.2162	4.4030	4.6076
	[12]	3.5810	3.8581	3.9948	4.0964	4.2759	4.4711

Table 2 Effect of the power law exponent p on the buckling load and the natural frequency FG plate (1-2-1; $c/2a = 0.3$, $d/2a = 0.25$, $h/2a = 0.1$, $E_0 = 1$ GPa, $\rho_0 = 1$ kg/m^3)

Type	$\widehat{N}_{cr}$, Λ	$p = 0$	$p = 0.5$	$p = 1$	$p = 5$	$p = 10$	$p = 100$
A	$\widehat{N}_{cr}$	9.915	7.205	5.966	3.905	3.560	3.245
	Λ	2.375	2.080	1.919	1.599	1.537	1.477
B	$\widehat{N}_{cr}$	5.864	4.841	4.381	3.754	3.685	3.625
	Λ	1.902	1.773	1.709	1.624	1.619	1.617

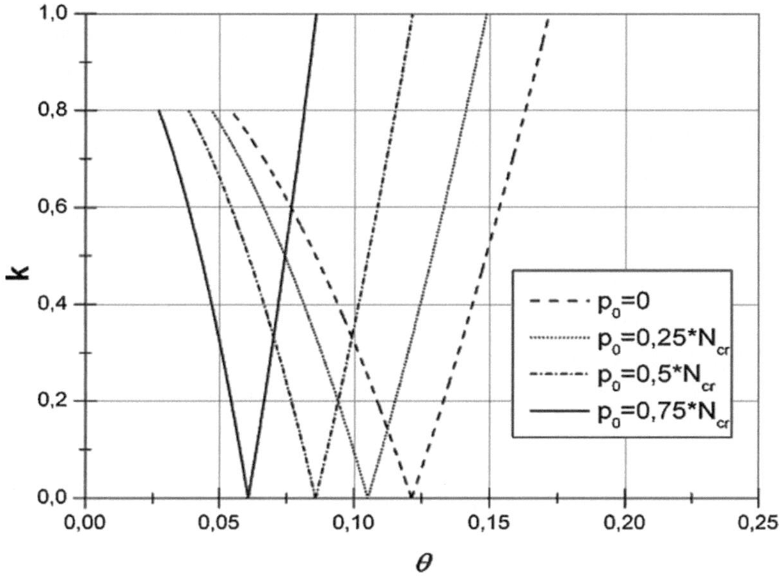

Fig. 2 Zones of dynamic instability (Type A)

Figure 2 shows zones of dynamic instability of the plate (Type A) for different values of the static load. The ratio of layers thickness is taken as $h^{(1)} - h^{(2)} - h^{(3)} = 1 - 2 - 1$, power low exponent $p = 1$.

From Fig. 2 one can observe that static constituent of the load p_{st} influences essentially on placement and size of the instability regions. Increase of this parameter causes to a shift of the instability domains toward the smaller values of exciting frequency.

Figure 3 shows backbone curves for the plate (Type B) with the same ratio of layers thickness, power law exponent $p = 1$. The effect of the static component of the load on the behavior of backbone curves is investigated. It can be concluded that all curves have the monotonically increasing character. The bigger values of the static component of the compression load causes the bigger increasing the value of ratio of nonlinear frequency to linear one.

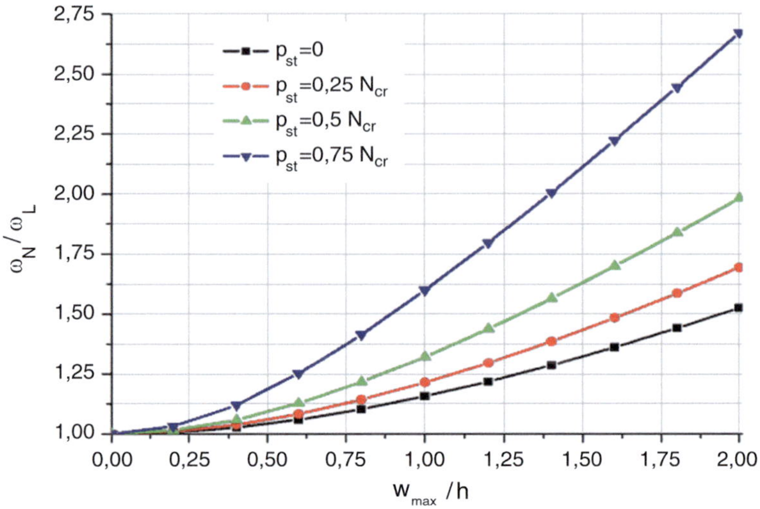

Fig. 3 Backbone curves of the plate (Type B)

5 Conclusions

The first time the R-functions theory and variational Ritz's method are used and developed to study buckling and parametric vibrations of three-layered FGM plates with an arbitrary geometry. Considered specific examples of sandwich plates with complex planform and different arrangement of layers compressed by load in the middle surface demonstrate the possibility of application of the R-functions theory to such class of problems. Effect of the layers thickness and arrangement of the material and gradient index in power law on the buckling critical load and frequencies of the plates are investigated. Instability regions and backbone curves are presented for different value of compressing load.

Conflict of Interest The authors declare that they have no conflict of interest.

References

1. Swaminathan, K., Naveenkumar, D.T., Zenkour, A.M., Carrera, E.: Stress, vibration and buckling analyses of FGV plates—a state-of-art review. J. Compos. Struct. **120**, 10–31 (2015)
2. Thai, H.-T., Kim, S.-E.: A review of theories for the modelling and analysis of plates and shells. J. Compos. Struct. **128**, 70–86 (2015)
3. Kumar, Y.: The Rayleigh-Ritz method for linear dynamic, static and buckling behavior of beams, shells and plates: a literature review. J. Vib. Control. **24**(7), 1205–1227 (2017). https://doi.org/10.1177/1077546317694724

4. Li, Q., Iu, V.P., Kou, K.P.: Three-dimensional vibration analysis of functionally graded material sandwich plates. J. Sound Vib. **311**, 498–515 (2008)
5. Lei, Z.X., Zhang, L.W., Liew, K.M.: Buckling analysis of CNT reinforced functionally graded laminated composite plates. J. Compos. Struct. **152**, 62–73 (2016)
6. Rvachev, V.L., Sheiko, T.I.: The R-functions in boundary value problems in mechanics. Appl. Mech. Rev. **48**(4), 151–188 (1995)
7. Awrejcewicz, J., Kurpa, L., Mazur, L.: Dynamical instability of laminated plates with external cutout. Int. J. Non Linear Mech. **81**, 103–114 (2016)
8. Awrejcewicz, J., Kurpa, L., Shmatko, T.: Analysis of geometrically nonlinear vibrations of functionally graded shallow shells of a complex shape. Lat. Am. J. Solids Struct. **14**, 1648–1668 (2017)
9. Reddy, J.N.: Mechanics of Laminated Composite Plates and Shells. Theory and Analysis, 2nd edn. CRC Press, Boca Raton (2004)
10. Zencour, A.M.: A comprehensive analysis of functionally graded sandwich plates: part 2—buckling and free vibration. Int. J. Solids Struct. **42**, 5243–5258 (2005)
11. Meiche, N.E., Tounsi, A., Ziane, N., Mechab, I., Bedia, A.A.: A new hyperbolic shear deformation theory for buckling and vibration of functionally graded sandwich plate. Int. J. Mech. Sci. **53**(4), 237–247 (2011)
12. Neves, A.M.A., Ferreira, A.J.M., Carrera, E., Cinefra, M., Jorge, R.M.N., Soares, C.M.M.: Buckling analysis of sandwich plates with functionally graded skins using a new quasi-3D hyperbolic sine shear deformation theory and collocation with radial basis functions. J. Appl. Math. Mech. **92**, 749–766 (2012)

Nonlinear Oscillation of a FG Cylindrical Shell on a Discontinuous Elastic Foundation

Frederico M. A. Silva, Patrícia C. Rodrigues, and Paulo B. Gonçalves

Abstract The aim of the present work is to investigate the influence of the elastic foundation discontinuities on the nonlinear response of a functionally graded cylindrical shell with internal flowing fluid. The fluid is assumed to be inviscid and incompressible and the flow, isentropic and irrotational. Nonlinear Donnell shallow shell theory is used to obtain the nonlinear partial differential equations of motion and the elastic foundation is described by the Pasternak model. Resonance curves are obtained to identify the nonlinear forced behavior of the FG cylindrical shell conveying fluid and surrounded by a discontinuous elastic foundation. Different configurations of discontinuities of elastic foundation and axial fluid flow velocities are considered, showing the influence of these parameters on the stable paths of resonance curves.

Keywords Cylindrical shell · FGM · Partial elastic foundation · Nonlinear vibrations

1 Theoretical Formulation

Typically, functionally graded (FG) cylindrical shells are composed of metallic and ceramic materials varying through the shell thickness. In the last two decades, several research works have been published on the dynamic behavior of FG structures. Recently, a complete review of various theories for modeling and analysis of FG shells has been presented [1]. In the literature, few research works evaluates the influence of discontinuous elastic foundation in FG shells. Kim [2] investigated the behavior of a FG cylindrical shell partially resting on elastic foundation at

F. M. A. Silva (✉) · P. C. Rodrigues
School of Civil and Environmental Engineering, Federal University of Goiás, Goiânia, Brazil
e-mail: silvafma@ufg.br

P. B. Gonçalves
Department of Civil Engineering, Pontifical Catholic University, Rio de Janeiro, Brazil
e-mail: paulo@puc-rio.br

W. Lacarbonara et al. (eds.), *New Trends in Nonlinear Dynamics*,
https://doi.org/10.1007/978-3-030-34724-6_9

an oblique angle. The parametric analysis focused on the influence of the elastic foundation stiffness and the oblique angle on the results. In another work Kim [3] studied a FG cylindrical shell with discontinuous elastic foundation, exploring the discontinuities along the circumferential angle and along the shell length [3]. However, both works were focused on the linear analysis. Thus, there is a lack of information on the influence of a discontinuous elastic medium on the nonlinear shell response.

Consider a simply supported circular cylindrical shell of radius R, length L, and thickness h. The shell material is assumed to be elastic and isotropic with Young's modulus E, Poisson ratio ν, and density ρ varying its properties along the shells thickness. The axial, circumferential, and radial coordinates are denoted by x, θ, and z, respectively, and the corresponding mid-surface displacements are denoted by u, v, and w, as shown in Fig. 1. The shell is resting on an elastic Pasternak foundation with reaction P_{k} and subjected to a time-dependent lateral pressure p and a hydrodynamic internal fluid pressure P_{H}.

The nonlinear Donnell shallow shell theory is used to obtain the nonlinear partial differential equation of motion, which is given in terms of the transversal displacement $w(x,\theta)$ and Airy stress function $f(x,\theta)$ by:

$$\rho_t\ddot{w}+2\eta_1\rho_t h\omega\dot{w}+\eta_2\frac{E_{\mathrm{M}}h^3}{12(1-\nu_{\mathrm{M}}{}^2)}\nabla^4\dot{w}+\frac{1}{R}f_{,xx}+\frac{1}{R^2}\left[f_{,\theta\theta}w_{,xx}-2f_{,x\theta}w_{,x\theta}+f_{,xx}w_{,\theta\theta}\right]$$
$$-\frac{2}{R}M_{x\theta,x\theta}-M_{x,xx}-\frac{1}{R^2}M_{\theta,\theta\theta}+P_{\mathrm{H}}-p+P_k=0 \quad (1)$$

where ρ_t is the average density, η_1 and η_2 are, respectively, the linear viscous damping and the viscoelastic material damping coefficients.

Airy stress function reduces a system with three equations to one single equation with two variables, $w(x,\theta)$ and $f(x,\theta)$. So it is necessary to obtain a second equation to solve the problem. The compatibility and continuity conditions are obtained using geometric compatibility equation for a plane deformation state. Using the relations $f_{,xx}=N_\theta, f_{,\theta\theta}=R^2N_x$ and $f_{,x\theta}=-RN_{x\theta}$, it takes the form:

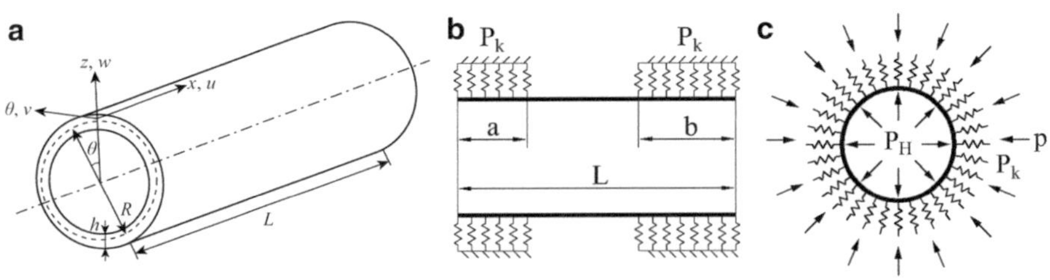

Fig. 1 (**a**) Geometry and system coordinates of a circular cylindrical shell. (**b**) Discontinuity elastic foundation along the length. (**c**) External applied loads and the internal hydrodynamic pressure

$$\left(\frac{2A_{12}}{R^2c_1}+\frac{1}{R^2A_{66}}\right)f^P_{,xx\theta\theta}+\frac{A_{11}}{c_1}f^P_{,xxxx}+\frac{A_{22}}{R^4c_1}f^P_{,\theta\theta\theta\theta}=-\frac{1}{R}w_{,xx}\left(w_{,\theta\theta}-1\right)+\frac{1}{R^2}w^2_{,x\theta}$$
$$+\frac{A_{12}B_{11}-A_{11}B_{12}}{c_1}w_{,xxxx}+\frac{A_{11}A_{22}B_{66}-A_{11}A_{66}B_{22}-A^2_{12}B_{66}+2A_{12}A_{66}B_{12}-A_{22}A_{66}B_{11}}{R^2A_{66}c_1}w_{,xx\theta\theta}$$
$$+\frac{A_{12}B_{22}-A_{22}B_{12}}{R^4c_1}w_{,\theta\theta\theta\theta} \tag{2}$$

where c_1 is a constant defined as $c_1=A_{11}A_{12}-A_{12}$ and A_{ij}, B_{ij} and C_{ij} $(i, j=1, 2, 6)$ are, respectively, the extensional, coupling and bending stiffness defined as:

$$\left\{A_{ij},B_{ij},C_{ij}\right\}=\int_{-h/2}^{h/2}Q_{ij}\left\{1,z,z^2\right\}dz \quad \text{with} \quad Q_{11}=\frac{E}{1-\nu^2},\ Q_{22}=Q_{11},$$
$$Q_{12}=\frac{E\nu}{1-\nu^2} \quad \text{and} \quad Q_{66}=\frac{E}{2\left(1+\nu\right)} \tag{3}$$

The moment resultants for a FG cylindrical shell are obtained from stress-strain relations and can be written as a function of displacements as Eq. (4). These moments are used to evaluate the equation of motion in Eq. (1).

$$\begin{Bmatrix}M_x\\M_\theta\\M_{x\theta}\end{Bmatrix}=\begin{bmatrix}B_{11}&B_{12}&0&C_{11}&C_{12}&0\\B_{21}&B_{22}&0&C_{21}&C_{22}&0\\0&0&B_{66}&0&0&C_{66}\end{bmatrix}\begin{Bmatrix}u_{,x}+w_{,x}{}^2/2\\\left(v_{,\theta}+w\right)/R+w_{,\theta}{}^2/2R^2\\u_{,\theta}/R+v_{,x}+\left(w_{,x}w_{,\theta}\right)/R\\-w_{,xx}\\-w_{,\theta\theta}/R^2\\-w_{,x\theta}/R\end{Bmatrix} \tag{4}$$

In the present work, the variation of constituent materials across the FGM shell thickness is given by a power law distribution, written as [5]:

$$V_{\mathrm{M}}(z)=\left[\left(1-\frac{2z}{h}\right)\left(1+\frac{2z}{h}\right)\right]^{2N+1} \qquad V_{\mathrm{C}}(z)=1-V_{\mathrm{A}}(z) \tag{5}$$

where V_{M} and V_{C} are, respectively, the volume fraction of metal and ceramic; z is the distance from the shell middle surface, and N is the volume fraction index. The Young's modulus E, Poisson ratio ν, density ρ varies with the volume fraction as $P=(P_{\mathrm{M}}-P_{\mathrm{C}})V_{\mathrm{M}}(z)+P_{\mathrm{C}}$, where P_{M} and P_{C} are the material properties of the metal and ceramic, respectively.

Pasternak model is used to evaluate the reaction of the surrounding medium P_{k} on the cylindrical shell, described as $P_{\mathrm{k}}=(K_{\mathrm{W}}\,w+K_{\mathrm{P}}\,\nabla^2 w)\,H_x$, where K_{W} and K_{P} are the radial and transverse stiffness of the spring layer, respectively; ∇^2 is the Laplace operator and H_x is the Heaviside step function in x direction, expressed as $H_x=H(x)-H(x-a)+H(x-b)-H(x-L)$. For $H_x=1$, elastic foundation becomes complete.

The hydrodynamic pressure is described by [4]:

$$P_{\mathrm{H}} = \rho_{\mathrm{F}} I_n(qr) \left(\ddot{w} + 2U \dot{w}_{,x} + U^2 w_{,xx} \right) / q I_n'(qr) \tag{6}$$

where U is the mean axial flow velocity, ρ_{F} is the fluid density; I_n is the modified Bessel function of the first kind of order n; I_n' is the derivative of I_n with respect to the argument and $q = m\pi/L$.

2 Numerical Results

Consider a simply supported cylindrical shell of radius $R = 2$ m, length $L = 2$ m, and thickness $h = 0.01$ m. The FGM shell is composed of nickel and silicon nitride, the metal and ceramic material, respectively. The nickel properties are: $E_{\mathrm{M}} = 2.051 \times 10^{11}$ N/m^2, $\rho_{\mathrm{M}} = 8900$ kg/m^3 and $\nu_{\mathrm{M}} = 0.3$ while the silicon nitride are: $E_{\mathrm{C}} = 3.222 \times 10^{11}$ N/m^2, $\rho_{\mathrm{C}} = 2370$ kg/m^3 and $\nu_{\mathrm{C}} = 0.24$. The shell is subjected by a time-dependent lateral pressure with the same shape as the fundamental vibration mode and is given by:

$$p(t) = P_{\mathrm{L}} \sin(m\pi x/L) \cos(n\theta) \cos(\omega t) \tag{7}$$

The nonlinear partial differential equation of motion, Eq. (1), is discretized using a transversal displacement field which is obtained by applying a perturbation technique [6] and describes the main modal couplings due to quadratic and cubic nonlinearities. The obtained modal solution has five degrees of freedom and is expressed by:

$$\begin{aligned} w_{\mathrm{p}} = &\overline{W}_{11}(t) h \cos(n\theta) \sin(m\pi x/L) + \overline{W}^{\mathrm{c}}_{11}(t) h \sin(n\theta) \sin(m\pi x/L) \\ &+ \overline{W}_{13}(t) h \cos(n\theta) \sin(3m\pi x/L) + \overline{W}^{\mathrm{c}}_{13}(t) h \sin(n\theta) \sin(3m\pi x/L) \\ &+ \overline{W}_{02}(t) h \left[(3/4) - \cos(2m\pi x/L) + (1/4) \cos(4m\pi x/L) \right] \end{aligned} \tag{8}$$

where m and n are the number of longitudinal half-waves and the number of circumferential waves, respectively. Also, due to discontinuity of elastic foundation along of longitudinal direction, additional linear vibration modes are considered to better discretize the displacement field generated by this discontinuity in the axials direction:

$$w_{\mathrm{d}} = \sum_{j=2,4,6}^{8} \overline{W}_{1j}(t) h \cos(n\theta) \sin(jm\pi x/L) + \overline{W}^{\mathrm{c}}_{1j}(t) h \sin(n\theta) \sin(jm\pi x/L) \tag{9}$$

The obtained modal solution satisfy the following boundary conditions for a simply supported cylindrical shell:

$$w = M_x = 0 \quad \text{at} \quad x = 0, L. \tag{10}$$

A convergence analysis of the nonlinear response is conducted considering the initial expansion described by Eq. (8) and an increasing number of modes in Eq. (9) up to $j = 8$, as show in Table 1. The complete modal expansion corresponds to Model X5. Adding the companion modes, the transversal displacement field has 13 degrees of freedom and is given by $w = w_\text{p} + w_\text{d}$.

Substituting the chosen modal expansion in Table 1 into the compatibility equation, Eq. (2), and solving it analytically, a compatible stress function is obtained. Substituting the obtained stress function and transversal displacement expansion into Eq. (1) and employing the standard Galerkin method, a system of ODEs in time domain with is obtained. The resonance curves are obtained employing the continuation software "AUTO" [7]. The resonance curve relates the magnitude of permanent nonlinear time response, for each modal amplitude, with external excitation frequency.

The numerical results are obtained considering the following parameters: foundation stiffnesses $K_\text{W} = 2.7 \times 10^7$ N/m^2 and $K_\text{P} = 2.7 \times 10^5$ N, volume fraction index $N = 0.1$, fluid velocity $U = 20$ m/s, magnitude of the lateral load $P_\text{L} = 5000$ N/m and damping coefficients $\eta_1 = 0.003$ and $\eta_2 = 0.0001$. It is important to notice that all these chosen values for geometrical parameters, material properties, and foundation stiffness were used as academic purposes. However these used values allow us to evaluate qualitatively the influence of discontinuous elastic foundation on resonance curves of FG cylindrical shells. Figure 2 illustrates the four types of discontinuous elastic foundation considered in the analysis. The lowest natural frequency for Cases A and D occurs for wave numbers $(m, n) = (1, 7)$ and, for Cases B and C, for wave numbers $(m, n) = (1, 8)$.

Table 1 Proposed modal expansions for transversal displacements field

Model	Modal amplitudes present in modal expansion	Number of degree-of-freedom (K)
Model X1	$W_{11}(t)$, $W_{11c}(t)$, $W_{13}(t)$, $W_{13c}(t)$, $W_{02}(t)$	5
Model X2	$W_{11}(t)$, $W_{11c}(t)$, $W_{13}(t)$, $W_{13c}(t)$, $W_{02}(t)$, $W_{12}(t)$, $W_{12c}(t)$	7
Model X3	$W_{11}(t)$, $W_{11c}(t)$, $W_{13}(t)$, $W_{13c}(t)$, $W_{02}(t)$, $W_{12}(t)$, $W_{12c}(t)$, $W_{14}(t)$, $W_{14c}(t)$	9
Model X4	$W_{11}(t)$, $W_{11c}(t)$, $W_{13}(t)$, $W_{13c}(t)$, $W_{02}(t)$, $W_{12}(t)$, $W_{12c}(t)$, $W_{14}(t)$, $W_{14c}(t)$, $W_{16}(t)$, $W_{16c}(t)$	11
Model X5	$W_{11}(t)$, $W_{11c}(t)$, $W_{13}(t)$, $W_{13c}(t)$, $W_{02}(t)$, $W_{12}(t)$, $W_{12c}(t)$, $W_{14}(t)$, $W_{14c}(t)$, $W_{16}(t)$, $W_{16c}(t)$, $W_{18}(t)$, $W_{18c}(t)$	13

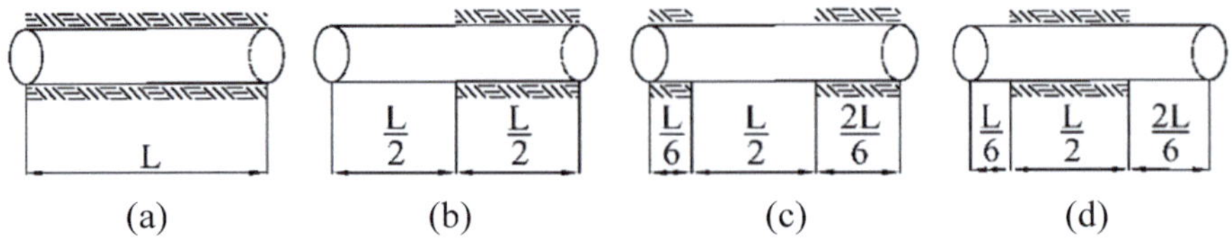

Fig. 2 Variation of elastic foundation along the longitudinal direction. (**a**) Case A. (**b**) Case B. (**c**) Case C. (**d**) Case D

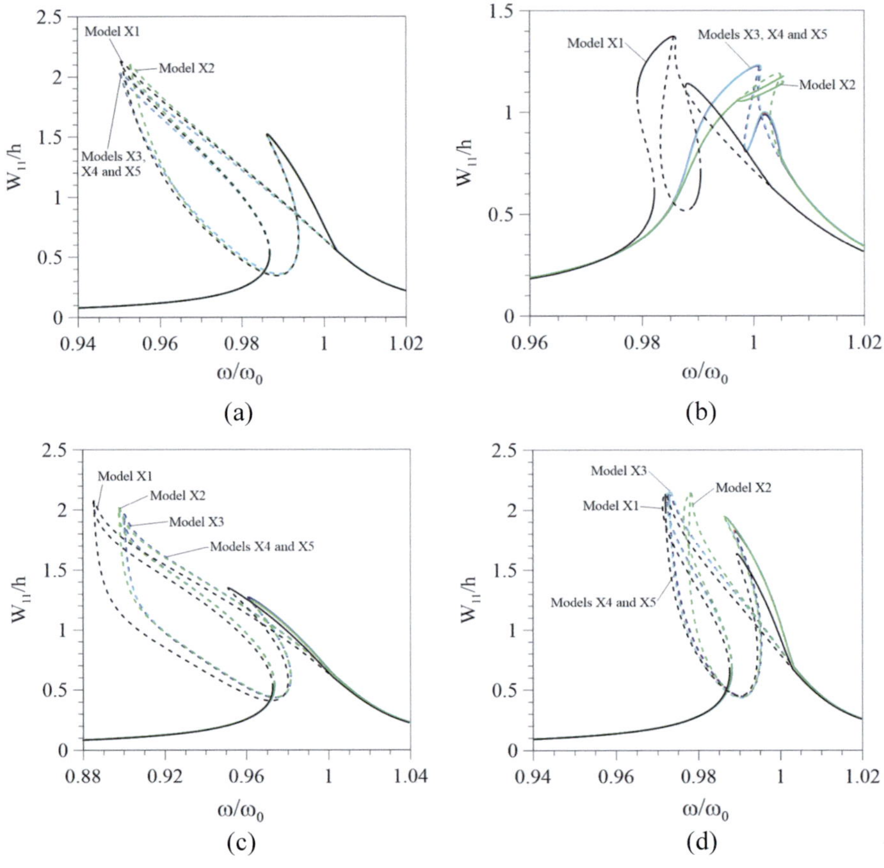

Fig. 3 Convergence of resonance curves for (**a**) Case A, (**b**) Case B, (**c**) Case C and (**d**) Case D

Figure 3, where the normalized maximum modal amplitude (W_{11}/h) is plotted as a function of the normalized frequency parameter (ω/ω_0), exhibits the convergence of the resonance curves for each elastic foundation configuration as the number of modes in Table 1 increases. It is shown Model X5 guarantees the convergence of the response in all cases up to large vibration amplitudes. The participation of the asymmetric terms on the nonlinear response is analyzed by plotting their maximum

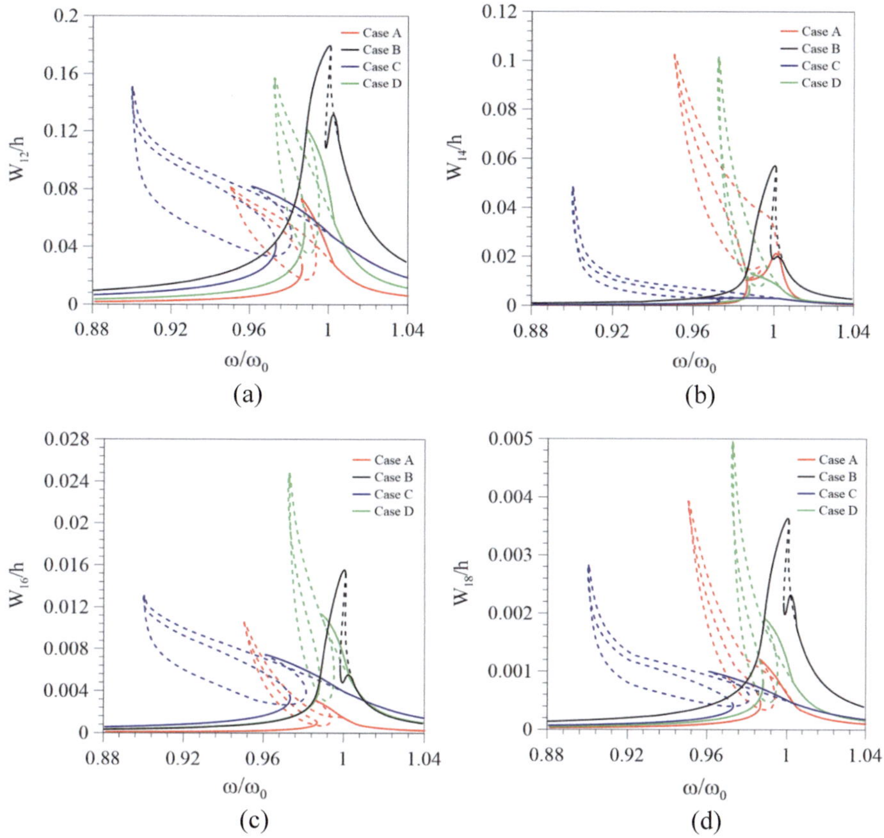

Fig. 4 Resonance curves for all cases of configuration of elastic base to linear vibration modes (**a**) $W_{12}(t)$, (**b**) $W_{14}(t)$, (**c**) $W_{16}(t)$, and (**d**) $W_{18}(t)$. (Model X5)

vibration amplitude as a function of the frequency parameter (ω/ω_0) in Fig. 4. Their influence decreases as the number of linear modes increase. The maximum vibration amplitude $W_{12}(t)$, considering all four cases of elastic foundation configuration, is lower than 20% of the shell thickness, while the maximum vibration amplitude $W_{14}(t)$, $W_{16}(t)$, and $W_{18}(t)$, reached approximately 10%, 2.5%, and 0.5% of the shell thickness, respectively. Thus, model MX5 is used in all subsequent analyses.

Figure 5 presents in detail the obtained resonance curves of Fig. 4, using Model X5, identifying clearly the stable and unstable paths, where LP means a limit points bifurcation and BP, a flip bifurcation. The right column in Fig. 5 shows the frequency range associated with each solution branch and where coexistent stable solutions occur.

Cases A, C, and D display a softening behavior with two frequency limit points. Along the resonant branch, for low values of the vibration amplitude a stable flip bifurcation occurs and the resonant branch becomes unstable for higher vibration

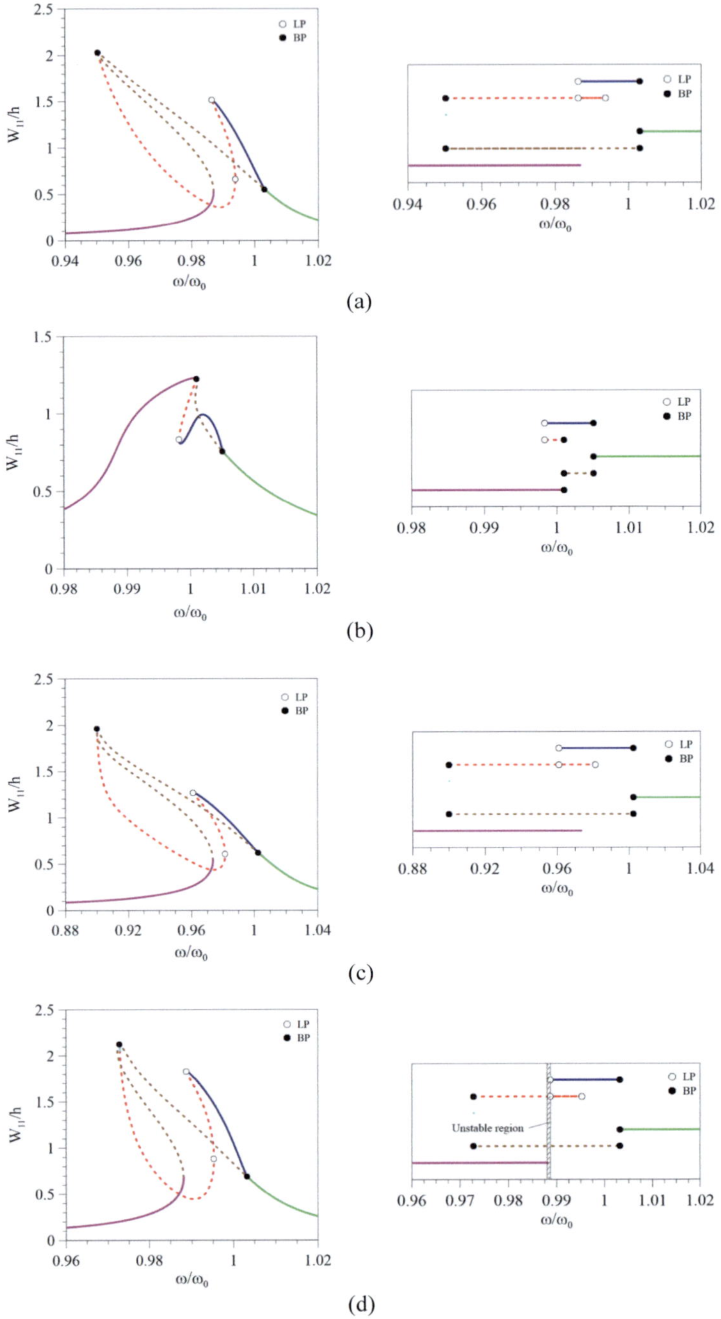

Fig. 5 Resonance curves and their respective diagrams for each stable and unstable paths for (**a**) Case A, (**b**) Case B, (**c**) Case C, and (**d**) Case D of discontinuity of elastic base. (Model X5)

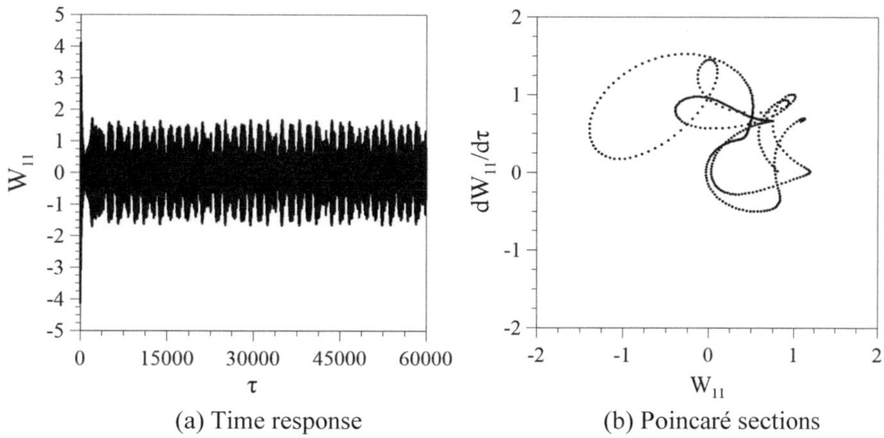

Fig. 6 (**a**) Time response for Case D in unstable region ($\omega/\omega_0 = 0.9883$) and (**b**) the correspondent Poincaré sections

amplitudes. The ensuing stable path (in blue) becomes unstable due to a saddle node bifurcation with the subsequent unstable path connecting this point to the upper saddle node bifurcation along the original resonance curve. This unstable branch also exhibits a frequency limit point. Thus there are two points where the slowly evolving system jumps to another attractor. Case B shows a slight hardening behavior with the same bifurcation sequence. Case D, Fig. 5d, presents a small frequency range where low order periodic orbits are not found by the continuation algorithm, indicated as "unstable region." Figure 6 shows the time response and Poincaré sections for a frequency in this region ($\omega/\omega_0 = 0.9883$), where a nonperiodic solution is identified, Fig. 6a, as corroborated by the Poincaré section, Fig. 6b.

3 Conclusions

In this work, the nonlinear behavior of FG cylindrical shells with internal flowing fluid, resting on a partial elastic foundation, is investigated. Applying Donnell's nonlinear shallow shell theory, the resultant nonlinear partial differential equation of motion is discretized as a set of nonlinear differential equations in time domain by the application of standard Galerkin method. Convergent low order models are obtained for different discontinuous foundation geometries, allowing a detailed nonlinear bifurcation analysis of the shell in the main resonance region. The results show that the foundation discontinuities have a palpable influence on the nonlinear resonance curves and bifurcation sequences. The competing coexisting solutions lead to dynamic jumps that may lead to stress peaks and shell damage. Thus a detailed parametric analysis of this problem is necessary, especially in the design of partially buried shells conveying fluid.

References

1. Thai, H.-T., Kim, S.-E.: A review of theories for the modelling and analysis of functionally graded plates and shells. Compos. Struct. **128**, 70–86 (2015)
2. Kim, Y.W.: Free vibration analysis of FGM cylindrical shell partially resting on Pasternak elastic foundation with an oblique edge. Compos. B Eng. **70**, 263–276 (2015)
3. Kim, Y.W.: Effect of partial elastic foundation on free vibration of fluid-filled functionally graded cylindrical shells. Acta Mech. Sin. **31**, 920–930 (2015)
4. Païdoussis, M.P., Denise, J.P.: Flutter of thin cylindrical shells conveying fluid. J. Sound Vib. **20**, 9–26 (1972)
5. Silva, F.M.A., Montes, R.O.P., Gonçalves, P.B., Del Prado, Z.J.G.N.: Nonlinear vibrations of fluid-filled functionally graded cylindrical shell considering a time-dependent lateral load and static preload. Proc. Inst. Mech. Eng. C J. Mech. Eng. Sci. **230**, 102–119 (2015)
6. Gonçalves, P.B., Silva, F.M.A., Del Prado, Z.J.G.N.: Reduced order models for the nonlinear dynamic analysis of shells. Proc. IUTAM. **19**, 118–125 (2016)
7. Doedel, E.J., Champneys, A.R., Fairgrieve, T.F., Kuznetsov, Y.A., Sandstede, B., Wang, X.: AUTO 97: Continuation and Bifurcation Software for Ordinary Differential Equations (with HomCont). Concordia University, Montreal (1998)

Nonlinear Fracture Dynamic Analysis of Double Cantilever Beam Sandwich Specimens

Vyacheslav N. Burlayenko, Tomasz Sadowski, and Svetlana D. Dimitrova

Abstract Double cantilever beam interlaminar fracture toughness sandwich specimens under different kind of dynamic loading and loading rates are investigated. The nonlinear dynamic response of those sandwich specimens being fractured during the test is numerically examined using the finite element method within the ABAQUS™ code. In such virtual tests, the dynamic displacement and stress fields are calculated and the components of the dynamic stress intensity factor are extracted from the two-dimensional finite element solutions by means of the interaction integral method. Cohesive finite elements are used for simulating the dynamic fracturing of the specimens. The influence of inertia on interfacial crack propagation in the specimens is evaluated as a direct outcome of the finite element analysis.

Keywords Double cantilever beam · Face sheet/core debonding · Dynamic fracture · Finite element analysis

1 Introduction

The interfacial debonding between the core and the face sheets arising due to either manufacturing or in-service reasons is a damage often encountered in sandwich composites. The debonding problem under dynamic loading is an important issue in engineering applications of sandwich structural components [1]. It has already been proved by both experimental and theoretical studies that due to debonding the vibration characteristics of sandwich composites are altered [2] and nonlinear effects

---

V. N. Burlayenko (✉) · T. Sadowski
Lublin University of Technology, Lublin, Poland

National Technical University 'KhPI', Kharkiv, Ukraine
e-mail: v.burlayenko@pollub.pl; burlayenko@yahoo.com

S. D. Dimitrova
National Technical University 'KhPI', Kharkiv, Ukraine

W. Lacarbonara et al. (eds.), *New Trends in Nonlinear Dynamics*,
https://doi.org/10.1007/978-3-030-34724-6_10

appear in their dynamics [3]. Moreover, dynamic loads may provoke a debonding propagation. That is dynamic failure of sandwich structures by debonding may occur [4].

Since the debonding poses a threat to the structural integrity of a whole sandwich construction, a considerable attention of the engineering community has been paid to fracture mechanics methods. The latter are able to validate sandwich composites in terms of damage tolerance and possible failure. In doing so, interlaminar fracture toughness specimens are used to supply necessary information regarding the fracture resistance of the material. One of the most commonly used experimental methods for studying the mode I interlaminar fracture is a double cantilever beam (DCB) sandwich specimen [5]. The fracture parameters such as stress intensity factors (SIFs) or strain energy release rate (ERR) controlling a fracture process are inferred from the test method. A literature search showed that experimental studies on the analysis of interface cracks in DCB sandwich specimens are mostly limited to quasi-static cases, because of a lack of reliable test set-ups for proper generating and, then, accurate tracking a fast crack propagation [6]. Some existing analytic results are confined by simple loading cases, e.g., in [7].

The aim of this research is to present an accurate and efficient finite element model of the DCB sandwich specimen for predictions of a nonlinear dynamics of sandwich composites in virtual fracture tests. The influence of dynamic loading on the overall dynamic response and strength of the DCB sandwich specimen based on this model is studied and discussed to gain a conceptual understanding of the dynamic fracture phenomenon.

2 Formulation of the Model

A general dynamic framework of the finite element method (FEM) with cohesive elements is used to formulate a model in the present work. Let us consider a two-dimensional (2-D) continuum composed of two or more dissimilar materials. The continuum contains a crack at one or more of material interfaces. At a time instant t, the medium occupies a space V with an external boundary ∂V, and the presence of crack is associated with two internal surfaces $\partial V_c^+ \cup \partial V_c^- = \partial V_c$ of the continuum. We assume that the cracked medium is subjected to infinitesimal deformations. Then, neglecting body forces and accounting for cohesive and contact forces along ∂V_c, the principle of virtual work can be stated as [4]:

$$\begin{aligned} &\int_{V\backslash\partial V_c} (\boldsymbol{\sigma} : \nabla\delta\mathbf{u} + \rho\ddot{\mathbf{u}}\cdot\delta\mathbf{u})\, dV + \int_{\partial V_c} \mathbf{T}\cdot\delta\boldsymbol{\Delta} dA + \\ &\int_{\partial V_c} (t_N\delta g_N + \mathbf{t}_T\cdot\delta\mathbf{g}_T)\, dA - \int_{\partial V_t} \bar{\mathbf{t}}\cdot\delta\mathbf{u} dA = 0 \end{aligned} \tag{1}$$

for all kinematically admissible displacement fields $\delta\mathbf{u}$ and with prescribed displacements $\bar{\mathbf{u}}$ at a boundary ∂V_u and traction $\bar{\mathbf{t}}$ at a boundary ∂V_t, where $\partial V_u \cup \partial V_t = \partial V$. The Cauchy stress $\boldsymbol{\sigma}$ caused by the displacement field $\mathbf{u}$ is related to strains in accordance with the linear elastic Hooke's relations, whereas at the crack tip within the surface ∂V_c with a normal $\mathbf{n}_c$, the cohesive zone reflecting fracture due to material damage accumulation is invoked. The latter can be expressed by a bilinear constitutive relation between the components of displacement jump $\boldsymbol{\Delta}$ and the cohesive traction $\mathbf{T} = \boldsymbol{\sigma} \cdot \mathbf{n}_c$ across ∂V_c, which are associated with single-mode loading ($i = I, II$) for a 2-D case as follows [8]:

$$T_i = \begin{matrix} k_i \Delta_i, & \Delta_i \le \Delta_i^0 \\ (1 - D_i) k_i \Delta_i, & \Delta_i^0 \le \Delta_i \le \Delta_i^f \\ 0, & \Delta_i \ge \Delta_i^f, \end{matrix} \tag{2}$$

where $D_i = \left(\Delta_i^f (\Delta_i - \Delta_i^0)\right) / \left(\Delta_i (\Delta_i^f - \Delta_i^0)\right)$ is a damage variable for each fracture mode. At the beginning of this traction-separation law (TSL), a linear response is defined by a penalty stiffness k_i; the softening part starts from the value Δ_i^0, where the traction reaches a maximum normal or shear cohesive value T_i^0 and, then, it evolves linearly till Δ_i^f, where complete failure occurs. In the case of mixed modes, an effective displacement jump $\Delta_m = \sqrt{\sum_{j=1}^{M} \Delta_j^2}$, where $M = I, II$ is introduced. Consequently, all the fracture parameters of the TSL are to be expressed in terms of equivalent mixed-mode values instead the pure mode components [8]. In doing so, we suppose the quadratic stress criterion for the fracture onset, and the Benzeggagh–Kenane (B–K) criterion for the fracture evolution [9]. Nevertheless, the knowledge of the pure mode components at the crack tip is a prerequisite to the understanding of the material strength.

The near-tip stress field of an interface bi-material crack can be quantified by the complex SIF, $K = K_1 + iK_2 = |K| e^{i\psi}$ with K_1 and K_2 used instead of K_I and K_{II} adopted for homogeneous materials. This emphasizes a coupling between tensile and shear stresses at the crack tip, which also oscillate depending on a distance r from the crack tip and the bi-material constant ϵ because of the term $r^{i\epsilon}$, [10]. The SIF components can be evaluated using the interaction integral method (IIM):

$$K_M = \frac{H}{2K_M^{aux}} I_M, \tag{3}$$

where $H = (2\cosh^2 \pi\epsilon)/(1/\bar{E}_1 + 1/\bar{E}_2)$ with $\bar{E}_k = E_k$ for in plane stress and $\bar{E}_k = E_k/(1 - \nu_k)$ for in plane strain, $k = 1, 2$; '*aux*' stands for auxiliary factors known from the asymptotic Williams type' solutions; the 2-D domain interaction integral I_M for a straight crack growing in x_1-axis direction is presented as [11]:

$$I_M = \int_A \left\{ \left[\boldsymbol{\sigma} \cdot \frac{\partial \mathbf{u}_M^{aux}}{\partial x_1} + \boldsymbol{\sigma}_M^{aux} \cdot \frac{\partial \mathbf{u}}{\partial x_1} \right] : \frac{\partial q_1}{\partial \mathbf{x}} - \boldsymbol{\sigma} : \boldsymbol{\varepsilon}_M^{aux} \frac{\partial q_1}{\partial x_1} + \rho \ddot{\mathbf{u}} \cdot \frac{\partial \mathbf{u}_M^{aux}}{\partial x_1} q_1 \right\} dA \tag{4}$$

where A is a domain enclosed by the contour around the crack tip; q_1 is a weighting function with values 0 and 1 within this domain.

In addition to the cohesive fracture, contact and friction along ∂V_c are also parts of the solution. The normal $\mathbf{t}_N = t_N \mathbf{n}_c$ and tangential $\mathbf{t}_T$ components of the contact traction are interrelated with appropriate normal g_N and tangential $\mathbf{g}_T$ gap functions of the displacement jump field $\boldsymbol{\Delta}$, [12]. Then, the impenetrability and friction constraints imposed on ∂V_c^+ and ∂V_c^- are expressed in terms of these variables as the Karush–Kuhn–Tucker conditions:

$$t_N \leq 0,\ g_N \geq 0,\ t_N g_N = 0,\ \|\mathbf{t}_T\| \leq \tau_{crit},\ \|\mathbf{g}_T\| \geq 0,\ (\|\mathbf{t}_T\| - \tau_{crit})\,\|\mathbf{g}_T\| = 0. \tag{5}$$

Coulomb friction model assumes, $\tau_{crit} = \mu t_N$ and μ is the coefficient of friction.

In the context of the FEM, Eq. (1) is to be transformed to the discrete system of equations with respect to nodal displacements $\{U\}$ at a time instant t as:

$$[M]\{\ddot{U}\}_t + \{R_{int}\}_t + \{R_{coh}\}_t + \{R_{cont}\}_t = \{R_{ext}\}_t\,, \tag{6}$$

where $\{R_{int}\}$, $\{R_{ext}\}$, $\{R_{coh}\}$, and $\{R_{cont}\}$ are nodal internal, external, cohesive, and contact forces, respectively; $[M]$ is the mass matrix. Either central difference explicit or Hilber–Hughes–Taylor (HHT) implicit time-stepping schemes available in ABAQUS [9] are used for solving the system (6).

3 Results of Simulations

A DCB sandwich specimen of length of $L = 270$ mm and width of $b = 10$ mm with a pre-existing debonding of length of $a = 90$ mm is considered. The specimen consists of laminated face sheets of thicknesses of 2.4 mm and a foam core of thickness of 50 mm which are bonded with a very thin glue layer. The specimen is subjected to bending moments, M_1 and M_2 at neutral lines of the specimen arms. The properties of the specimen constituents are summarized in Table 1.

Table 1 Material properties of the DCB sandwich specimen

Constituents	Material constants
GFRP face sheet	$E_x = E_z = 16.5$ GPa; $E_y = 3.8$ GPa; $G_{xy} = G_{xz} = 1.3$ GPa; $G_{yz} = 6.6$ GPa; $\nu_{xy} = 0.05$; $\nu_{xz} = \nu_{yz} = 0.25$; $\rho = 1650\,\mathrm{kgm}^{-3}$
PVC H 100 foam core	$E_x = E_y = E_z = 105$ MPa; $G_{xy} = G_{xz} = G_{yz} = 78$ MPa; $\nu_{xy} = \nu_{xz} = \nu_{yz} = 0.325$; $\rho = 100\,\mathrm{kgm}^{-3}$
G-VE/H 100 interface	$k_I = k_{II} = k_{III} = 100$ GPa; $G_{Ic} = 400\,\mathrm{Jm}^{-2}$; $G_{IIc} = G_{IIIc} = 500\,\mathrm{Jm}^{-2}$; $T_I^0 = 10$ MPa; $T_{II}^0 = T_{III}^0 = 20$ MPa

A 2-D finite element model of the DCB specimen was developed using 8-node reduced integration plane strain finite elements (CPE8R) available in ABAQUS. The mesh contained a refinement of the near-tip region as shown in Fig. 1. Quarter-point singular elements were placed around the crack tip to model square-root singularity. The implicit dynamic analysis has been carried out. The 'hard contact' model and frictionless conditions, associated with the constraints (5) have been tackled by the penalty contact algorithm. For a stationary debonding, the dynamic SIFs were computed by ABAQUS built-in IIM, while cohesive elements have been used to simulate nonlinear dynamics of debonding growth.

First, the specimen is subjected to impact loading in the form: $M_i = M_i^0 H(t)$, where the moment amplitudes $M_1^0 = 1.512\,\text{Nm}$ and $M_2^0 = -0.076\,\text{Nm}$ inducing the static ERR about $400\,\text{J/m}^2$. Figure 2 shows the dimensionless DSIFs separated by the IIM. A strong mixed-mode stress state due to a bi-material nature of the interface crack is clearly seen. Also, the instantaneous loading leads to dynamic effects in the DCB due to generated stress waves. The DSIFs exceed their static counterparts, especially at the beginning of loading. Herewith, the mode II state prevails for the first few milliseconds because the shear stress waves arrive at the crack tip faster than the longitudinal ones, but later the major response is the mode I dominated state. The picks and dips in the DSIF curves are related to the effects of reflected and scattered waves interacting with the crack tip.

Second, the specimen under harmonic loading $M_i = M_i^0 \sin \Omega t$ with various driving frequencies is considered. A long-term behaviour of the DSIFs is illustrated

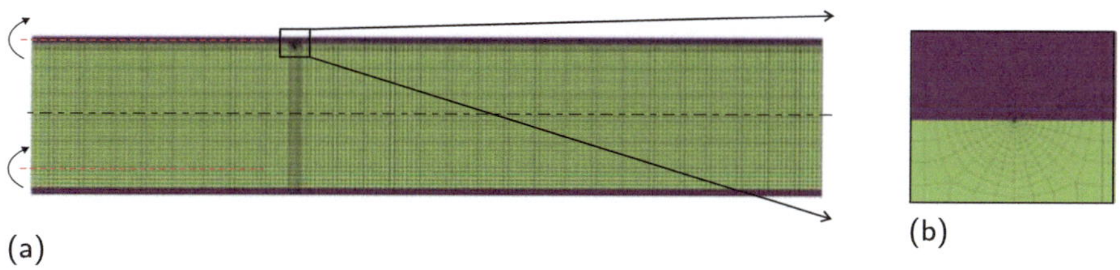

Fig. 1 FE model of the DCB specimen: (**a**) FE mesh; (**b**) mesh refinement

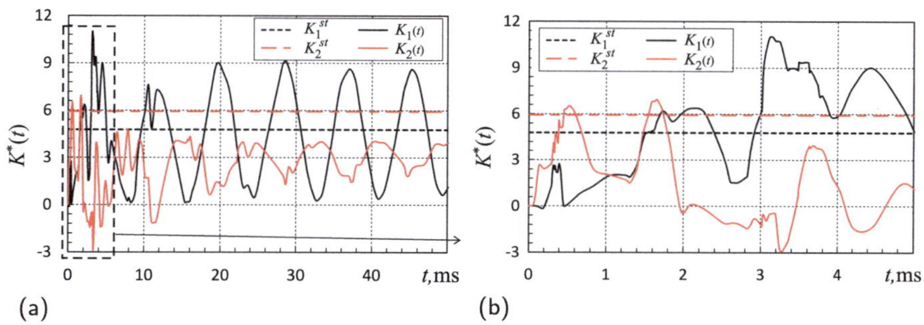

Fig. 2 DSIFs, $K_i^*(t) = K_i(t)b^2\sqrt{a}/M_1^0$ due to step loading: (**a**) analysis time; and (**b**) a zoom

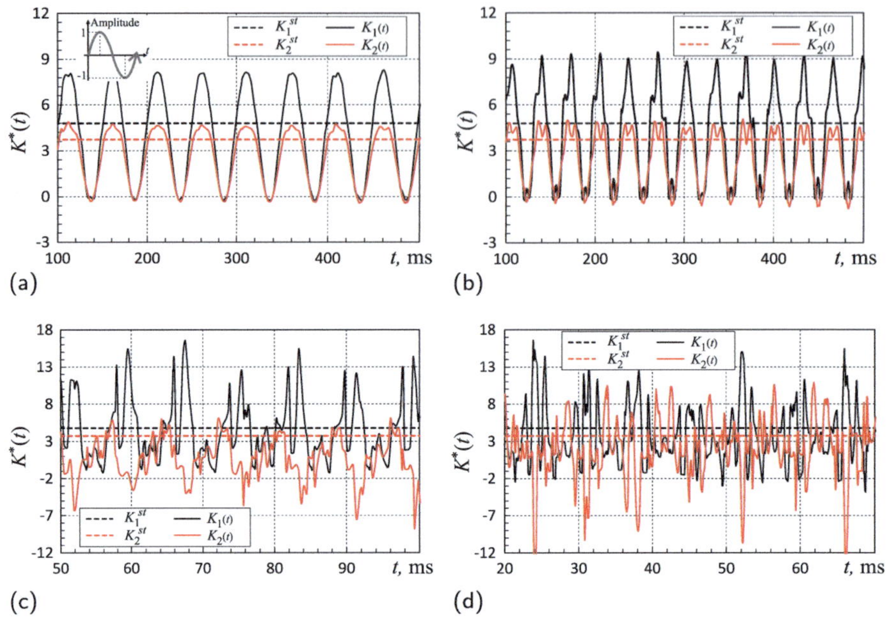

Fig. 3 DSIFs, $K_i^*(t) = K_i(t)b^2\sqrt{a}/M_1^0$ due to harmonic loading at a driving frequency: (**a**) $\zeta = 1/6$; (**b**) $\zeta = 1/2$; (**c**) $\zeta = 3/2$; (**d**) $\zeta = 2$

in Fig. 3, where the unsteady oscillations have been discarded. One can see that the behaviour of the DSIFs highly depends on the driving frequency, accepted as a fraction $\zeta = \Omega/\omega_1$ (1/6, 1/2, 3/2, 2) of the fundamental frequency ω_1 of the DCB. At the higher driving frequencies, the DSIFs have bigger amplitudes and more complicated responses than at the lower ones. This could be attributed to an additional contribution of stress waves generated from irregular contacts within the debonded region as studied, e.g., in [13, 14].

The effect of impulsive loading on the transient DSIFs of the DCB is demonstrated in Fig. 4, where different forms and rates of the loading impulses with the same load amplitudes are examined. It follows from the plots that both the pulse duration and the pulse form affects the DSIFs. All the pulses of very short durations induced the near-tip stresses not exceeding those in the case of static loading. The same situation has been revealed for the triangular and sinusoidal pulses of relatively long durations. For these pulses, only the loading during 1 and 10 ms results in the increasing of the dynamic K_2 and K_1, respectively, over the analogous static SIFs. However, the rectangular pulse produces the DSIFs which are bigger than analogous static ones and identical to the step load with increasing the pulse duration. Thus, such pulses with abrupt ascending and descending paths of loading are more dangerous in terms of the dynamic strength.

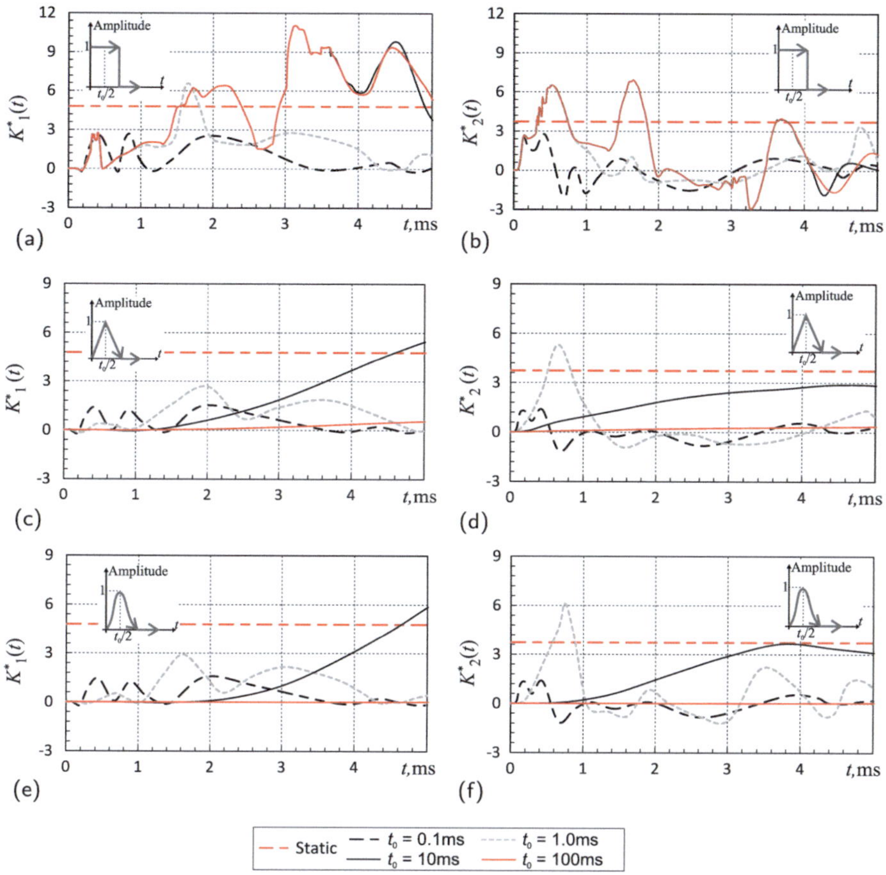

Fig. 4 DSIFs, $K_i^*(t) = K_i(t)b^2\sqrt{a}/M_1^0$ due to: (**a**–**b**) rectangular pulse; (**c**–**d**) triangular pulse; and (**e**–**f**) sinusoidal pulse

Finally, to simulate the nonlinear fracture dynamics of the specimen, four-node cohesive elements (COH2D4) satisfying the TSL (2) were inserted into the finite element mesh of the DCB model, Fig. 1. Explicit dynamic analyses with the frictionless hard contact model, handled by the kinematic contact algorithm [9] have been performed. The debonding propagation under a step load with a linear ascending path (a ramp time t_0) of different slops is shown in Fig. 5a. It is seen that the debonding increases with decreasing the ramp time, and for the shortest one a disintegration of the specimen occurs. In the cases without fracture, the debonding was extending with a relatively constant speed after a short time of fast growing. Unlike this relatively stable debonding growth, the stick-slip debonding propagation has been observed in the DCB under harmonic loading with a driving frequency ratio, $\zeta = 3/2$ as illustrated in Fig. 5b. Also, the simulations revealed that the debonding was intensively growing when a mode II dominated regime took place at

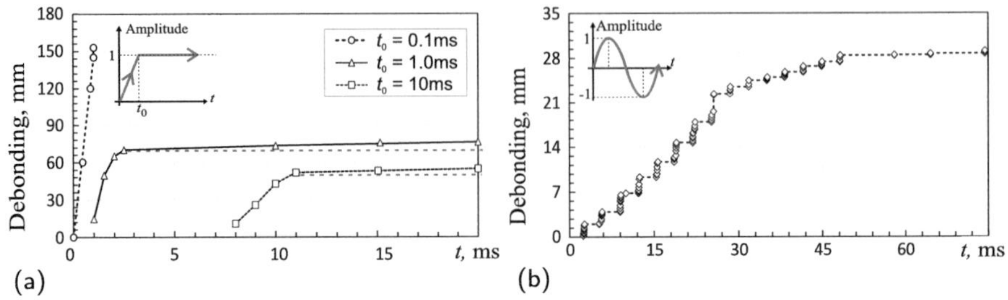

Fig. 5 Debonding propagation under: (**a**) impulsive loading; (**b**) harmonic loading

the crack tip of the DCB. Such the dynamic fracture behaviour during harmonic loading could be explained by the generation of additional stress waves due to contact. This is unlike fracture under impulse loading, where the stress wave effect disappears after a while.

4 Conclusions

The FEM calculations revealed that there is a large dynamic effect in dynamic DCB tests, primarily due to stress waves from both the loading and crack face contact. Such waves interact with the crack tip and strongly affect the fracture parameters and the debonding behaviour of the DCB sandwich specimens.

Acknowledgements The research has been supported by the POLONEZ 2 programme of the National Science Centre of Poland, grant agreement No UMO-2016/21/P/ST8/00790 co-financed by the EU's Horizon 2020 Marie Skłodowska-Curie programme, grant agreement No 665778.

References

1. Abrate, S., Castanié, B., Rajapakse, Y.D.S. (eds.): Dynamic Failure of Composite and Sandwich Structures. Solid Mechanics and its Applications, vol. 192. Springer, Dordrecht (2013)
2. Burlayenko, V.N., Sadowski, T.: Dynamic analysis of debonded sandwich plates with flexible core: numerical aspects and simulation. In: Altenbach, H., Eremeyev, V.A. (eds.) Shell-like Structures. Advanced Structured Materials, vol. 15, pp. 415–440. Springer, Heidelberg (2011)
3. Qu, Y., Meng, G.: Nonlinear vibro-acoustic analysis of composite sandwich plates with skin-core debondings. AIAA J. **55**(5), 1723–1733 (2017)
4. Burlayenko, V.N., Sadowski, T.: Simulations of post-impact skin/core debond growth in sandwich plates under impulsive loading. J. Appl. Nonlin. Dyn. **3**(4), 369–379 (2014)
5. Carlsson, L., Matteson, R., Aviles, F., et al.: Crack path in foam cored DCB sandwich fracture specimens. Compos. Sci. Technol. **65**, 2612–2621 (2005)

6. Ravi-Chandar, K.: Experimental challenges in the investigation of dynamic fracture of brittle materials. In: Bouchaund E. et al. (eds.) Physical Aspects of Fracture, vol. 32, pp. 323–342. Springer, Dordrecht (2001)
7. Lü, N.-C., Cheng, Y.-H., Li, X.-G., Chen, J.: Analytical solutions of dynamic mode I crack under the conditions of displacement boundary. Nonlinear Dyn. **67**, 2197–2205 (2012)
8. Camanho, P.P., Dávila, C.G., de Moura, M.F.: Numerical simulation of mixed-mode progressive delamination in composite materials. J. Compos. Mater. **37**(16), 1415–1438 (2003)
9. ABAQUS User's manual, ver. 2016. Dassault Systémes. Simulia Corp., Providence (2016)
10. Hutchinson, J.W., Suo, Z.: Mixed mode cracking in layered materials. Advances in Applied Mechanics, vol. 29, pp. 63–191. Elsevier, Amsterdam (1991)
11. Nakamura, T., Kushner, A., Lo, C.Y.: Interlaminar dynamic crack propagation. Int. J. Solids Struct. **32**(17–18), 2657–2675 (1995)
12. Burlayenko, V.N., Sadowski, T.: Linear and nonlinear dynamic analyses of sandwich panels with face sheet-to-core debonding. Shock Vib. **2018**, 5715863 (2018)
13. Elmalich, D., Rabinovitch, O.: On the effect of inter-laminar contact on the dynamics of locally delaminated FRP strengthened walls. Int. J. Nonlinear Mech. **77**, 141–157 (2015)
14. Saito, A.: Nonlinear resonances of chains of thin elastic beams with intermittent contact. ASME J. Comput. Nonlinear Dyn **13**(8), 081005 (2018)

Nonlinear Vibration Analysis of a Sandwich Beam and Assessment of the Dynamic Behavior

Airton Nabarrete, Eduardo Francisco Rocha de Araujo, Jose Manoel Balthazar, and Angelo Marcelo Tusset

Abstract The aim of the current work is to discuss the dynamic buckling of a sandwich plate, to identify the response signals, and to find the relevant frequencies and amplitudes. The analyses consider the thin face-sheets of the sandwich plate which are subject to in-plane compressive load lower than the estimated critical value for the local buckling, but the plate is also subject to vibration in specific frequency causing variation of the load amplitude. The dynamic variation of the in-plane stress stiffening changes the natural frequencies representing the out-of-plane modes dynamically. The nonlinear variation of the response for the out-of-plane displacements is analyzed with the continuous wavelet transform (CWT) method for characterizing this behavior.

Keywords Nonlinear vibration · Sandwich beam · Dynamic buckling · Dynamic behavior

1 Introduction

Light and very light slender structures are present in different industries, for instance, in aerospace and aeronautic industries, since a high ratio of strength and low weight are a necessity. Components as helicopter blades, satellite panels, airplane control surface skins, upper and lower wing panels, leading edges, trailing edges are among these light structures. These components are subject to different nature of dynamic loading. One may realize a rectangular plate under in-plane compressive load with value insufficient to cause a local instability of buckling type. An additional compressive load due to vibration may yield the panel to buckle

A. Nabarrete (✉) · E. F. R. de Araujo
Department of Aerospace Engineering, Instituto Tecnológico de Aeronáutica, São José dos Campos, SP, Brazil
e-mail: nabarrete@ita.br

J. M. Balthazar · A. M. Tusset
Department of Electronic Engineering, UTFPR, Ponta Grossa, SP, Brazil

W. Lacarbonara et al. (eds.), *New Trends in Nonlinear Dynamics*,
https://doi.org/10.1007/978-3-030-34724-6_11

periodically. Thus, many structure parts may be subject to this dynamic compressive loading. Furthermore, some natural frequencies change with the vibration. This is the scenario to be studied.

The estimative of parameters that characterize this state and the answer to vibration loading can help to prevent situations with loss of stability of the structure. This phenomenon requires nonlinear dynamic analyses. It is observed that there are equations with analytical solutions for some classical nonlinear dynamic problems, usually for few degrees of freedom (DOFs). It is a challenge to consider a large number of DOFs even for the solution or post-processing the nonlinear results. Simplified tools as phase portraits diagrams and Poincaré Maps are not well suited to evaluate these results.

1.1 Nonlinear Behavior

Sandwich plates composed by a honeycomb structure for the core and metallic or laminated composites for the face-sheets are usually analyzed using simplified theories or by the finite element method with accurate load predictions. The use of soft core materials implies additional difficulties to the analytical analysis due to large differences between face-sheet and core stiffness. In these cases, finite element analyses with solid elements are usually used, but large number of elements must be employed for a model with thin face-sheets and thicker core, since the aspect ratio constraint raises the cost of the numerical solutions [1].

The model of a structural element is simplified to a sandwich plate with soft core subject to in-plane compressive loads, as depicted in Fig. 1. The finite element formulation is chosen for stability analyses in order to put some complexities in a

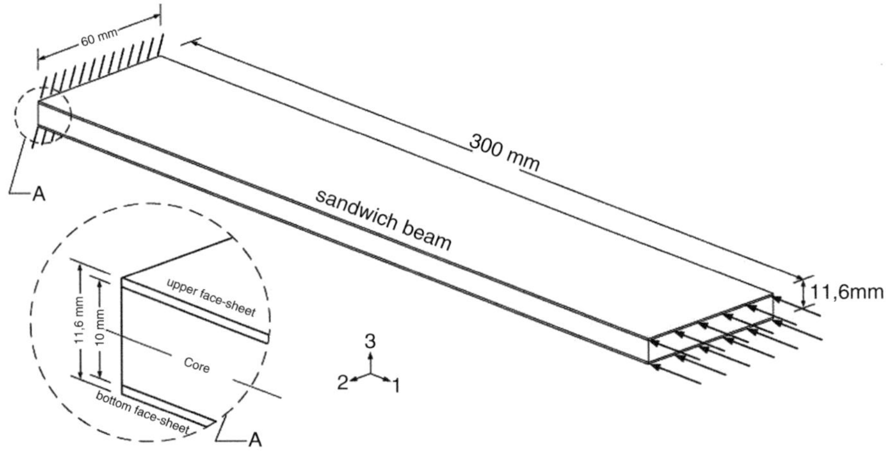

Fig. 1 Specimen sandwich clamped at face edge and free in the others

near future. Analytical models yet published in the literature are available [2]. The in-plane compressive load in one direction and the small width of the symmetric plate may allow the result comparison with analytical formulations of sandwich beams [3, 4]. The compressive load is composed by a major value representing the static in-plane equilibrium of the structural system. The vibration of the loaded system is done by in-plane harmonic loads applied in the model [5]. Different load values are applied in separated analyses.

The compressive load yields stress stiffening of the sandwich structure. The estimative of parameters through linear analyses uses the static in-plane equilibrium to characterize the level of stability, but it cannot prevent the dynamic response with loss of stability for the structure [6]. The behavior of the structure when low values of out-of-plane natural frequencies are achieved claims for a nonlinear stability analysis and depending on the dynamic load level a nonlinear response is expected.

1.2 Model Description

A structural finite element model was built with hexahedral solid elements C3D8R [7] to perform a dynamic simulation. The sandwich plate dimensions, constraints, and loading are depicted in Fig. 1 and characterize it as a sandwich beam. Clamped-free constraint conditions are considered for both ends of the sandwich beam with constraints applied to top and bottom face-sheets. The dynamic axial loading $N(t)$ is applied on the free end of the beam, distributed on top and bottom faces. The mechanical properties of the face-sheets, laminated with Carbon Epoxy AGP370-5H/2501-6s, as well as, the core constructed with rigid foam Divinycell® H-200 are tabulated by Daniel and Ishai [8].

1.3 Equation of Motion

Equation (1) expresses the equation of motion. $[M]$, $[D]$, and $[K]$ are mass, damping, and stiffness matrices which are constant for a linear analysis. $\{N(t)\}$,$\{u\}$,$\{\dot{u}\}$,$\{\ddot{u}\}$ are the in-plane dynamic load, displacement, velocity, and acceleration vectors.

$$[M]\{\ddot{u}\} + [D]\{\dot{u}\} + [K]\{u\} = \{N(t)\} \tag{1}$$

Damping $[D]$ is considered as proportional to the stiffness $[D] = \beta\,[K]$. Stiffness $[K]$ is constant for the linear elastic problem. This is the case for small displacements and small strain. In this chapter, the stress stiffening and the curvature after buckling are taken into account. Thus, $[K]$ may be written as Eq. (2), where $[K_{\text{elast}}]$,

$[K_{large}]$, $[K_G]$ are elastic, large displacement, and geometric stiffness matrices, respectively.

$$[K] = [K_{elast}] + [K_G] + \left[K_{large}\right] \tag{2}$$

1.4 Buckling Problem

The variational formulation of the linear buckling of a sandwich plate requires the minimization of a functional $\boldsymbol{\Pi} = \boldsymbol{U} - \boldsymbol{\Delta W}$, where $\boldsymbol{U}$ is the sum of the face-sheet and core strain energies. $\boldsymbol{\Delta W}$ is the change of the potential energy due to the applied in-plane forces. The stress stiffening can be calculated for linear buckling analysis and strongly affects the vibration frequencies of the sandwich structure with thin face-sheets. Equation (3), represents the linear buckling problem, where $[K_G]$ is the geometric matrix referring to a coupling between in-plane stress and lateral displacements associated with bending [9].

$$[[K_{elast}] + \lambda\,[K_G]]\,\{u\} = 0 \tag{3}$$

Large compressive in-plane stress reduces the bending stiffness. One can consider the calculation of natural frequencies for unloaded plate, as well as, for stress-stiffened sandwich face-sheets. The buckling load is calculated as $P_{cr} = \lambda\, N_0$, where N_0 is the arbitrarily chosen in-plane static load applied for both sandwich face-sheets. This buckling load is used as reference for choosing the amplitude of the harmonic excitation function. The first global buckling mode is considered, i.e., in Eq. (3) the buckling load is calculated as $P_{cr} = 5433.2$ N.

The free vibration problem includes the stress stiffening. In Eq. (4), ω_i is the i-th undamped natural frequency and $\{\phi_i\}$ is the i-th modal shape.

$$\left([K]_{elas} + [K_G] - \omega_i^2\,[M]\right).\{\phi_i\} = 0 \tag{4}$$

The pretension load P is applied as proportional to P_{cr}. Thus, the ratio P/P_{cr} varies from 0.1 up to 1.0. Figure 2 shows the linear buckling analysis for the sandwich plate with variation of the three first natural frequencies as function of the ratio P/P_{cr}.

1.5 Dynamic Loading

The dynamic excitation $N(t)$ is described in terms of the ratio P/P_{cr}. Initially, the compressive load varies from zero to 0.8 for P/P_{cr}, during a half second of time. For the next half second, the load is kept constant ($P/P_{cr} = 0.8$). After that, a

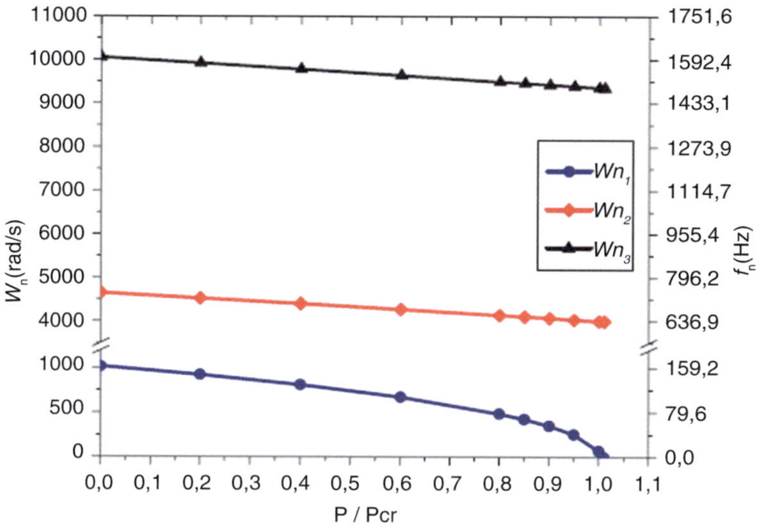

Fig. 2 Natural frequency value variation against axial compression loading

sinusoidal load with amplitude $P/P_{\rm cr} = 0.1$ is added for more 2 s. The total time of the numerical simulation is $t = 3$ s for this chapter. From one test to other the excitation frequency changes looking at different dynamic responses. The dynamic excitation is expressed by Eqs. (5), (6), and (7).

$$N(t) = 3.2\, t\ P_{\rm cr}, \qquad t \leq 0.5 \text{ s} \tag{5}$$

$$N(t) = 0.8\ P_{\rm cr}, \qquad 0.5 < t \leq 1.0 \text{ s} \tag{6}$$

$$N(t) = 0.8\ P_{\rm cr} + 0.1\ P_{\rm cr} \sin{(2\pi f t)}\,, \qquad t > 1.0 \text{ s} \tag{7}$$

It is interesting to note that the natural frequency of the pretensioned sandwich plate is $f = 72.2$ Hz considering the ratio $P/P_{\rm cr} = 0.9$.

2 Results

Time responses are obtained from a nonlinear transient analysis of the finite element model with the dynamic loading described above. The transverse displacement U_3 of a sandwich plate node on the upper face-sheet is used in this chapter.

The buckling load of the sandwich plate, as well as, the undamped natural frequencies for unloaded conditions and pretensioned vibration are calculated using

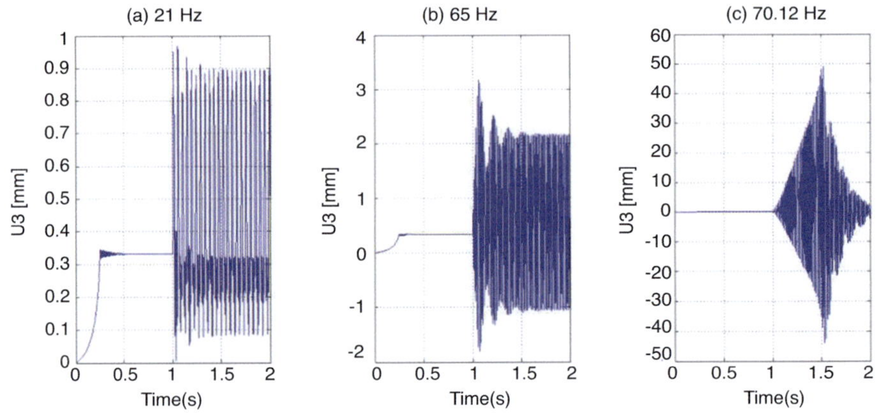

Fig. 3 Transverse displacement U_3: (**a**) $f = 21$ Hz; (**b**) $f = 65$ Hz; (**c**) $f = 70.12$ Hz

the finite element solver at Abaqus 6.5 [7]. In this chapter, an explicit time integration scheme is used for the transient analysis. The nonlinear transient analysis takes into account a nonconstant stiffness.

Figure 3 shows the transverse displacement U_3 on the sandwich plate tip for frequencies $f = 21$ Hz, $f = 65$ Hz and $f = 70.12$ Hz.

The responses in Fig. 3 show different behaviors of the sandwich plate depending on the excitation frequency. It is important to note that for the three responses the applied excitation has the same loading scheme with different harmonic frequencies. The high level compressive in-plane load reduces the bending stiffness promoting a transverse response with expressive amplitude. It is observed that the transverse displacement is not always stationary after damping the transient response. Variable amplitude is observed, in special for $f = 70.12$ Hz, as depicted in Fig. 3(c). In this case, a large amplitude value is achieved. One may observe that the excitation frequency is very close to the natural frequency for the pretensioned sandwich beam. Also, a nonstationary response indicates possibly variable frequencies, characterizing nonproportionality, i.e., nonlinearity.

The stress stiffening of the sandwich plate is considerable high and varies with the dynamic loading. Therefore the bending stiffness reduction is variable depending on the loading scheme, which implies in a variation of the natural frequencies of the sandwich beam. As a matter of fact, this problem is related to the parametric excitation equated by Mathieu–Hill [10].

The observation of the frequency contents of the time response is very useful in vibration analysis. In this chapter, the understanding of frequency content is necessary for a stability evaluation of the sandwich plate. The Fast Fourier Transform (FFT) is a very used tool to verify the frequency components of the time response. The magnitude and phase is obtained considering the frequency resolution. This allows the identification of the frequency content of the response considering a linear combination of harmonic functions with magnitude and phase

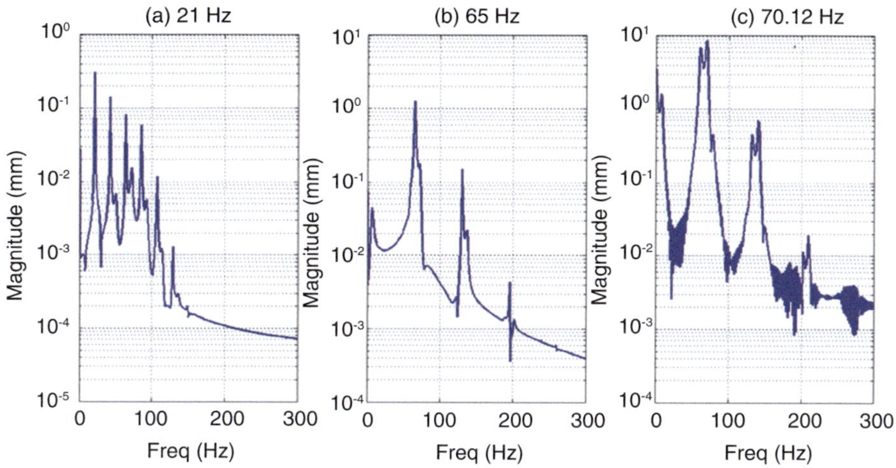

Fig. 4 Frequency response: (**a**) $f = 21$ Hz; (**b**) $f = 65$ Hz; (**c**) $f = 70.12$ Hz

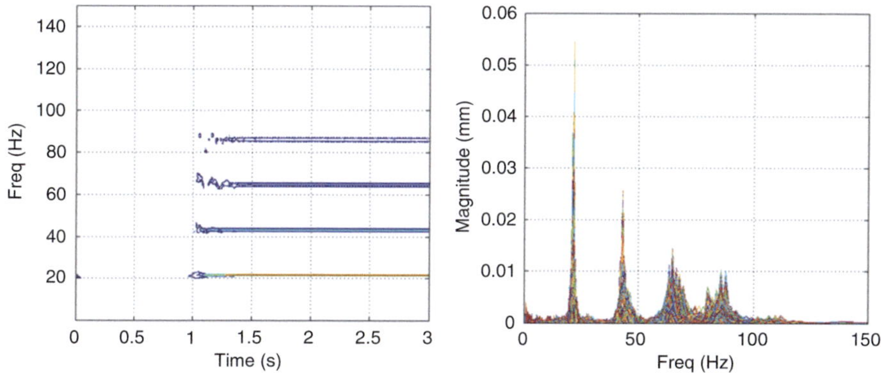

Fig. 5 CWT analysis for $f = 21$ Hz

in order to reproduce the original response. Figure 4 depicts the FFT for the transverse displacement U_3 on the sandwich plate tip for frequencies $f = 21$ Hz, $f = 65$ Hz and $f = 70.12$ Hz.

Another method used to obtain the frequency contents of the time response is the continuous wavelet transform (CWT) [11]. CWT is a function expansion, analogous to a Fast Fourier Transform. Meanwhile FFT is based on a combination of harmonic functions, and from these functions, amplitudes and phases are obtained, CWT is based on combination of continuous time and space functions, not necessarily harmonic ones. It allows an accurate local description of the signal of plate response in this chapter, and it allows the separation of the signal characteristics. Also in this chapter, the time response having transient and stationary behaviors is transformed to obtain the frequency contents. These results are depicted in Figs. 5, 6, and 7.

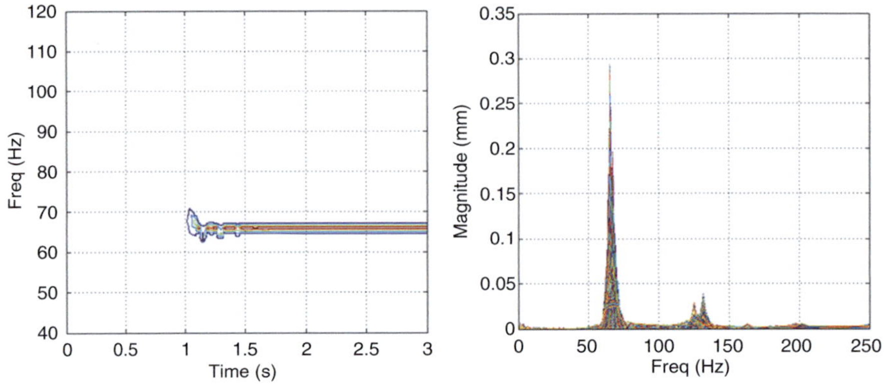

Fig. 6 CWT analysis for $f = 65$ Hz

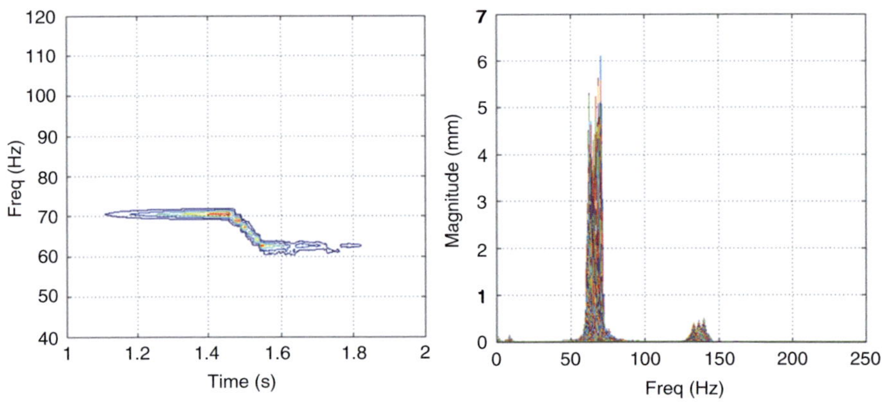

Fig. 7 CWT analysis for $f = 70.12$ Hz

Table 1 Summary of frequencies calculated from FFT and CWT

Freq. contents	FFT 21 Hz	FFT 65 Hz	FFT 70.1 Hz	CWT 21 Hz	CWT 65 Hz	CWT 70.1 Hz
f_1 [Hz]	21.49	64.94	69.9	21.5	65.98	70.54
f_2 [Hz]	42.5	129.9	140	43.26	131.7	139.8
f_3 [Hz]	64	195.8	210	64.3	NaN	NaN
f_4 [Hz]	85.5	NaN	NaN	87	NaN	NaN

Table 1 presents a summary of frequency results obtained from FFT and CWT. The excitation frequencies are 21, 65, and 70.12 Hz. The frequencies f_1 as well as the subharmonics f_2, f_3 and f_4 are present. The response amplitudes for f_3 and f_4 are very small in some cases if compared with the amplitudes for f_1 and f_2 (negligible).

3 Discussion and Conclusions

A refined finite element model of the sandwich beam is constructed for getting the necessary accuracy on nonlinear analysis response. In the analysis the response of a clamped-free sandwich plate excited with compressive vibration loading is evaluated. Transient results as displacements are obtained from the finite element analysis. Due to stress stiffening, changes on the sandwich natural frequencies are observed in Fig. 4 as function of the dynamic loading. Variable amplitude and frequencies are observed, characterizing nonlinear behavior. A large change in amplitude is observed for 70.12 Hz. Fast Fourier Transform (FFT) and continuous wavelet transform (CWT) are used for characterizing this behavior [11], as showed in Fig. 6. CWT and FFT transforms are used to obtain relevant frequencies. One may observe different responses: harmonic frequency, as well as, subharmonic frequencies, presented in parametric excited plate. Response frequencies were captured by both FFT and CWT method. CWT provides additionally the time frequency displacement visualization as shown in Fig. 7 that helps to identify abrupt changes in frequency along the time event.

References

1. Nabarrete, A., Almeida, S.F.M., Hansen, J.: Sandwich-plate vibration analysis: three-layer quasi-three-dimensional finite element model. AIAA J. **41**(8), 1547–1555 (2003)
2. Abou-Rayan, A.M., Nayef, A.H., Mook, D.T., Nayfeh, M.A.: Nonlinear response of a parametrically excited buckled beam. Nonlinear Dyn. **4**(5), 499–525 (1993)
3. Sokolinsky, V.S., Nutt, S.R., Frostig, Y.: Boundary condition effects in free vibrations of higher-order soft sandwich beams. AIAA J. **40**(6), 1220–1227 (2002)
4. Yomayusa Agredo, C.E.: Análise estática de estabilidade e dinâmica não linear de vigas sanduiche pelo método dos elementos finitos (In Portuguese). M.Sc. Thesis, Instituto Tecnológico de Aeronaútica (2009)
5. Yaffe, R., Abramovich, H.: Dynamic buckling of cylindrical stringer stiffened shells. Comput. Struct. **81**(8–11), 1031–1039 (2003)
6. Carbonara, W.: A theoretical and experimental investigation of nonlinear vibrations of buckled beams. M.Sc. Thesis, Virginia Polytechnic Institute and State University (1997)
7. Dassault System Abaqus, release 6.5, Finite Element Software (2005)
8. Daniel, I.M., Ishai, O.: Engineering Mechanics of Composite Material. Oxford University Press, New York (2006)
9. Cook, R.D.: Finite Element Modeling for Stress Analysis. Wiley, New York (1995)
10. Dwivedy, S.K., Sahu, K.C., Babu, S.K.: Parametric instability regions of three layered soft-cored sandwich beam using higher order theory. J. Sound Vib. **304**, 326–344 (2007)
11. Burrus, C.S., Gopinath, R., Guo, H.: Introduction to Wavelets and Wavelet Transforms. Prentice Hall, Upper Saddle River (1998)

Dynamic Buckling of FGM Cylindrical Shells Under Torsional Impact Loads

Jinghua Zhang, Shuai Chen, and Like Chen

Abstract According to the classical shell theory and considering torsional stress wave, buckling of functionally graded material cylindrical shells under torsional impact load are studied by the symplectic method. Considering the radial, circumferential, and axial displacements of the shells, the original variables and the dual variables are established. Then the symplectic method is introduced, which converts the problem into obtaining the eigenvalues and eigenvectors in Hamilton system. After that, the corresponding buckling loads and buckling modes are obtained respectively relevant to the eigenvalues and eigensolutions via the bifurcation conditions. Finally, the influences of material gradient and parameters of structural geometry on buckling loads are analyzed and discussed.

Keywords Functionally graded materials · Cylindrical shell · Torsional impact · Dynamic buckling

1 Introduction

As composite materials, the properties of functionally graded materials (FGM) change continuously and smoothly in a specific direction [1, 2]. The analyses of the mechanical behaviors of FGM structures are more difficult than conventional homogeneous structures. Up to now, considerable research works on the buckling of the FGM structures have been published, but most of them are only limited to static problems. For example, thermal buckling and post-buckling for the FGM plates were investigated in [3–5]. Li et al. [6] and Zhang et al. [7] researched the buckling and post-buckling of FGM Timoshenko beams and imperfect FGM plates.

J. Zhang (✉) · L. Chen
Department of Engineering Mechanics, Lanzhou University of Technology, Lanzhou, China

S. Chen
Faculty of Vehicle Engineering and Mechanics, Dalian University of Technology, Dalian, China

W. Lacarbonara et al. (eds.), *New Trends in Nonlinear Dynamics*,
https://doi.org/10.1007/978-3-030-34724-6_12

The research on dynamic buckling of the FGM structures is much fewer compared to it on static buckling. The works of [2, 8, 9] investigated the dynamic buckling of FGM thin cylindrical shells under axial compression, a longitudinal constant velocity or thermal load. Additionally, two-dimensional analysis of FGM partial annular disks subjected to radial thermal shock was presented by Mehrian and Naei [10]. All of the above research uses traditional methods, such as the finite element method, Galerkin method, etc. However, it is difficult to solve the complex partial differential equations of the dynamic buckling problems using these methods. In contrast, based on the symplecticity method in the Hamilton system [11], the equations of structural stability will be easily solved by variable separation approach and symplectic eigenfunction expansion. Xu et al. [12] studied the buckling and post-buckling behaviors of homogeneous cylindrical shells based on the symplecticity method. Sun et al. [13] studied the static buckling behaviors of FGM cylindrical shells combined thermal and compressive loads by the symplecticity method, too. They showed this method is very efficient and accurate in solving the problem of structural stability.

To the best of the authors' knowledge, few researches study on impact buckling of FGM structures by the symplecticity method. Therefore, it is meaningful to exam the torsional impact buckling of FGM cylindrical shells in Hamilton system. In symplectic space, a canonical equation will be established, then buckling mode equations and bifurcation conditions will be solved by analytical methods. Finally, critical buckling loads and buckling modes will be obtained and discussed.

2 Mathematical Formulas

The thin-walled FGM cylindrical shell with length l, mid-surface radius R, and thickness h is considered, which is fixed at one end and subjected to a torsional impact load at the other end. A coordinate system (x, θ, z) is referred, in which the x-axis coincides with the generatrix of the middle surface, measured from the left end. θ is in the circumferential direction and z is in the transverse direction. The corresponding displacements in the mid-surface are designated as u, v, and w. No initial displacement or velocity exists at any point.

For the FGM cylindrical shell [14], the linear rule of mixtures is used to describe the variations in material properties P [15, 16]; and the power law function is used to describe the variation of volume fractions V_1 and V_2, expressed as

$$P = P_1 V_1 + P_2 V_2, \quad V_1 = \left(\frac{h - 2z}{2h} \right)^k, \quad V_2 = 1 - V_1 \tag{1}$$

where k is the power law index used to quantify the inhomogeneous properties of FGM. Since Poisson's ratio μ does not significantly vary in material gradient direction, so $\mu(z)$ is taken as a constant μ.

2.1 Fundamental Equations

Based on the classical thin shell theory, the strains $\varepsilon_{xx}^{(z)}, \varepsilon_{\theta\theta}^{(z)}, \varepsilon_{x\theta}^{(z)}$ at any point are expressed as

$$\varepsilon_{xx}^{(z)} = \varepsilon_{xx} + z\kappa_{xx}, \quad \varepsilon_{\theta\theta}^{(z)} = \varepsilon_{\theta\theta} + z\kappa_{\theta\theta}, \quad \varepsilon_{x\theta}^{(z)} = \varepsilon_{x\theta} + z\kappa_{x\theta} \tag{2}$$

where strains $\varepsilon_{xx}, \varepsilon_{\theta\theta}, \varepsilon_{x\theta}$ and curvatures $\kappa_{xx}, \kappa_{\theta\theta}, \kappa_{x\theta}$ on the middle surface are

$$\varepsilon_{xx}=\frac{\partial u}{\partial x}, \quad \varepsilon_{\theta\theta}=\frac{1}{R}\frac{\partial v}{\partial \theta}-\frac{w}{R}, \quad \varepsilon_{x\theta}=\frac{1}{R}\frac{\partial u}{\partial \theta}+\frac{\partial v}{\partial x}, \quad \kappa_{xx}=-\frac{\partial^2 w}{\partial x^2}, \quad \kappa_{\theta\theta}=-\frac{1}{R^2}\frac{\partial^2 w}{\partial \theta^2},$$
$$\kappa_{x\theta}=-\frac{1}{R}\frac{\partial^2 w}{\partial x \partial \theta} \tag{3}$$

Considering linear elastic deformations, the constitutive equations are expressed as:

$$\sigma_{xx}=\frac{E}{1-\mu^2}\varepsilon_{xx}^{(z)}+\frac{\mu E}{1-\mu^2}\varepsilon_{\theta\theta}^{(z)}, \quad \sigma_{\theta\theta}=\frac{\mu E}{1-\mu^2}\varepsilon_{xx}^{(z)}+\frac{E}{1-\mu^2}\varepsilon_{\theta\theta}^{(z)}, \quad \sigma_{x\theta}=\frac{E}{2(1+\mu)}\varepsilon_{x\theta}^{(z)} \tag{4}$$

where σ_{xx} and $\sigma_{\theta\theta}$ are normal stresses, $\sigma_{x\theta}$ is shear stress.

2.2 Canonical Equations

Taking the geometric equations into account, the density of strain energy $\overline{U}$ of the shell is expressed as

$$\begin{aligned}
\overline{U} = {} & \frac{A}{2}\left(\frac{\partial u}{\partial x}+\frac{1}{R}\frac{\partial v}{\partial \theta}-\frac{w}{R}\right)^2-(1-\mu)\left[A\frac{\partial u}{\partial x}\left(\frac{1}{R}\frac{\partial v}{\partial \theta}-\frac{w}{R}\right)+C\frac{\partial^2 w}{\partial x^2}\frac{1}{R^2}\frac{\partial^2 w}{\partial \theta^2}\right] \\
& +\frac{C}{2}\left(\frac{\partial^2 w}{\partial x^2}+\frac{1}{R^2}\frac{\partial^2 w}{\partial \theta^2}\right)^2-B\frac{\partial u}{\partial x}\frac{\partial^2 w}{\partial x^2}-\frac{B}{R^2}\left(\frac{1}{R}\frac{\partial v}{\partial \theta}-\frac{w}{R}\right)\frac{\partial^2 w}{\partial \theta^2}-\mu B\frac{\partial u}{\partial x}\frac{1}{R^2}\frac{\partial^2 w}{\partial \theta^2} \\
& -\mu B\left(\frac{1}{R}\frac{\partial v}{\partial \theta}-\frac{w}{R}\right)\frac{\partial^2 w}{\partial x^2}-2(1-\mu)\left[\frac{B}{R}\left(\frac{1}{R}\frac{\partial u}{\partial \theta}+\frac{\partial v}{\partial x}\right)-C\left(\frac{1}{R}\frac{\partial^2 w}{\partial x \partial \theta}\right)^2\right]
\end{aligned}$$

in which $A=\int_{-h/2}^{h/2}\frac{E}{1-\mu^2}\mathrm{d}z$, $B=\int_{-h/2}^{h/2}\frac{Ez}{1-\mu^2}\mathrm{d}z$, and $C=\int_{-h/2}^{h/2}\frac{Ez^2}{1-\mu^2}\mathrm{d}z$ are stiffness coefficients. Assuming the shells under torsional impact loading, it is expressed as

$$N_T=\begin{cases} T & 0 \le x \le C_{\mathrm{e}}t,\ t \le L/C_{\mathrm{e}} \\ 0 & C_{\mathrm{e}}t \le x \le L,\ t \le L/C_{\mathrm{e}} \end{cases} \tag{5}$$

where $C_e \approx \sqrt{\frac{(1-\mu)A}{2I_0}}$ is wave velocity; $I_0 = \int_{-h/2}^{h/2}\rho(z)\mathrm{d}z$ is mass per unit area; $x_e = C_e t$ is elastic transverse wavefront position. Then the density of Lagrange function can be expressed as

$$L = \frac{1}{2}I_0\left(\frac{\partial u}{\partial t}\right)^2 + \frac{1}{2}I_0\left(\frac{\partial v}{\partial t}\right)^2 + \frac{1}{2}I_0\left(\frac{\partial w}{\partial t}\right)^2 - \overline{U} - N_T\frac{\partial w}{\partial x}\frac{\partial w}{\partial \theta} - N_T\frac{\partial v}{\partial x}w \tag{6}$$

The torsional wave equation can be obtained by variation with respected to v

$$\frac{\partial^2 v}{\partial t^2} - C_e^2\frac{\partial^2 v}{\partial x^2} = 0 \tag{7}$$

Define dimensionless variables $X = \frac{x}{R}$, $W = \frac{w}{R}$, $\alpha = \frac{AR^2}{C}$, $\beta = \frac{BR}{C}$, $T = \frac{C_e t}{R}$, $T_{cr} = \frac{N_T R^2}{C}$. Introduce the original variables $\mathbf{q} = \begin{Bmatrix} q_1 \\ q_2 \end{Bmatrix} = \begin{Bmatrix} W \\ \psi \end{Bmatrix}$ and the dual variables $\mathbf{p} = \begin{Bmatrix} p_1 \\ p_2 \end{Bmatrix} = \frac{\partial L}{\partial \dot{\mathbf{q}}} = \begin{Bmatrix} -\dddot{q}_1 - \dot{q}_1'' \\ q_1'' - \dot{q}_2 + \beta q_1 \end{Bmatrix}$, where $\dot{\mathbf{q}} = \frac{1}{R}\frac{\partial \dot{\mathbf{q}}}{\partial \theta}$, $\psi = -\dot{q}_1$, $\mathbf{q}' = \frac{\partial \mathbf{q}}{\partial X}$.

The canonical equation can be obtained by the Hamiltonian variational principle

$$\begin{Bmatrix} \dot{q}_1 \\ \dot{q}_2 \\ \dot{p}_1 \\ \dot{p}_2 \end{Bmatrix} = \begin{bmatrix} 0 & -1 & 0 & 0 \\ -\partial_x^2 - \beta & 0 & 0 & 1 \\ -2\partial_x^4 - [2(1+\mu)\beta]\partial_x^2 - \beta^2 - \alpha & -T_{cr}\partial_x & 0 & \partial_x^2 + \beta \\ -T_{cr}\partial_x & 0 & 1 & 0 \end{bmatrix} \begin{Bmatrix} q_1 \\ q_2 \\ p_1 \\ p_2 \end{Bmatrix} \tag{8}$$

Introducing a state vector $\dot{\boldsymbol{\varphi}} = \{q_1, q_2, p_1, p_2\}^{\mathrm{T}}$, then the canonical equation becomes

$$\dot{\boldsymbol{\varphi}} = \mathbf{H}\boldsymbol{\varphi} \tag{9}$$

The solution of Eq. (9) is written as the following variable separation form

$$\boldsymbol{\varphi}(X,\theta) = \boldsymbol{\varphi}_n(X)\mathrm{e}^{\lambda_n \theta} \tag{10}$$

where λ_n is the eigenvalue of the function; φ_n is the eigenvector, and they satisfy the following eigenequation

$$\mathbf{H}\boldsymbol{\varphi} = \lambda_n\boldsymbol{\varphi} \tag{11}$$

For the FGM cylindrical shell, the eigenvalue can be obtained from the sealing condition $\varphi(X, 0) = \varphi_n(X) = \varphi(X, 2\pi) = \varphi_n(X)e^{2\lambda_n \pi}$, expressed as

$$\lambda_n = in \tag{12}$$

where $i = \sqrt{-1}$, the values of n are 0, ±1, ±2.... For $\lambda_n \neq 0$, the eigensolutions $\varphi_n(X)$ are nonzero. Solving Eq. (11) gives the eigenvector φ

$$\varphi_n = c_1 \begin{bmatrix} e^{\lambda_1 X} \\ -ine^{\lambda_1 X} \\ in\xi e^{\lambda_1 X} \\ \xi_1 e^{\lambda_1 X} \end{bmatrix} + c_2 \begin{bmatrix} e^{\lambda_2 X} \\ -ine^{\lambda_2 X} \\ in\xi e^{\lambda_2 X} \\ \xi_2 e^{\lambda_2 X} \end{bmatrix} + c_3 \begin{bmatrix} e^{\lambda_3 X} \\ -ine^{\lambda_3 X} \\ in\xi e^{\lambda_3 X} \\ \xi_3 e^{\lambda_3 X} \end{bmatrix} + c_4 \begin{bmatrix} e^{\lambda_4 X} \\ -ine^{\lambda_4 X} \\ in\xi e^{\lambda_4 X} \\ \xi_4 e^{\lambda_4 X} \end{bmatrix} \tag{13}$$

where $c_1 \sim c_4$ are constants; $\xi_j = \lambda_j^2 + \beta + n^2$; λ_j ($j = 1$–4) are solutions of equation

$$\lambda^4 + 2\left(\beta\mu + i^2 n^2\right)\lambda^2 + 2Tin\lambda + \alpha - i^4 n^4 = 0 \tag{14}$$

2.3 *Bifurcation Conditions*

Since the shell is considered to be fixed at one end, there are no displacement and rotation, i.e., $w = 0$, $\frac{\partial w}{\partial \theta} = 0$. By introducing the boundary conditions and continuity conditions into Eq. (10), homogeneous algebraic equations are obtained. The condition for buckling is the equations have nonzero solutions, and the determinant of coefficient equals to zero

$$\begin{vmatrix} 1 & 1 & 1 & 1 \\ \lambda_1 & \lambda_2 & \lambda_3 & \lambda_4 \\ e^{\lambda_1 X_e} & e^{\lambda_2 X_e} & e^{\lambda_3 X_e} & e^{\lambda_4 X_e} \\ \lambda e^{\lambda_1 X_e} & \lambda e^{\lambda_2 X_e} & \lambda e^{\lambda_3 X_e} & \lambda e^{\lambda_4 X_e} \end{vmatrix} = 0 \tag{15}$$

Using the above bifurcation conditions Eq. (15), the torsional buckling loads T_{cr} for FGM shell buckling can be determined. After obtaining the buckling loads, the corresponding buckling modes can be solved by Eq. (13).

3 Numerical Results and Discussions

In this chapter, FGM cylindrical shells are made from ceramic SiC and metal Ni. The material properties of the constituents can be found in [17]. The Poisson's ratio is $\mu = 0.3$.

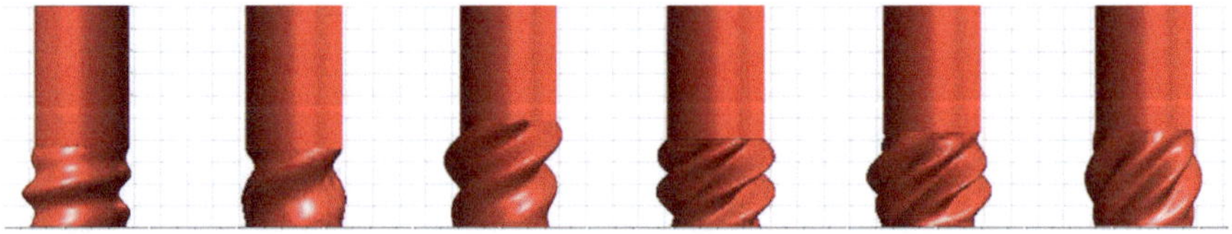

Fig. 1 Buckling modes of the shells with different circumferential order ($m = 1$)

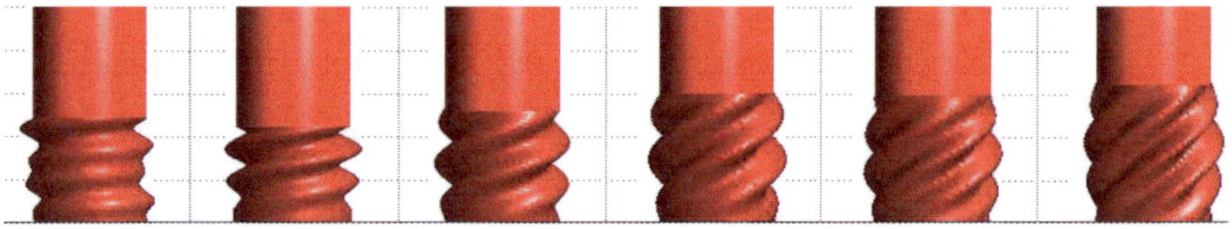

Fig. 2 Buckling modes of the shell with different circumferential order ($m = 2$)

Table 1 Variation of torsional critical loads with circumferential order

n	1	2	3	4	5	6
Present	612.5	317.3	195	131.8	92.3	65.1
[18]	600	303.8	179.8	130.8	97.6	71.6

The torsional shock buckling modes of the shells with different wavefronts are shown in Figs. 1 and 2. It can be seen that the buckling modes are different for the wavefronts, the more the axial wavenumber$m$ is, the larger the wavefront is. At the same time, they are all local buckling that occurs in areas disturbed by stress waves. Comparing with the torsional buckling modes of homogeneous cylindrical shell in [18], it is found that the buckling modes of the homogeneous FGM cylindrical shells are identical, that is, the material changes do not affect the mode pattern.

Let $k = 0$, FGM reduce to homogeneous material. Table 1 shows comparisons between the torsional buckling loads of the homogeneous cylindrical shell calculated in this chapter and the corresponding results in [18]. It can be found the present results are very close to the corresponding results in the literature, indicating that the theoretical derivation and numerical calculation are correct and reliable. Since the critical loads are obtained by numerically solving bifurcation conditions through the Newton iterative method, some differences arise.

If not specified, the geometries of the cylindrical shells are chosen to be $h = 0.05\,\text{m}$, $R = 1\,\text{m}$ in the subsequent calculation. Table 2 gives the first sixth-order buckling loads for different wavefronts. It can be seen from the table that as the axial mode order increases, the buckling loads increase. And the longer the wavefront is selected, the smaller the buckling load is.

Table 3 lists the first sixth-order critical loads of FGM shells with different k when the circumferential wave is $n = 4$ and wavefront is $T = 4$. Figures 3 and 4 further show the variations of the first sixth-order circumferential and the

Table 2 First sixth-order buckling loads for different wavefronts ($n = 1$)

m	1	2	3	4	5	6
$T = 0.9$	804	827	1570	1603	2835	2850
$T = 1.3$	687	720	1100	1112	1730	1752
$T = 3.2$	516	523	558	570	632	638
$T = 4.8$	509	513	526	531	557	559

Table 3 Buckling loads of FGM cylindrical shells with different axial waves (unit: N)

m	1	2	3	4	5	6
SiC	4.69×10^8	4.7×10^8	4.91×10^8	4.96×10^8	5.36×10^8	5.46×10^8
$k = 0.5$	3.87×10^8	3.91×10^8	4.11×10^8	4.19×10^8	4.56×10^8	4.62×10^8
$k = 1$	3.47×10^8	3.49×10^8	3.72×10^8	3.77×10^8	4.14×10^8	4.23×10^8
$k = 2$	3.08×10^8	3.09×10^8	3.33×10^8	3.35×10^8	3.74×10^8	3.8×10^8
$k = 10$	2.51×10^8	2.53×10^8	2.71×10^8	2.75×10^8	3.05×10^8	3.11×10^8
Ni	2.26×10^8	2.27×10^8	2.37×10^8	2.39×10^8	2.59×10^8	2.64×10^8

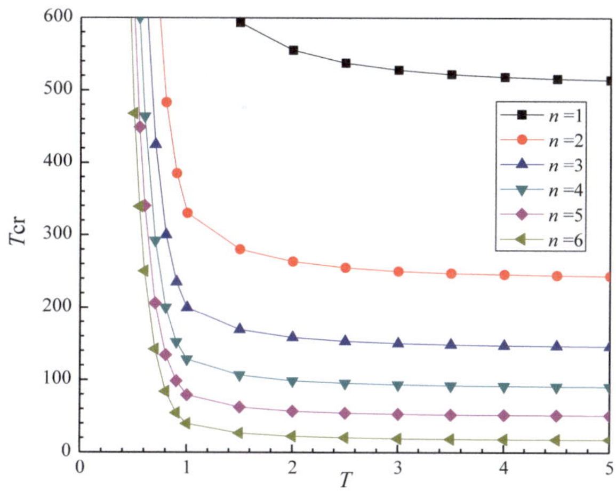

Fig. 3 Variations of the first sixth-order circumferential buckling loads with the wavefront ($m = 1$)

first eighth-order radial buckling loads with the wavefront. It can be seen that the torsional buckling loads decrease as the order of the buckling modes increase. Additionally, the buckling loads drop when the power law index k increases, i.e., the ability of the shells to withstand dynamic torsional loads decreases. This is due to the fact that the constituents of the ceramics decrease with the increasing k. And the longer the wavefront of the action is selected, the smaller buckling loads are.

Finally, the variations of critical buckling loads are presented in Table 4 with the changes of k and some specified ratios γ representing ratios of radius to thickness. When the wavefront is $T = 4$, the circumferential and axial orders are $n = 4$ and $m = 1$, respectively. It shows that the critical buckling loads generally decrease as the ratios of radius to thickness γ increase. This is because that the bending stiffness decreases as γ increases.

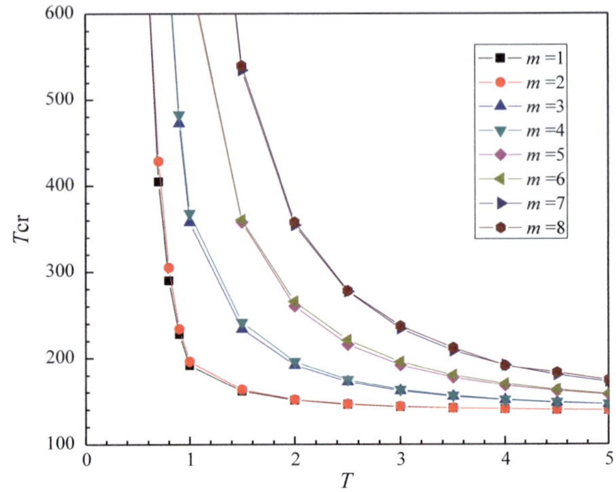

Fig. 4 Variations of the first eighth-order axial buckling loads with the wavefront ($n = 3$)

Table 4 The critical buckling loads for FGM shells with some specified ratio γ (unit: N)

γ	16	20	24	28	32
SiC	4.91×10^8	4.69×10^8	4.58×10^8	4.44×10^8	4.28×10^8
$k = 0.5$	3.82×10^8	3.87×10^8	3.8×10^8	3.68×10^8	3.51×10^8
$k = 1$	3.4×10^8	3.47×10^8	3.43×10^8	3.32×10^8	3.17×10^8
$k = 2$	2.99×10^8	3.08×10^8	3.04×10^8	2.91×10^8	2.84×10^8
$k = 5$	2.62×10^8	2.7×10^8	2.65×10^8	2.54×10^8	2.48×10^8
$k = 10$	2.46×10^8	2.51×10^8	2.47×10^8	2.42×10^8	2.3×10^8
Ni	2.37×10^8	2.26×10^8	2.21×10^8	2.14×10^8	2.06×10^8

4 Conclusions

In this chapter, the torsional impact buckling of the ceramic-metal FGM cylindrical shell is investigated in Hamilton system by symplectic method. The canonical equations are established, and a complete buckling mode space is given. The relationship between the critical loads and the eigenvalues has been revealed, so has it between the buckling modes and eigensolutions. It is found that the gradient properties of FGM have significant effect on the buckling loads. The buckling loads of the metal/ceramic FGM shells are intermediate to those of metal and ceramic shells, and decrease monotonously with the increasing of power law index, furthermore, the ratio of radius to thickness also have great influences on the critical buckling loads.

Acknowledgments This work was supported by the National Natural Science Foundation of China [grant numbers 11662008, 11862012].

References

1. Zhang, J.H., Li, G.Z., Li, S.R.: Analysis of transient displacements for a ceramic–metal functionally graded cylindrical shell under dynamic thermal loading. Ceram. Int. **41**, 12378–12385 (2015)
2. Gao, K., Gao, W., Wu, D., et al.: Nonlinear dynamic buckling of the imperfect orthotropic E-FGM circular cylindrical shells subjected to the longitudinal constant velocity. Int. J. Mech. Sci. **138–139**, 199–209 (2018)
3. Javaheri, R., Eslami, M.R.: Thermal buckling of functionally graded plates. AIAA J. **40**, 162–169 (2002)
4. Ma, L.S., Wang, T.J.: Nonlinear bending and post-buckling of a functionally graded circular plate under mechanical and thermal loadings. Int. J. Solids Struct. **40**, 3311–3330 (2003)
5. Liew, K.M., Yang, J., Kitipornchia, S.: Post-bucking of piezoelectric FGM plates subject to thermo-electro-mechanical loading. Int. J. Solids Struct. **41**, 3869–3892 (2003)
6. Li, S.R., Zhang, J.H., Zhao, Y.G.: Thermal post-buckling of functionally graded material Timoshenko beams. Appl. Math. Mech. **27**(6), 803–810 (2006)
7. Zhang, J.H., Pan, S.C., Chen, L.K.: Dynamic thermal buckling and postbuckling of clamped–clamped imperfect functionally graded annular plates. Nonlinear Dyn. **95**, 565–577 (2019)
8. Mirzavand, B., Eslami, M.R., Reddy, J.N.: Dynamic thermal postbuckling analysis of shear deformable piezoelectric FGM cylindrical shells. J. Therm. Stresses. **36**, 189–206 (2013)
9. Han, Q., Huang, H.W.: Nonlinear dynamic buckling of functionally graded cylindrical shells subjected to time-dependent axial load. Compos. Struct. **92**, 593–598 (2010)
10. Mehrian, S.M.N., Naei, M.H.: Two-dimensional analysis of functionally graded partial annular disk under radial thermal shock using Hybrid Fourier-Laplace transform. Appl. Mech. Mater. **436**, 92–99 (2013)
11. Lim, C.W., Xu, X.S.: Symplectic elasticity: theory and applications. Appl. Mech. Rev. **63**, 50802 (2010)
12. Xu, X.S., Ma, J.Q., Lim, C.W., Chu, H.: Dynamic local and global buckling of cylindrical shells under axial impact. Eng. Struct. **31**, 1132–1140 (2009)
13. Sun, J.B., Xu, X.S., Lim, C.W.: Buckling of functionally graded cylindrical shells under combined thermal and compressive loads. J. Therm. Stresses. **37**(3), 340–362 (2014)
14. Strozzi, M., Pellicano, F.: Nonlinear vibrations of functionally graded cylindrical shells. Thin-Walled Struct. **67**, 63–77 (2013)
15. Alijani, F., Amabili, M., Karagiozis, K., Bakhtiari-Nejad, F.: Nonlinear vibrations of functionally graded doubly curved shallow shells. J. Sound Vib. **330**, 1432–1454 (2011)
16. Alijani, F., Amabili, M., Bakhtiari-Nejad, F.: Thermal effects on nonlinear vibrations of functionally graded doubly curved shells using higher order shear deformation theory. Compos. Struct. **93**, 2541–2553 (2011)
17. Zhang, J.H., Li, S.R.: Dynamic buckling of FGM truncated conical shells subjected to non-uniform normal impact load. Compos. Struct. **92**, 2979–2983 (2010)
18. Xu, X.S., Ma, J.Q., Lim, C.W., et al.: Dynamic torsional buckling of cylindrical shells. Comput. Struct. **88**, 322–330 (2010)

Part II
MEMS/NEMS and Energy Harvesters

Nonlinear Dynamic Modeling for High Temperature Superconductivity in Nanocluster Topological Structures on Solid Surface

Sergei M. Arakelian, Igor Yu. Chestnov, Alexander V. Istratov, Timur A. Khudaiberganov, and Oleg Ya. Butkovskiy

Abstract We studied laser-induced nanocluster structures of different types in both topology and the element compositions due to the nonlinear interaction of laser radiation with the condensed matter taking into account the correlations in nanoparticle ensemble by quantum states. The problem of both optical response and high temperature superconductivity, due to topological surface structures with correlated states, is under our consideration in the frame of nonlinear dynamic modeling resulting, e.g., in the electronic Cooper pairs appearance. Random temporal and spatial variations in selected topological parameters may result in large variations of such functional properties. The analogy with nonlinear dynamics of system under external noise takes place in the case. Quantum mobility of electrons over different trajectories in the spatially inhomogeneous structures/nanocluster systems is presented in accordance with the path integral-theory approach.

Keywords High temperature superconductivity · Nonlinear dynamics · Nanoclusters · Topology structures

1 Introduction

The achievements of modern nonlinear dynamics determine both the behavior and universal processes in highly excited trigger-like systems of different types [1]. In particular, they are used for predictive modeling of the development of phenomena in many branches of physics and its applications [2]. Especially, this concerns

S. M. Arakelian (✉) · A. V. Istratov · T. A. Khudaiberganov · O. Y. Butkovskiy
Department of Physics and Applied Mathematics, Vladimir State University, Vladimir, Russia
e-mail: arak@vlsu.ru

I. Y. Chestnov
Department of Physics and Applied Mathematics, Vladimir State University, Vladimir, Russia

Institute of Natural Sciences, Westlake Institute for Advanced Study, Zhejiang Province, China

W. Lacarbonara et al. (eds.), *New Trends in Nonlinear Dynamics*,
https://doi.org/10.1007/978-3-030-34724-6_13

the space-time description and forecast of consequences due to development of complex nonlinear threshold interactions under laser radiation [3]. In fact, e.g., the transformation of topological charge of optical singularities in both amplitude and polarization characteristics is under intensive study at present time (see e.g., [4]).

In this aspect, for such specific objects as quantum units, the development of nonlinear wave processes may be described in analogy with nonlinear dynamics [5]. Therefore, it is of considerable interest to analyze the fundamental quantum phenomena and their macroscopic manifestations in real experiments. We explore these possible effects, in particular in electrophysics and optics, for nanostructured systems with controlled topology induced by laser radiation.

However, in practical aspect, this requires a prior consideration of the technology of synthesizing in a given direction such nanostructures in a laser experiment [6]. Our chapter is devoted to such type of interdisciplinary direction on the basis of nonlinear dynamics achievements. We briefly review as well the nonlinear dynamic modeling for necessary conditions in a high temperature superconductivity problem for the thin film nanocluster topological structures on solid surface in the above complex procedure. We are talking about the following approach being implemented.

The physical properties of nanocluster systems are very sensitive to the form, size, and distance/spatial distribution between their components. The fact is well known for any material in general, but to change these parameters and carry out the stable conditions for ordinary monolith solid objects we need to put the objects under extremal conditions (cf. [7]). In contrast, nanocluster structures can be easily modified in the necessary way in femto-nanophotonics laser experiments (cf. [6]). They may be presented as a grid ensemble of a single-electron system (with a certain effective mass m_{eff}) demonstrating a nonlinear quantum dynamic behavior, and the nonlinear spectroscopic effects manifestation around the forbidden gap occur. In superconductor problem, the main question is how to fabricate the coupling states for charged particles being responsible for electroconductivity. For a cluster system, we discuss some alternative mechanisms of electronic coupling (in equilibrium states), but not via a standard Cooper phonon coupling only. In our early experiments, we used a multilayer film formation by two-stage dynamic nonlinear process: first, laser-induced nanoparticles arise in colloidal solution by ms/ns-laser pulses (10^6–10^7 W/cm^2) acting on the solid target in liquid; and second, the creation of thin films due to laser deposition of nanoparticles from colloid on a solid surface under cw-radiation. By such technique we obtained a dramatic enhancement (in several orders) of electroconductivity due to the variation of topological peculiarities of an induced nanocluster thin film system (cf. [8]). The process may be interpreted as a nonlinear phase transition in topological structure with bifurcations in functional properties.

Such hopping/bifurcation behavior in electrical conductivity is extremely important for applications. Indeed, it allows one to realize two different states with a controlled transition between them in a nonlinear system. This results in development of new physical principles for creating of topological elements and devices for modern photonics and nanoelectronics.

2 Physical Basis and Reasonable Models

We obtained nanostructures and thin films with controllable topology under the development of different nonlinear processes in the system: thermodiffusion, gas-dynamic evaporation in pore-like structures with bubbles, ablation products, and ballistic movement of the particles in liquid [6]. All these processes can be analyzed in accordance with different nonlinear hydrodynamic regimes. We discuss both some possible nonlinear mechanisms being responsible for a high electroconductivity and the features of obtaining a hopping conductivity in such inhomogeneous thin film surface structures (with the thickness of up to 100 nm) when the charged particles are propagating along the boundary surface of granulated structures.

In terms of discussion, first, it is obvious that electrophysics strongly depends on the topology of nanostructured films on the solid surface. Such nonlinear electrical transport properties are due to the quantum correlated states resulting in the tunnel and hopping electrical conductivity. There is a competition between the bulk and surface electrical conductivity contributions, controlled (to a great extent) by a deposited cluster topology/temperature [9].

Second, optical spectra of thin films (produced layer by layer) result in some new physical states of the system, particularly, in optical response (cf. [10]).

Below, we discuss the problems in different aspects with more details for the nanocluster topology models mentioned above.

3 The Topology Cluster Shape Variations and Functional Characteristics of the Objects vs Different Key Parameters

Different topological structures for nanoobjects, obtained by computer simulation in arbitrary units in spherical coordinate system (R, θ, ϕ), were modeled by us.

From the very beginning, the shape of the body was spherical (radius R_0) which is perturbed by variation of some key parameters: the ratio of the values of both azimuthal (k_1) and zenithal (k_2) distortions ($0 \leq k_1 \leq 1$, $0 \leq k_2 \leq 1$, respectively); as well as the number of distortions for the same parameters ($p_1 = 0, 1, 2, \ldots$; $p_2 = 0, 1, 2, \ldots$).

In the case, a surface deformation model is defined by:

$$R(\theta,\phi) = \begin{cases} R_0 \left[\begin{array}{l} (1+k_1 \cos(\theta)\cos(p_1\phi)) + \\ + k_2 \left((-1)^{\frac{p_2-1}{2}} \operatorname{mod}(p_2, 2) \sin(p_2\theta) + \operatorname{mod}(p_2+1, 2)\cos(p_2\theta) \right) \end{array} \right], & p_1 \neq 0, \\ R_0 k_2 \left((-1)^{\frac{p_2-1}{2}} \operatorname{mod}(p_2, 2) \sin(p_2\theta) + \operatorname{mod}(p_2+1, 2)\cos(p_2\theta) \right), & p_1 = 0. \end{cases} \quad (1)$$

We analyzed the following cases for the deformation parameters: $k_1 = k_2 = 0.4$ and $p_2 = 0, 1, 2, \ldots$; $p_1 = 0, 1, 2$. Soft Mathcad programming has been used for the modeling.

Several selected results in accordance with formula (1) are shown in Fig. 1a.

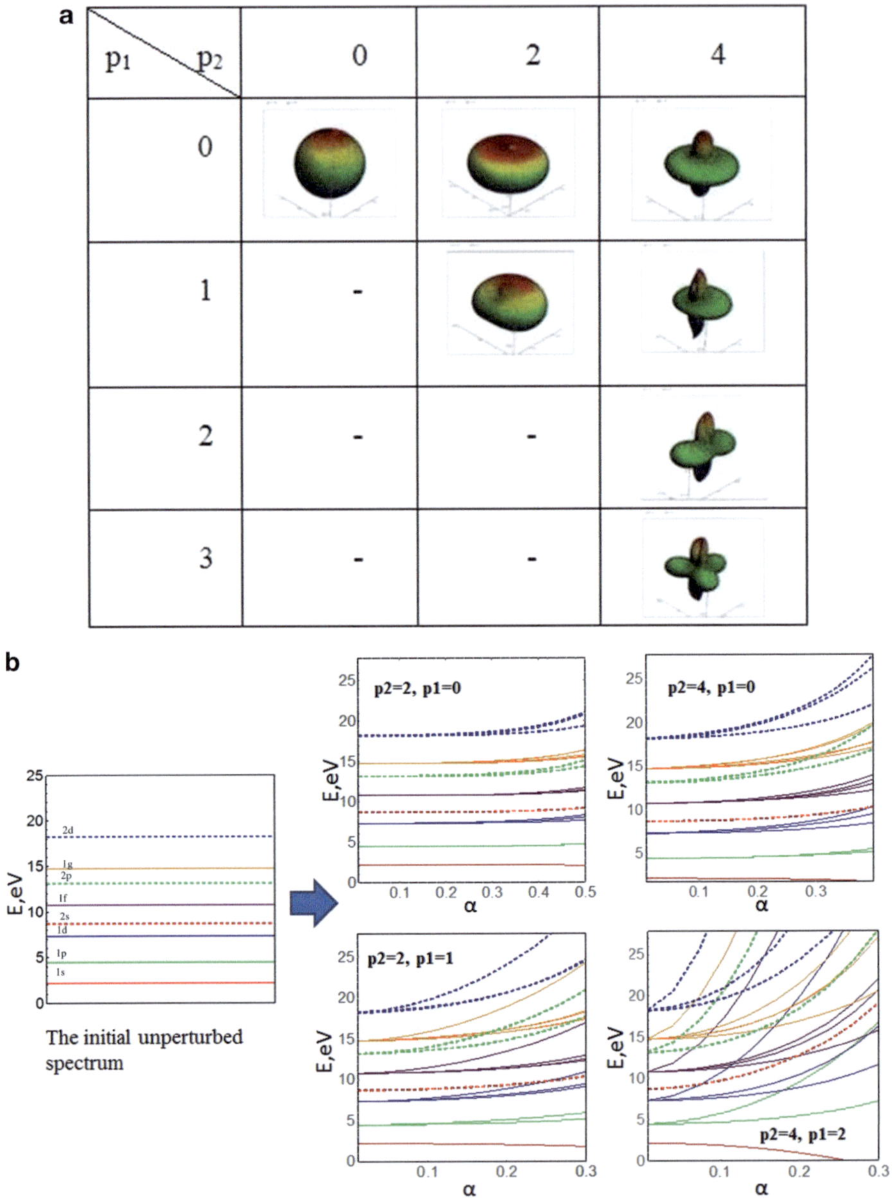

Fig. 1 (**a**) Several topological structures controlled by the key parameters (see text Eq. (1)). (**b**) Spectra of deformed nanoclusters. The split energy levels are shown for different distortions when we fixed $k_1 = k_2 = \alpha$ and variable magnitudes of p_1, p_2 (see Eq. (1)). Numerical parameters are: valent electron density $N_e = 40$; Wigner–Seitz parameter $r_s = 2.1$ (being $\sim 1/N_e^{1/2}$); electron effective mass $m^*_{\rm eff} = 1.4m_e = 0.708$ MeV

The energetic parameters, relying on the fact that the obtained nanoobject complex structures result in the modification of the potential energy distribution for quantum wells, can be discussed in a standard quantum mechanics approach.

In this case, the energy levels are split as follows (see Fig. 1b). We used the simplest case in approximation of the first order linearization in the small deformations vs geometric parameters which result in perturbation of the energy levels as $\delta E_{n,l,m} = \int_0^{2\pi}\int_0^{\pi}\int_0^{R_0} \psi^*_{n,l,m}(r,\theta,\phi)\,\delta H\,\psi_{n,l,m}(r,\theta,\phi)\,r^2 \sin\theta dr d\theta d\phi$, where *n, l, m* are the standard quantum numbers in the atomic spectrum of quantum system, δH—perturbation of Hamiltonian, ψ—wave function.

We note that the transformations of the energy spectra (determined by depth and width of the potential wells) become more significant for electrons especially at higher energy levels. Therefore, the processes of their going beyond the cluster boundaries of the potential wells occur. Then, the problem should be considered as the travelling of electrons through the ensemble of potential barriers with different parameters. However, it is now of fundamental importance for us, that the movement of electrons in a nanostructured system under external voltage leads to electrical conductivity (see Fig. 2a). Thus, the problem can be reduced to a dynamic model of the electrons motion, being nonlinear in a general case (e.g., with a hopping behavior) but we should latter discuss the problem in accordance with the path integrals over quantum trajectories (cf. [11])—see below division 4.

Transmission optical spectra *T* for different conditions are shown in Fig. 2b.

4 Enhancement Electroconductivity Physics: Nonlinear Dynamic Model

In the electrical conductivity for bound (Cooper) electrons (*N*), a spectral picture determines the opportunity of the superconducting state, in principle [12]. Therefore, it is possible that the two alternative processes be realized (the pair production and their decay) with some certain ratios of the energy parameters for such systems. As for semiconductors, we are speaking about the generation (production)—g_N of pairs and their recombination (decay)—r_N in the presence of a forbidden band.

This problem can be reduced, e.g., to a well-known nonlinear dynamic Ferhulst model of the growth of biological population [1, 2]. Thus, the nonlinear dynamics of the emergency of the Cooper pair density (*N*) may be used in the frame of such model.

The main recognized mechanism for the emergence of superconductivity is considered to be the electron–phonon interaction [13], when the attraction arises between two electrons under the influence of lattice vibrations. But in cluster system the topological parameters may be responsible for the coupling [8].

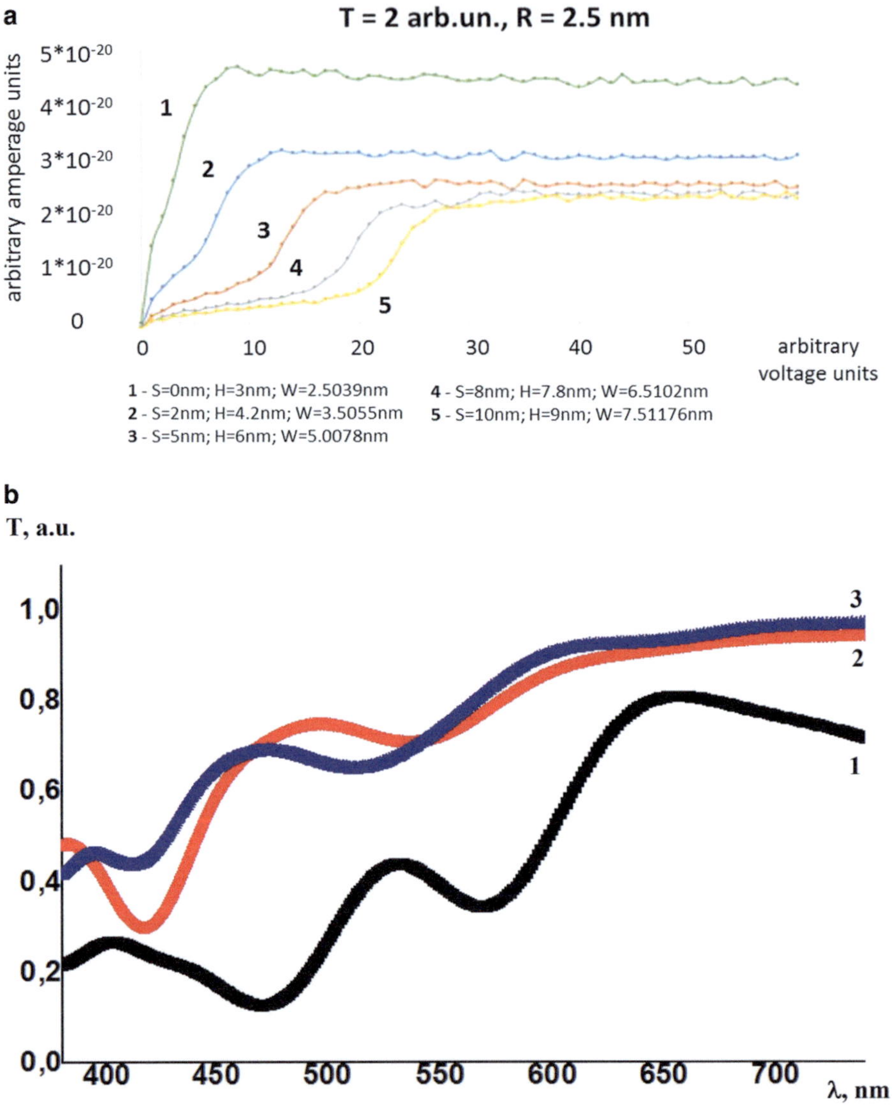

Fig. 2 (**a**) Volt-ampere characteristics of thin films vs the distance (S) between the conducting islands and radius (R) of deposited particles; τ—time of laser deposition; H—thickness of the thin film; W—distance between microcontacts. (**b**) Modeling the optical transmission spectra of deposited bimetallic films for noble metal compositions Au:Ag (concentration 1:1): (1) $2R = 50$ nm, one layer, the distance between the particles (gap)—5 nm; (2) $2R = 10$ nm, five layers, a gap—4 nm; (3) $2R = 10$ nm, five layers, a gap—2 nm. The FDTD method has been used for the modeling

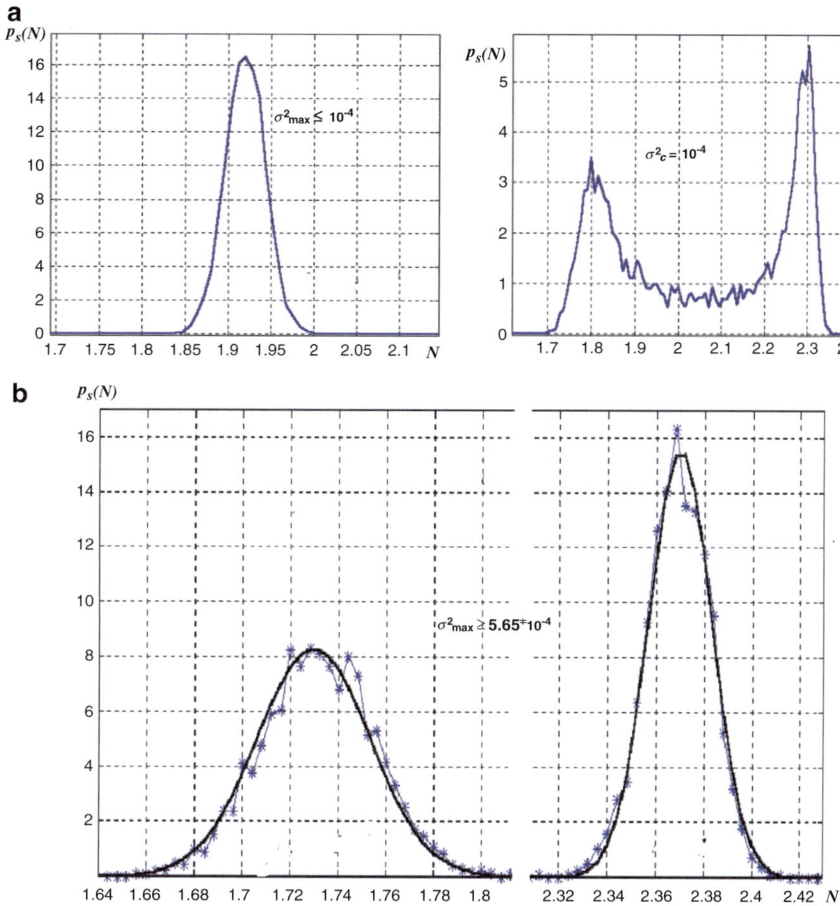

Fig. 3 A stationary probability density for coupled pairs $p_s(N)$—initial distribution function when the fluctuation intensity $\sigma^2_{max} < \sigma^2_c$ (**a**); $p_s(N)$ at the critical point $\sigma^2_c = 10^{-4}$ (**b**); $p_s(N)$ after the bifurcation point when the current value $\sigma^2_x > \sigma^2_c$ (**c**). The numerical modeling is carried out while $p_s(N) = 1/\pi\sqrt{(N-1.8)(2.3-N)}$

For studying a bifurcation in the system behavior under presence of fluctuations, as in [2], the condition $N > 0$ occurs in respect of the some critical noise value σ^2_c—see Fig. 3. In our case, it is a question of a specific topology structure in nanocluster system being controlled by laser radiation.

5 Electroconductivity in a Granulated Metallic Film in Terms of the Path Integral-Theory Approach

Quantum mobility of electrons over different trajectories in a spatially inhomogeneous structure, i.e., in a nanocluster network system, may be presented in accordance with the path integral-theory approach [11].

We analyzed the problem in frame of standard model of harmonic oscillator (with frequency ω and quantum numbers l, m) in an external electric field E with an offset $(x - x_{l\mathrm{m}})$ in the one-dimensional case.

Propagator K for the moving of charge particles between two space-localized states with normal coordinates $(q' - q'')$ at the corresponding time points $(t' - t'')$ are written in the form:

$$K = \left\langle q'' \right| U\left(t'', t'\right) \left| q' \right\rangle = \int \exp\left(\frac{i}{\hbar} \int\limits_{t'}^{t''} L dt \right) Dq, \tag{2}$$

with U as an evolution operator, $Dq = \frac{1}{A} \prod_{j=1}^{n-1} \frac{1}{A} dq_{\mathrm{j}}\left(t_{\mathrm{j}}\right)$ is a path integral measure, and Lagrangian L is represented in quadratic form:

$$L = \frac{1}{2} \sum_{l,m} \left\{ \frac{x\prime^2}{2N} - \omega^2 (x - x_{lm})^2 + Ex \right\}, \tag{3}$$

where A—cross-section of the structure, N—number of the time sections due to fragmentation of $(t' - t'')$-interval (transfer matrix approach [14]).

Finally, our results for electrical conductivity in the simple model of Kronig–Penney [7] are presented in Fig. 4. It depends on relationship between electron energy (E) and Fermi-energy (E_{F}). Region 1—shown on a small scale—is a dependence in accordance with Ohm's law. With a positive charge shift to the right microcontact relative to the left one due to the applied voltage, the level of Fermi energy (E_{F}) on the left corresponds to the resonant level. This leads to a significant increase in current (region 2). With a larger value of the charge bias, the current stops flowing when the E_{F} level falls below the edge of the width of the conduction band. The result is a noticeable decrease in the current value with increasing voltage, which corresponds to the emergence of a region of negative differential resistance and/or a mode with significant current suppression (region 3). With a larger value of the charge displacement, the current increases again as the charge particles (electrons) acquire sufficient kinetic energy (region 4).

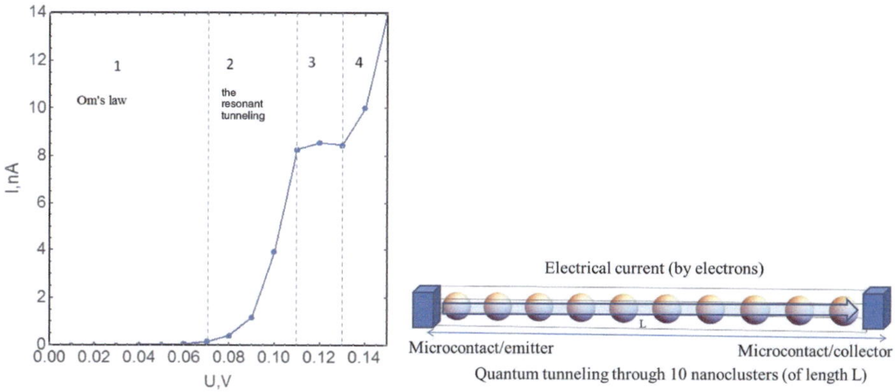

Fig. 4 Different regimes of electroconductivity in cluster systems (for Kronig–Penney model of superlattice—see on the right). Vertical axis—current I, horizontal axis—voltage U. The values of the numerical parameters are the same as in Fig. 1b; $E_F = 6$ eV

6 Conclusions

The main problem is how to determine the specific numerical values of the control parameters of a nanocluster system relying on these general relations for the achievement of the desired final result in the experiment. This requires a detailed research (both in theory and in experiment) of the influence of nanocluster topological parameters on the functional properties of the substances used. It is a very complicated problem, and is still far from being solved.

Nevertheless, the observed phenomena give us an opportunity to establish the basis of new physical principles to create the functional elements for topological photonics in hybrid set-up (optics + electrophysics) being controlled by nonlinear dynamic processes (cf. [6]). In fact, we have obtained in the experiment [8, 9] several unusual effects, both in optical characteristics and in electroconductivity. The study of the subject in the future should introduce to the analysis the methods of the stochastic nonlinear dynamic approach based on exact data obtained from experiment in the aspect of connection of topological parameters of nanocluster system (in both space and time domains) with observed functional properties.

Acknowledgements The chapter was prepared within the framework of the state task of VlSU No 16.1123.2017/4.6, and was partially supported by the Ministry of Science and Higher Education of Russia – Agreement No. 075-15-2019-1838.

References

1. Horsthemke, W., Lefever, R.: Noise-Induced Transitions. Theory and Application in Physics, Chemistry and Biology, vol. 15, p. 332. Springer Series in Synergetics (1984)
2. Van Kampen, N.G.: Stochastic Processes in Physics and Chemistry, p. 470. North-Holland Personal (1984)
3. Klimontovich, Y.L.: The Statistical Theory of Open Systems, p. 624. M.: Janus LLP (1995)
4. Grigoriev, K.S., Makarov, V.A., Perezhogin, I.A.: Polarization singularities in a sumfrequency light beam generated by a bichromatic singular beam in the bulk of an isotropic nonlinear chiral medium. Phys. Rev. A. **92**, 023814 (2015)
5. Kravtsov, Y.A., Bilchinskaya, S.G., Butkovskii, O.Y., et al.: Prebifurcational noise rise in nonlinear systems. Russ. J. Exp. Theor. Phys. **93**(6), 1323–1329 (2001)
6. Arakelian, S.M., Kucherik, A.O., Prokoshev, V.G., et al.: Introduction to the Femto-nanophotonics, Fundamental Principles and Methods of Laser Diagnostics and Control of Nanostructured materials, p. 744. M.: Logos (2015)
7. Lifshitz, E.M., Pitaevsky, L.P.: Theoretical Physics. Statistical Physics. Part 2: Theory of Condensate State, vol. IX, p. 440. M.: Fizmatlit (2015)
8. Arakelian, S.M., Kucherik, A.O., Kutrovskaya, S.V., et al.: Verification of quantum dimension effects in electrical conductivity with different topology of laser-induced thin-film structures. J. Phys.: Conf. Ser. **951**, 012018.1–012018.5 (2018)
9. Kavokin, A.V., Kutrovskaya, S.V., Kucherik, A.O., et al.: Crossover between tunnel and hopping conductivity in granulated films of noble metals. Superlattice. Microst. **111**, 335–339 (2017)
10. Kutrovskaya, S.V., Arakelian, S.M., Kucherik, A.O., et al.: Synthesis of hybrid gold-silicon nano particles in a liquid. Sci. Rep. **7**, 10284.6 (2017)
11. Feinmann, R.P., Hibbs, A.R.: Quantum Mechanics and Path Integrals, p. 384. McGraw—Hill Book Company, New York (1965)
12. Kresin, V.Z., Ovchinnikov, Y.N.: Shell structure and strengthening of superconducting pair correlation in nanoclusters. Phys. Rev. **B74**, 024514 (2006)
13. Abrikosov, A.A.: Fundamentals of the Theory of Metals, p. 600. M.: Fizmatlit (2009)
14. Christian, J.: Accuracy of transfer matrix approaches for solving the effective mass Schrodinger equation. IEEE J. Quantum Electron. **45**, 1059–1067 (2009)

Nonlinear Dynamic Processes in Laser-Induced Transitions to Low-Dimensional Carbon Nanostructures in Bulk Graphite Unit

Kirill Khorkov, Dmitriy Kochuev, Ruslan Chkalov, Valery Prokoshev, and Sergei Arakelian

Abstract Development of nonstationary technique for the laser-induced functional elements synthesis based on micro- and nanostructures in graphite samples is under study. Carbon nanostructures such as graphene, nanopeaks, and crystals have been obtained in our experiments. The nonlinear formation mechanisms of nanostructures and microcrystals under femtosecond laser radiation for graphite in liquid nitrogen are analyzed. Femtosecond laser pulses with high power allow achieving the local transient conditions for the nonstationary material processing resulting in ablation, sufficient modification of the structure, and/or changing of the phase composition of the materials. Liquid nitrogen as a cryogenic and/or reaction liquid contributes to additional fast cooling and stabilization of the fabricated micro- and nanostructures.

Keywords Femtosecond lasers · Nanostructures · Graphene · Carbon microcrystals · Liquid nitrogen · Qualitative model

1 Introduction

Development of synthesis methods of functional elements based on micro- and nanostructures is one of the most important and promising areas of modern technologies. One of the manifestations of unique properties of carbon is the presence of a large number of its crystal modifications, ranging from superhard forms to various variations of graphite. Carbon nanostructures such as graphene, nanotubes, and nanodiamonds are widely used in various fields of photonics, micro-, and optoelectronics as the detectors of individual gas molecules [1, 2], the material of supercapacitors electrodes with a high specific surface area and internal capacity [3], the saturated absorbers [2, 4], etc.

Using intense laser radiation with ultrashort duration allows to reaching the local conditions in the area of impact and sufficiently change the phase composition

K. Khorkov (✉) · D. Kochuev · R. Chkalov · V. Prokoshev · S. Arakelian
Vladimir State University, Vladimir, Russia

W. Lacarbonara et al. (eds.), *New Trends in Nonlinear Dynamics*,
https://doi.org/10.1007/978-3-030-34724-6_14

of the material, including dielectrics, semiconductors, and metals [5–7]. The use of ultrashort-duration laser radiation is widely used for nanostructuring [8–11], modification and processing of various materials [12, 13], thin films deposition, production of colloidal solutions, and many others [14, 15]. Research and development of new laser synthesis methods of nanostructures and microcrystals in liquid nitrogen is an actual task of creating a modern element base of photonics, micro-, and optoelectronics.

2 Experimental Technique

In our experiments, we used two femtosecond laser systems: Ytterbium laser system (wavelength $\lambda = 1030$ nm, radiation pulse duration $\tau = 280$ fs, pulse repetition rate $f = 10$ kHz, pulse energy $E_{max} = 150$ µJ) and Titanium-sapphire laser system (wavelength $\lambda = 800$ nm, radiation pulse duration $\tau = 50$ fs, pulse repetition rate $f = 1$ kHz, pulse energy $E_{max} = 1$ mJ). Surface processing of graphite was carried out in liquid nitrogen, which covered it with a layer thickness of 10 mm. The laser spot diameter was up to 100 µm. As the samples we used target of highly oriented pyrolytic graphite HOPG-1,7-10 $\times$ 10 $\times$ 1 and glassy carbon GC-2000.

For experiments of processing and micro-nanostructuring of materials by ultra-short laser pulses, a hardware–software laser complex was developed, combining a number of required technical components and having a single interface for controlling the component elements of the setup. Also the software was developed and the scheme of precision controlled treatment by femtosecond laser radiation of materials for the formation of micro- and nanostructures was implemented [16].

3 Physical Processes of Laser-Induced Carbon Structures Formation

All currently known methods of obtaining graphene can be divided into two groups: synthesis (chemical vapor deposition, obtaining graphene in the electric arc, the epitaxial growing on the metal surface, etc.) and separation (micromechanical exfoliation of graphite, liquid-phase exfoliation of graphite the oxidation of graphite, etc.). Today, there are many methods of liquid-phase graphene separation. Proposed method is based on the intercalation of foreign atoms or molecules into the interlayer space of graphite. As a result of the introduction of impurities weakens the energy of interaction between the layers by increasing the distance between them [17].

Development the laser-induced graphene exfoliation method at the interaction of femtosecond laser radiation with carbon samples in liquid nitrogen allows to eliminate two serious drawbacks of existing implementations of the liquid-phase separation method: duration of treatment and contamination by impurities. For graphene exfoliation in this method, laser photomechanical effects are used, which are more pronounced, the shorter pulse duration of laser radiation. It is known that rapid laser heating leads to appearance of strong thermoelastic stresses, so that the front surface of the irradiated sample can spall in the area of femtosecond laser radiation [6, 18]. Some thermal effects, accompanying laser exposure, prevent the direct formation of nanostructures with clear boundaries, as shown by researches in the field of interaction of laser radiation with carbon materials [19], including nanostructuring. A strong smoothing effect has a slow cooling process of the irradiated area and the impact of formed plasma plume.

In the interaction of ultrashort laser pulses with a graphite sample in a liquid nitrogen medium, several variants of the course of the processes of structural changes in the material can be identified (Fig. 1a). When the laser radiation power is insufficient for laser ablation processes to occur, the photon energy is absorbed by the carbon lattice and intercalated nitrogen. In view of the sharp increase in the lattice vibrations and nitrogen molecules, weak interplane π-bonds break. Cleavage of graphite layers occurs mainly on defects of the processed structure (cracks, chips, and scratches, including dissimilar lattice defects in the form of gaps) conducive to the penetration of liquid nitrogen molecules. Accordingly, increasing a number of nitrogen atoms in the lattice of a material contributes to a more intense exfoliation of graphene sheets.

Using intense laser radiation of ultrashort duration allows reaching the local conditions in the area of impact and sufficiently changing the phase composition of the material. Femtosecond laser radiation provides nonlinear ultrafast transfer of target energy—laser energy is localized and absorbed by the substance; the use

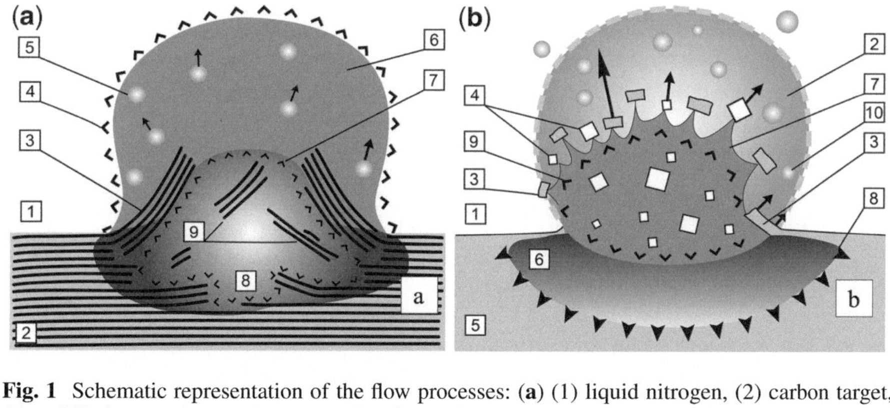

Fig. 1 Schematic representation of the flow processes: (**a**) (1) liquid nitrogen, (2) carbon target, (3) exfoliating graphene, (4) propagation front of vapor-gas region, (5) evaporated nitrogen, (6) vapor-gas area, (7) front of high-pressure region, (8) high-pressure area (the heating of intercalated nitrogen, laser ablation), (9) exfoliated graphene; (**b**) (1) liquid nitrogen, (2) vapor-gas area, (3) near-surface layer, (4) formed carbon crystals, (5) carbon target, (6) the area of heat removal into the volume of the material, (7) the region of subterranean overheating, (8) front of spreading/rising pressure, (9) front of temperature propagation due to thermal conductivity, (10) evaporated nitrogen

of liquid nitrogen as a cryogenic liquid accelerates the cooling process of sharply heated target material and contributes to the stabilization of crystalline forms.

Processes, accompanied by a rapid increase in temperature and pressure, as well as their rapid decrease, are highly unsteady. In this case, it is difficult to talk about the carbon phase diagram, which makes sense only for equilibrium states. Under the conditions of exposure accompanied by an additional laser-induced mechanism—subsurface overheating of the substance, the formation of an intermediate metastable state is possible. It is possible to form a local layer of liquid carbon at high pressure, which makes permissible to synthesize crystalline carbon structures. Subsurface heating up to several thousand degrees and the achievement of critical pressure values during the action of laser pulses train allow creating nonequilibrium conditions that initiate a phase explosion accompanied by the formation of single-crystal carbon micro- and nanostructures. Rapid cooling is necessary to stabilize the resulting forms.

Heating of the subsurface regions of the target from pyrolytic graphite and glass carbon was successfully used to produce graphene. To achieve conditions sufficient to change the phase composition of the material, experiments were carried out using a more powerful femtosecond Ti:Sapphire laser.

Based on the experiments results [20], we can conclude about a sharp increase in temperature and pressure, accompanied by a change in the phase composition at a certain depth of the sample, followed by the expansion of the formed bulk crystal structures to the surface. It can be assumed that the time of formation of crystal structures is much longer than the time of exposure to a single femtosecond laser pulse. The energy contribution from a single pulse is also insufficient for the growth of such a structure, consequently, there is an energy accumulation in the subsurface layer, stimulating the growth of crystals to a certain point (the expansion of structures at the opening of the surface as a result of overcoming the threshold of the pressure difference, a significant excess of stresses, etc.). It can be assumed that some calculations [18] made for nanosecond laser ablation are also valid in our case.

Figure 1b schematically shows the dynamics of physical processes at the femtosecond laser exposure at carbon sample in liquid nitrogen. During the action of a femtosecond laser pulse, energy is transferred to the electronic subsystem of the material. The duration of this process is comparable to the laser pulse duration. A photoelectron avalanche is formed, followed by energy transfer through electron–phonon interaction. The growth of the stress state of the crystal lattice is caused by the unsteady growth of temperature and pressure in the redistribution of electron–phonon energy. The process of expansion of this area develops uniformly in all directions relative to the conditional center, but this does not occur due to the highly nonequilibrium conditions of the system. The growth of this area deep into the material is limited by the "mobility" of the lattice due to the strong "cooling" of the sample volume. Heat removal, in turn, is limited by the thermal conductivity. The bulk expansion of the material into the liquid nitrogen medium is more intense due to the lower density of the medium. Thus, in the direction of the medium (liquid nitrogen), conditions are formed that contribute to the cooling of the sample surface (pressure drop, expansion of the material area into a less dense medium, heat exchange with the medium, etc.). In the direction of the volume of the main material, due to the small heat sink and the inability to expand deep into, the conditions for a significant increase in temperature and pressure are formed, which in turn leads to structural changes in the lattice of the material. The described mechanism of the processes accompanying the energy exchange in the system correlates well with the results of experimental studies.

4 Results

At the laser processing with Ytterbium system, the speed of movement of the laser beam along the surface of the target was from 10 to 100 cm/s. Accordingly, depending on the scanning speed of the beam, each region of the target surface was exposed from 1 to 10 femtosecond pulses. Graphene splitting occurs at a laser beam velocity of 25 cm/s, when the graphite surface is exposed to four laser pulses (overlapping—75%). With a larger overlap, a strong destruction of the surface of the target and at the same time an exfoliated carbon material begins [21].

Microscopic studies of the carbon material detached from the surface of HOPG under the influence of femtosecond laser radiation confirmed the production of graphene sheets of various shapes. Graphene tapes with a width of up to 50 μm and a length of more than 150 μm have been identified, as well as graphene plates of arbitrary shape with a characteristic size of more than 150 μm (Fig. 2b).

Graphene structures obtained by laser treatment of the surface of glassy carbon have a fundamentally different shape (Fig. 2d). Graphene sheets form lumps with a complex folded structure. Clumps with a characteristic size of the order of 1 μm are recorded.

At the laser processing with Ti:Sapphire system, the scanning electron microscope were revealed nanostructures and crystal structures of carbon in various forms (Fig. 3). It should be noted that when selecting a sample (HOPG or glassy carbon) similar types of nanostructures are formed. However, when processing a sample of glassy carbon, an ordered formation of nanopics is observed (Fig. 3a). Arrays

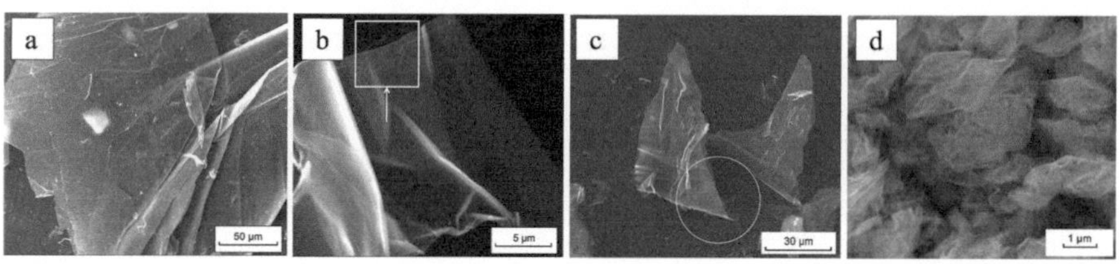

Fig. 2 SEM images of graphene, formed under the action of femtosecond laser radiation on carbon samples in liquid nitrogen: (**a**) graphene plates, (**b**) the edge region is highlighted and shown by an arrow, (**c**) edges of graphene sheets, (**d**) crumpled graphene

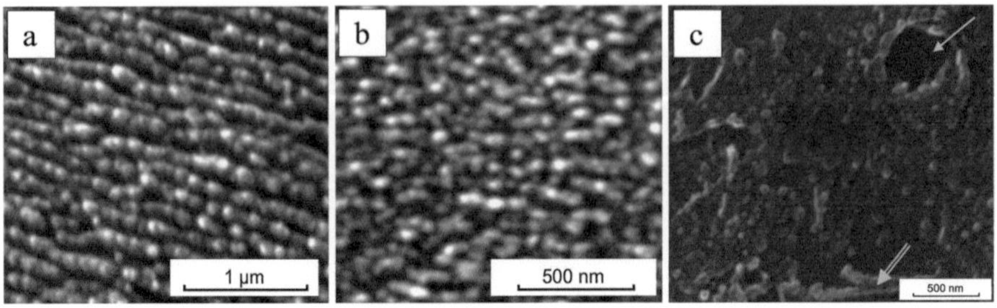

Fig. 3 SEM images of formed nanostructures on the surface of carbon samples: (**a**) glassy carbon, (**b**) HOPG, (**c**) formation of pores on the surface

Fig. 4 SEM images of different forms of carbon crystal structures formed under the influence of femtosecond laser radiation on carbon samples in liquid nitrogen

of nanostructures with characteristic sizes from 30 to 100 nm are obtained on the surface of pyrolytic graphite modified by laser radiation.

At the modes of influence accompanied by the subsurface overheating of the substance initiating phase explosion, formation of an intermediate liquid state is possible. The effect of laser radiation on the sample leads to the separation of sufficiently thick layers of pyrographite. Carbon nanostructures in the form of nanopeaks and similar to them are registered on the surface of the target, released as a result of the break of the surface layers. There were also found larger structures, clearly having a crystalline nature with dimensions of the order of 1÷10 microns (Fig. 4). The crystal structures registered after the laser-induced break of the surface layer of glassy carbon also have comparable dimensions. This is much larger than the size of nanodiamonds. Nevertheless, it can be assumed that the synthesis of both nanodiamonds and larger crystals is based on general principles. First of all, it is providing the necessary temperature and pressure. A more detailed examination of the sample separation areas allowed us to see the places where the remains of the surface layer cover some crystals and the space next to them. Removal of the surface layer cannot be associated with traditional laser ablation. In this case, the material must be removed from the entire surface within the laser spot. The remaining areas of the surface layer indicate that its removal is more nonstationary. A subsurface phase explosion can have this character [22].

The developed model of subsurface overheating allows making a qualitative assessment of the temperature distribution in the volume of the material. The temperature distribution as a result of energy transfer to the target material and the imposition of the thermal effect in the sample volume from a series of femtosecond laser pulses is simulated. The simulation results confirm the phenomenon of subsurface overheating. It is possible to adapt the model to different experimental conditions and phenomena under study in accordance with the observed experimental data [23, 24].

The use of an additional term describing the spatial region (Δx—the region near the boundary at which the heat exchange between the sample and liquid nitrogen occurs) in the simulation allows achieving such a balance of energies at which the subsurface overheating occurs. Δx-zone affects the temperature profile inside the target.

Thus, the final formula in the form of the sum of solutions of the base scenario and the summand of Δx-zone is as follows:

$$T(x, t-\tau) = T_0(x, t-\tau) + (2\gamma - 1)\frac{1}{2a\sqrt{\pi(t-\tau)}}\int_0^{+\infty} e^{-\frac{(x+x'+2\Delta x)^2}{4a^2(t-\tau)}}\varphi(x')\,dx' \tag{1}$$

where $\varphi(x) = (1-R)I_0\alpha e^{-\alpha x}$—the function forming a model initial condition, α—absorption coefficient, γ—parameter of the distribution profile, Δx—local area of interaction and thermoexchange.

Figure 5a shows the time dependence of the temperature at some depth within the sample. Figure 5b presents a graph of the temperature dependence on the depth of the sample under the influence of a series of femtosecond laser pulses, constructed in accordance with the proposed model.

In Fig. 5b digit 1 marked the temperature at which the possible phase transformation of carbon (melting). Upon reaching values of temperature 2 is the development phase of the explosion, the formation of carbon crystals in particular. The digit 3 indicates the location of the zones of possible phase transformations at some distance from the surface when the required temperature conditions are reached.

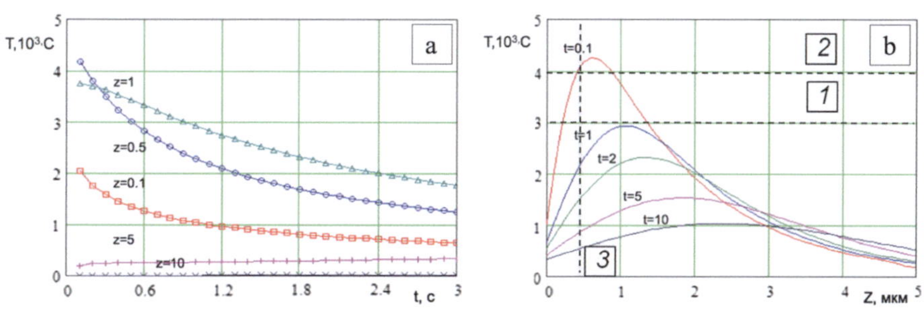

Fig. 5 Graphs of temperature versus time (**a**) and sample depth (**b**) when exposed to a series of femtosecond laser pulses

5 Conclusions

Formation of carbon nanostructures/microcrystals in liquid nitrogen under femtosecond laser beams, as a complicated nonlinear process, was under study in respect of both fundamental research and possible application in topological photonics. Obtained controllable low size structures have a perspective for creation of different functional elements. Physical processes are described and possible mechanisms of graphene layers splitting under the influence of femtosecond laser radiation in liquid nitrogen are proposed. The mechanisms of single crystal formation consisting in subsurface overheating under laser action on carbon samples in liquid nitrogen are described and confirmed by experimental data and simulation results.

Acknowledgements This work was performed as a part of Vladimir State University's State Task 3.5531.2017/8.9 GB-1106/17.

References

1. Schedin, F., Geim, A.K., Morozov, S.V., et al.: Nat. Mater. **6**(9), 652 (2007)
2. Hua-Qiang, W., Chang-Yang, L., Hong-Ming, L., He, Q.: Graphene applications in electronic and optoelectronic devices and circuits. Chin. Phys. B. **22**(9), 098106 (2013)
3. Liu, C., Yu, Z., Neff, D., et al.: Graphene-based supercapacitor with an ultrahigh energy density. Nano Lett. **10**(12), 4863–4868 (2010)
4. Bao, Q., Zhang, H., Wang, Y., et al.: Atomic-layer graphene as a saturable absorber for ultrafast pulsed lasers. Adv. Funct. Mater. **19**(19), 3077–3083 (2009)
5. Shugaev, M.V., Wu, C., Armbruster, O., et al.: Fundamentals of ultrafast laser–material interaction. MRS Bull. **41**(12), 960–968 (2016)
6. Ionin, A.A., Kudryashov, S.I., Samokhin, A.A.: Material surface ablation produced by ultrashort laser pulses. Physics-Uspekhi. **60**(2), 149 (2017)
7. Anisimov, S.I., Luk'yanchuk, B.S.: Selected problems of laser ablation theory. Physics-Uspekhi. **45**(3), 293–324 (2002)
8. Zabotnov, S.V., Golovan', L.A., Ostapenko, I.A., et al.: Femtosecond nanostructuring of silicon surfaces. JETP Lett. **83**(2), 69–71 (2006)
9. Korte, F., Serbin, J., Koch, J., et al.: Towards nanostructuring with femtosecond laser pulses. Appl. Phys. A. **77**(2), 229–235 (2003)
10. Zavestovskaya, I.N.: Laser nanostructuring of materials surfaces. Quantum Electron. **40**(11), 942–954 (2010)
11. Vorobyev, A.Y., Guo, C.: Femtosecond laser nanostructuring of metals. Opt. Express. **14**(6), 2164–2169 (2006)
12. Zhang, Y.L., Chen, Q.D., Xia, H., Sun, H.B.: Designable 3D nanofabrication by femtosecond laser direct writing. Nano Today. **5**(5), 435–448 (2010)
13. Martinez, A., Dubov, M., Khrushchev, I., Bennion, I.: Direct writing of fibre Bragg gratings by femtosecond laser. Electron. Lett. **40**(19), 1170–1172 (2004)
14. Makarov, G.N.: Laser applications in nanotechnology: nanofabrication using laser ablation and laser nanolithography. Physics-Uspekhi. **56**(7), 643 (2013)
15. Eason, R.: Pulsed Laser Deposition of Thin Films: Applications-Led Growth of Functional Materials. Wiley, Hoboken (2007)

16. Chkalov, R.V., Khorkov, K.S., Kochuev, D.A., et al.: Thin film elements design: software and possibilities of femtosecond laser techniques. J. Phys. Conf. Ser. **1109**(1), 012029 (2018)
17. Eletskii, A.V., Iskandarova, I.M., Knizhnik, A.A., Krasikov, D.N.: Graphene: fabrication methods and thermophysical properties. Physics-Uspekhi. **54**(3), 227–258 (2011)
18. Bulgakov, L.V., Bulgakova, N.M., Burakov, I.M., et al.: Synthesis of Nanosized Materials on Exposure of Matter to Intense Energy Fluxes. Inst. Teplofiz. Sib. Otd. Ross. Akad. Nauk, Novosibirsk (2009)
19. Georgakilas, V., Perman, J.A., Tucek, J., Zboril, R.: Broad family of carbon nanoallotropes: classification, chemistry, and applications of fullerenes, carbon dots, nanotubes, graphene, nanodiamonds, and combined superstructures. Chem. Rev. **115**(11), 4744–4822 (2015)
20. Abramov, D.V., Arakelyan, S.M., Galkin, A.F., et al.: Melting of carbon heated by focused laser radiation in air at atmospheric pressure and temperature below 4000 K. JETP Lett. **84**(5), 258–261 (2006)
21. Khorkov, K.S., Kochuev, D.A., Ilin, V.A., et al.: Mechanisms of graphene exfoliation under the action of femtosecond laser radiation in liquid nitrogen. J. Phys. Conf. Ser. **951**(1), 012014 (2018)
22. Khorkov, K.S., Abramov, D.V., Kochuev, D.A., et al.: Formation monocrystalline carbon micro-and nanostructures under femtosecond laser irradiation of graphite in liquid nitrogen. Phys. Procedia. **83**, 182–187 (2016)
23. Khorkov, K.S., Maleev, A.V., Chkalov, R.V., et al.: Investigation of carbon structures of single crystals obtained by laser synthesis. J. Surf. Invest.: X-Ray Synchrotron Neutron Tech. **12**, 392–394 (2018)
24. Khorkov, K.S., Zvyagin, M.Y., Kochuev, D.A., et al.: Model of the subsurface overheating of carbon samples upon laser impact in liquid nitrogen. Bull. Russ. Acad. Sci. Phys. **81**(12), 1433–1437 (2017)

Electromechanical Characterization of an Electrospun Piezoelectric Microfiber

Krishna Chytanya Chinnam, Arnaldo Casalotti, Edoardo Bemporad, and Giulia Lanzara

Abstract In this chapter, the influence of the electrical and mechanical coupling on the dynamic response of an electrospun piezoelectric microfiber is investigated. Particular focus is given to the fiber response dependence on the applied voltage. A frequency shift of the order of 5–9% in resonance frequency is observed when the microfibers are actuated with an alternating field. It is believed that this observed shift toward higher frequencies is related to the increasing alignment of polar crystallites that probably increases the polymer stiffness. This result highlights the possibility of making materials characterized by tunable stiffness and resonance frequency.

Keywords PVDF (polyvinylidene fluoride) · Microfiber · Actuation · Primary piezoelectric fiber resonance · Frequency shift · AC voltage excitation

1 Introduction

Piezoelectric sensors and actuators are widely used in several industrial fields (e.g., energy harvesting, structural health monitoring systems, morphing materials, stretchable sensors) thanks to their interesting electromechanical properties. While piezo-ceramics have the disadvantage of being stiff and brittle, their polymeric competitors are characterized by ultrahigh flexibility and lightness. For this reason, they are considered among the best candidates wherever large deformations coupled with low weight represent key application-driven requirements. Electrospun (PVDF) piezoelectric membranes represent the frontier in this field (see e.g., [1–7]).

While standard processes typically require a time-consuming polarization step that follows the material fabrication, in the electrospinning case this step is not

K. C. Chinnam · A. Casalotti · E. Bemporad · G. Lanzara (✉)
Department of Engineering, Roma Tre University, Rome, Italy
e-mail: giulia.lanzara@uniroma3.it

W. Lacarbonara et al. (eds.), *New Trends in Nonlinear Dynamics*,
https://doi.org/10.1007/978-3-030-34724-6_15

necessary. In this process the membranes are formed by nano/microfibers that are spilled-out of a needle under high electric fields. During this electric flight the fibers turn out to be stretched forming membranes that are readily intrinsically polarized as the non-electroactive α phase transforms into the electroactive β phase [8] demonstrated with the help of Fourier-transform infrared spectroscopy (FTIR) [9]. The remnant polarization (P_r) is a critical parameter which dictates the realistic piezo response of the material upon removal of the electric field. P_r was assessed with the help of well-known ferroelectric hysteresis curves [10]. These interesting PVDF membranes have mostly been actuated with DC input voltages; however, their usage under dynamic conditions (e.g., ultrasound sensors/actuators [11]) requires a thorough understanding of the material response under AC inputs.

PVDF membranes are often made of an aggregation of randomly distributed nano/microfibers formed during the electrospinning process. This implies the need to understand the response of the single fibers and then understand how these fibers interact with each other as an ensemble. The electromechanical properties of single electrospun PVDF fibers and of fiber bundles were characterized in a previous work by the authors for actuation applications [7, 8]. It was found that the response to an electrical stimulus on a bundle of fibers is a collective response of all fibers forming the bundle. An extensive experimental campaign was conducted in [12, 13] to study the actuation capability of single electrospun PVDF fibers fabricated with a direct-write method. Single poly[(vinylidenefluoride-co-trifluoroethylene] (PVDF-TrFE) fibers were also investigated with respect to their sensing capability and in terms of their intrinsic dielectric properties [14, 15]. Localized electromechanical interactions of single PVDF-TrFE fibers have also been investigated using a scanning probe microscope [16]. The mechanical and acoustic (sensing) properties of single PVDF fibers were also investigated [17, 18].

To the authors' best knowledge, despite the large effort in understanding the response of single PVDF fibers, no studies have been performed so far with respect to the actuation capability of a single fiber under dynamic AC inputs. In this work, the response of a single PVDF fiber under an alternating electric field is investigated.

2 Fabrication Process and Test Setup

Piezoelectric PVDF mats were fabricated via electrospinning. The electrospinning process parameters were tuned to control the diameter of the single fibers. A 21 wt%–25 wt% of PVDF in DMAc/Acetone, flow rate 0.08 mL/h and needle tip to collector distance 15 cm were adopted as process parameters. It is worth to note that the main parameter governing the fiber diameter is the solution concentration. In this case, the 21 wt% leads to a 65 μm, while the 25 wt% leads to a 75 μm fiber diameter.

Once the fiber mats were prepared, single fiber samples were manually extracted and installed on a testing rig that was specifically designed for this scope, see Fig. 1. The rig consisted of a highly insulating and robust Teflon plate comprising holes

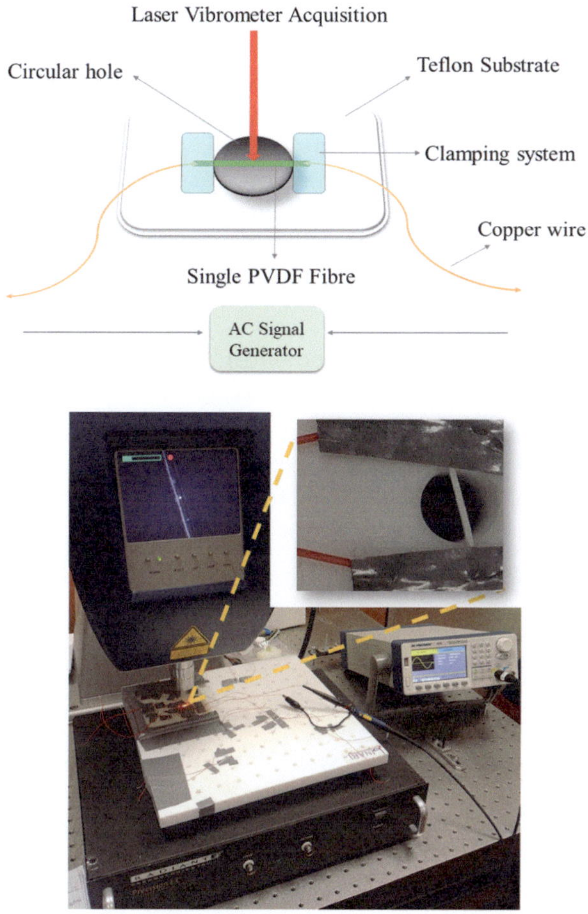

Fig. 1 Schematic of the test setup (top) and actual picture (bottom) to study the frequency response of a constrained PVDF fiber by actuating it via an alternating voltage signal

of different diameters. The fibers were placed across the diameter of the holes and then the electrical connections were realized using copper wires and conductive tape. The wires and the conductive tapes were carefully insulated using a duct tape. This reduced any loose connections and it very well supported the fiber from pre-stretching.

The free length of the fiber was defined by the hole of the Teflon substrate and in this case two different measures were selected: 3 mm and 10 mm for the 75 μm and 65 μm fibers, respectively. Moreover, the fixing system adopted for the electrical connections can at the same time provide the mechanical retaining system, resembling a clamped-clamped configuration.

As previously mentioned, the focus of the present work is to investigate the behavior of the single fibers under dynamic loading conditions, with the final objective of studying their sensitivity to applied voltages and frequencies. To this aim, the fiber is connected to an external signal generator, providing the AC input at selected voltages and frequencies. The top part of Fig. 1 depicts a schematic diagram of the test setup that was used for the electromechanical characterization, while the bottom part shows a picture of the actual setup. An electric field was applied across the fiber length that directly activates a bending motion with ensuing vibrations in the vertical direction. A laser scanning vibrometer allows to focus directly at the center of the fiber and detects the velocity of the point subject to the highest deformations. The following section discusses the results of the experiments in detail.

3 Results and Discussion

Fiber mats with a diameter of 7–8 cm were obtained by electrospinning for 3 min. Figure 2 shows the obtained mat (a) and its corresponding microscopic image (b). It is observed that the microstructure formed by the fibers is coherent with the literature [8, 10] where a similar fabrication process was adopted.

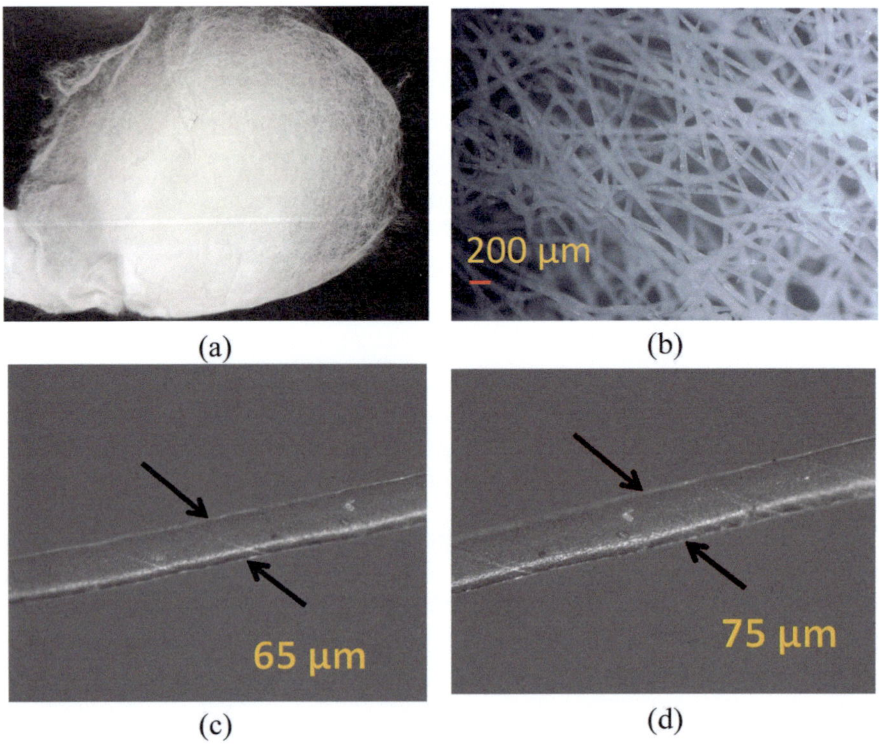

Fig. 2 Fabricated PVDF fiber mat (**a**) and its microscopic image (**b**). Microscopic images of the electrospun PVDF fibers of the two adopted diameters, 65 and 75 µm (**c**, **d**)

The tuned process parameters delivered fibers characterized by a diameter of 65 μm and 75 μm. These sizes were chosen to be compatible with the spot size of the laser in order to allow an accurate monitoring of the out-of-plane displacement. Single fibers were then extracted from the mat and analyzed under an optical microscope, see Fig. 2c, d, showing a good quality of the fibers without any significant imperfections. The study of the dynamic behavior of the manufactured microfibers is mainly focused on their flexural motion induced by the piezoelectric effect. A voltage is driven through the fiber length thus inducing flexural vibrations in the vertical plane in agreement with the fiber morphology. The main objective is to investigate the sensitivity of the fibers to the applied voltage and the selected frequency in ambient environment.

Preliminary tests are performed to identify the first flexural vibration mode of the microfiber. The resonance frequencies are identified by inducing a free vibration to the microfiber which was done by applying an impulsive load. The results are shown in Fig. 3, where the acquired time history is reported together with the Fast Fourier Transform (FFT) of the signal. The test is conducted on fibers with different lengths and diameter to highlight the effects that geometry can induce in the response. The first natural frequency, f_0, turned out to be ~ 2.3 kHz and 1.2 kHz for the 3 mm and 10 mm long fibers, respectively. The definition of the main natural frequency

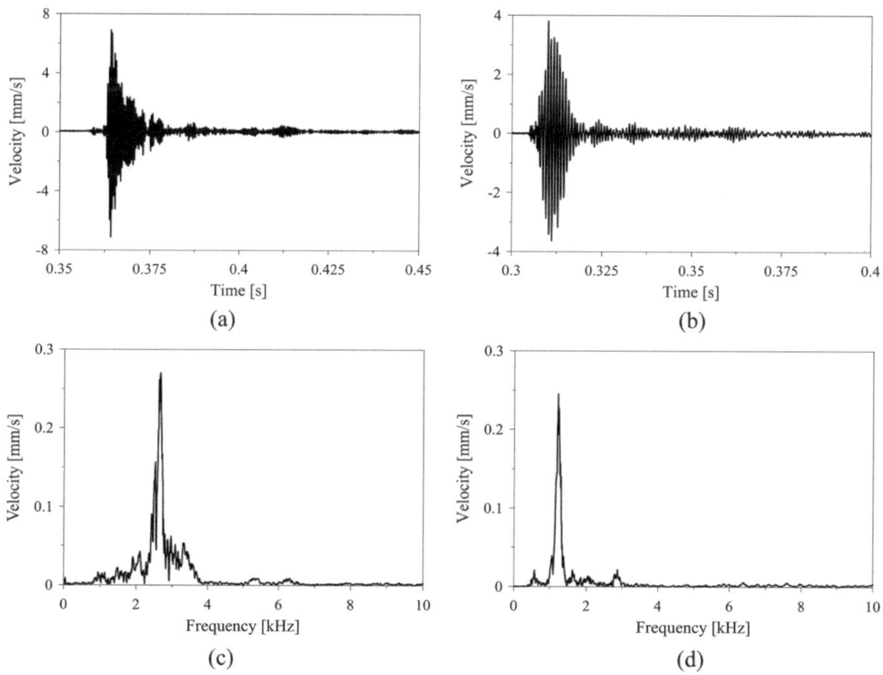

Fig. 3 Time history (**a**, **b**) of free vibration test and corresponding FFT (**c**, **d**) on the single PVDF fiber to obtain the resonance frequency of the fiber. Left part: 3 mm fiber. Right part: 10 mm

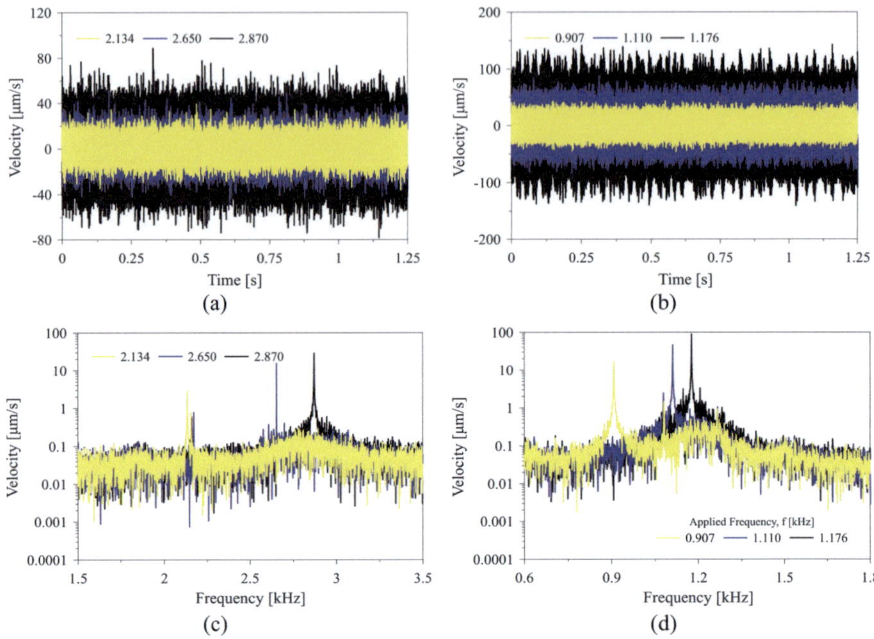

Fig. 4 Time history (**a**, **b**) and corresponding FFT (**c**, **d**) of forced vibration test on the single PVDF fiber, by applying sinusoidal voltage (20 V peak-to-peak) at three selected frequencies close to the resonance condition, namely, (2.134, 2.650, 2.870) kHz. Left part: 3 mm fiber. Right part: 10 mm fiber

allows to properly set the frequency bandwidth in which the fibers can be excited and experience a primary resonance when subject to an electromechanical external excitation.

As shown in Fig. 4, the fibers are subjected to a sine wave signal with fixed amplitude equal to 20 V (peak-to-peak) while the frequency is varied around the first natural frequency. Though a significant noise is affecting the quality of the signal, it can be noticed that when a sinusoidal voltage is applied to the fiber it experiences harmonic oscillations with an amplitude that is strictly related to the voltage, but also to the selected frequency. The results allow to observe that the response of the fiber experiences an increase in amplitude as the applied frequency, f, approaches its natural frequency, thus resulting in a primary resonance condition. It is observed that the frequency of oscillation is equal to that of the applied signal without any lags or losses in time domain while the electrical energy is being converted into the mechanical energy.

The test is extended by performing a refined frequency sweep at different applied voltages and the results are illustrated in Fig. 5 for the 3 mm fiber and in Fig. 6 for the 10 mm fiber, respectively. The selected voltages are set in the range 5–20 V (peak-to-peak) while the frequency bandwidth is extended to [0.75–1.30] f_0. As observed in the previous test, the response of the fibers increases when the applied frequency approaches f_0, giving rise to a primary resonance.

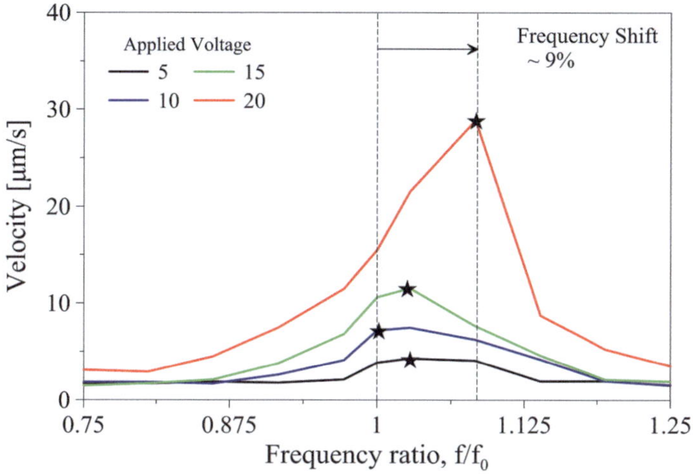

Fig. 5 Frequency response curves of the 3 mm long PVDF fiber by applying a sinusoidal voltage at different amplitudes (5–20 V)

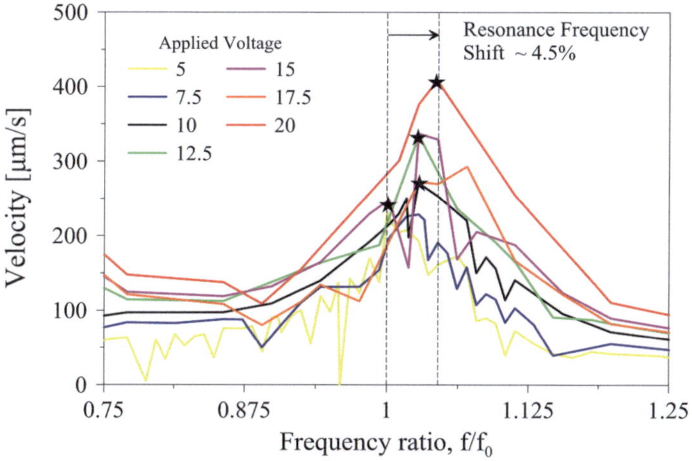

Fig. 6 Frequency response curves of the 10 mm long PVDF fiber by applying a sinusoidal voltage at different amplitudes (5–20 V)

Moreover, a considerable shift in the resonance frequency is also observed: the resonance frequency corresponds to f_0 at very low voltages, but as the voltage increases the resonance condition is achieved at different frequencies. As shown in Fig. 5, the 3 mm fiber experiences a slight decrease of the resonance frequency at 10 V and then shifts toward higher frequencies at larger voltages. The overall frequency shift range is around 9%. On the other hand, the response of the 10 mm fiber experiences directly an increase of the resonant frequency with a final shift of around 4.5% increasing applied voltage. This shift can be attributed to the increased orientation of polar crystallites due to the additional charges induced into the polymer when the amplitude of the AC input increases. Due to this effect the polymer probably stiffens and this is observed as a shift in the resonance frequency at higher fields.

There are a lot of factors that are affecting the whole measurement. The measurements were done in atmospheric conditions and it was observed that air flow, background noise, temperature, humidity, etc., have an influence on the fiber response. Further studies will be performed in controlled environment to avoid the influence of external factors on the microfiber response.

4 Conclusions

The results of the conducted experiments revealed a strong sensitivity of the PVDF fibers to the applied voltage. The increase in excitation amplitude causes a variation in the material response that leads to a shift in the resonance frequency. This response is probably related to the increasing changes in crystallites orientation with increasing voltage, which, as a result, increases the fiber rigidity.

References

1. Sengupta, D., Kottapalli, A.G.P., Chen, S.H., Miao, J.M., Kwok, C.Y., Triantafyllou, M.S., Warkiani, M.E., Asadnia, M.: Characterization of single polyvinylidene fluoride (PVDF) nanofiber for flow sensing applications. AIP Adv. **7**, 105205 (2017)
2. Sodano, H.A., Park, G., Inman, D.J.: An investigation into the performance of macro-fiber composites for sensing and structural vibration applications. Mech. Syst. Signal Process. **18**(3), 683–697 (2014)
3. Chuang, K.-C., Liou, H.-C., Ma, C.C.: Investigation of polyvinylidene fluoride (PVDF) films in identifying high-frequency vibration modes of flexible plates. IEEE Trans. Ultrason. Ferroelectr. Freq. Control. **61**(6), 1047–1058 (2014)
4. Goh, Y., Shakir, I., Hussain, R.: Electrospun fibers for tissue engineering, drug delivery, and wound dressing. J. Mater. Sci. **48**(8), 3027–3054 (2013)
5. Baozhang, L., Chengyi, X., Jianming, Z., Chunye, X.: Sensitivity of pressure sensors enhanced by doping silver nanowires. Sensors. **14**(6), 9889–9899 (2014)
6. Wang, X., Sun, F., Yin, G., Wang, Y., Liu, B., Dong, M.: Tactile-sensing based on flexible PVDF nanofibers via electrospinning: a review. Sensors. **18**, 330 (2018)
7. Fabriani, F., Chinnam, K., Iucci, G., Lanzara, G.: Characterization of electrospun PVDF fibres for sensing and actuation, ICEM 2018 July 1-5 Brussels (Belgium). MDPI Proc. **2**(8), 501 (2018)
8. Chinnam, K., Fabriani, F., Iucci, G., Lanzara, G.: Electrospun PVDF miniaturized muscles for bio inspired morphing materials. In: Paper No. SMASIS2018-8054, pp. V002T06A006. ASME 2018 SMASIS, San Antonio, Texas, USA, Sept. 10–12, 2018 (2018)
9. Chang, J., Dommer, M., Chang, C., Lin, L.: Piezoelectric nanofibers for energy scavenging applications. Nano Energy. **1**, 35–371 (2012)
10. Wang, S., Li, Q.: Design, synthesis and processing of PVDF-based dielectric polymers. IET Nanodielectr. **1**(2), 80–91 (2018)

11. Fan, F.R., Tang, W., Wang, Z.L.: Flexible nanogenerators for energy harvesting and self-powered electronics. Adv. Mater. **28**, 4283–4305 (2016)
12. Pu, J., Yan, X., Jiang, Y., Chang, C., Lin, L.: Piezoelectric actuation of a direct write electrospun PVDF fiber. In: 2010 IEEE 23rd International Conference on Micro Electro Mechanical Systems (MEMS), Wanchai, Hong Kong, Jan. 24–28, pp. 1163–1166 (2010)
13. Pu, J., Yan, X., Jiang, Y., Chang, C., Lin, L.: Piezoelectric actuation of direct-write electrospun fibers. Sens. Actuat. A. **164**, 131–136 (2010)
14. Lu, M.S., Lu, C., Li, M.H., Liu, C.L., Chen, W.C.: Nonvolatile organic transistor memory devices based on nanostructured polymeric materials. In: Proceedings of SPIE—The International Society for Optical Engineering, San Diego, California, United States, 2014, Aug 17-21 (2014)
15. Jiang, Y., Gong, L., Hu, X., Yong, Z., Chen, H., Feng, L., Zhang, D.: Aligned P(VDF-TrFE) nanofibers for enhanced piezoelectric directional strain sensing. Polymers **10**, 364 (2018)
16. Yonatan, C., Richard, A.W., Qingshen, J., Vijay, N., Soini, K.-N.: Localized electromechanical interactions in ferroelectric P(VDF-TrFE) nanowires investigated by scanning probe microscopy. APL Mater. **4**, 116106 (2016)
17. Xin, L., Hang, Q., Maksim, S.: Piezoelectric micro- and nanostructured fibers fabricated from thermoplastic nanocomposites using a fiber drawing technique: comparative study and potential applications. ACS Nano. **11**(2), 2103–2114 (2017)
18. Cha, S.N., Seo, J., Kim, S.M., Kim, H.J., Park, Y.J., Kim, S., Kim, J.M.: Sound-driven piezoelectric nanowire-based nanogenerators. Adv. Mater. **22**, 4726–4730 (2010)

On Modeling of Springless Electromagnetic Energy Harvesters

Ramy A. Mohamed, Ayman El-Badawy, Ahmed Moustafa, Andrew Kirolos, Mostafa Soliman, and Eihab M. Abdel-Rahman

Abstract We present an analytical model of an electromagnetic levitation energy harvester and validate it by comparing its result with FEM simulations and experimental measurements. The model is based on dipole-moment approximations of magnetic fields and interaction forces. The level of agreement of both models with measurements is discussed.

Keywords Energy harvesting · Electromagnetic

1 Introduction

Electromagnetic energy harvesting has been used to power a wide-range of devices, such as mobile electronics, wearable devices, and biomedical implants [1, 2]. Accurate modeling of energy transduction is essential for practical harvester design, but challenging because of the highly nonlinear nature of magnetic fields and

---

R. A. Mohamed
Department of Mechanical Engineering, Al-Azhar University, Cairo, Egypt

A. El-Badawy (✉)
Department of Mechanical Engineering, Al-Azhar University, Cairo, Egypt

Mechatronics Engineering Department, German University, Cairo, Egypt
e-mail: ayman.elbadawy@guc.edu.eg

A. Moustafa · A. Kirolos
Mechatronics Engineering Department, German University, Cairo, Egypt
e-mail: ahmad.abdelmegid@guc.edu.eg; andrew.nashed@guc.edu.eg

M. Soliman
Power Electronics Department, Electronics Research Institute, Cairo, Egypt
e-mail: mostafa.soliman@si-ware.com

E. M. Abdel-Rahman
Department of Systems Design Engineering, University of Waterloo, Waterloo, ON, Canada
e-mail: eihab@uwaterloo.ca

W. Lacarbonara et al. (eds.), *New Trends in Nonlinear Dynamics*,
https://doi.org/10.1007/978-3-030-34724-6_16

forces. In this paper, we assess two modeling approaches and compare them with experimental measurements from a fabricated electromagnetic energy harvester. The first approach utilizes the analytical point-dipole model of magnets widely adopted to describe magnetic fields. It allows us to calculate the interaction forces between the magnets as well as the coupling factor of the harvester. The second approach uses a finite element model and simulates the energy harvester in the commercial package COMSOL Multiphysics to calculate the magnetic interaction forces, seismic mass displacements, and output voltage.

2 Electromagnetic Energy Harvester

The energy harvester, depicted in Fig. 1, is composed of two identical magnets fixed at the top and bottom of a tubular track and a freely moving cylindrical magnet with radius r_m and height h_m acting as seismic mass m. The polarity of the three magnets is set such that the end magnets exert repulsive restoring forces on the moving magnet as it translates along the track. A coil made of copper wire with radius r_w is wound around the tube to make N_x turns in the axial direction and N_r turns in the radial direction.

As external vibrations $y(t)$ move the tube in a vertical direction aligned with gravity, the seismic mass moves by $x(t)$ inside the tube with respect to the center point along the tube axis at rest. This movement creates a varying magnetic field with respect to the coil, which in turn induces voltage across its terminals. The design parameters of the fabricated harvester are listed in Table 1.

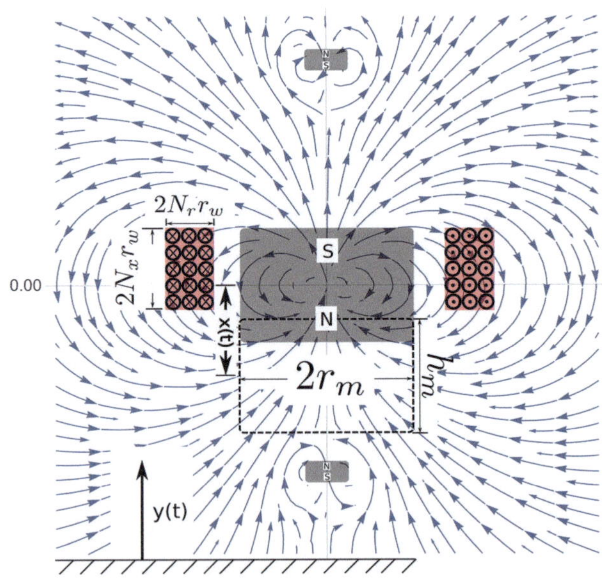

Fig. 1 A schematic of the electromagnetic energy harvester showing the field lines computed using the point dipole-moment model

Table 1 Harvester design parameters

Parameter	Value
Track length L	46 mm
Moving magnet radius r_m	9.525 mm
Moving magnet height h_m	12.7 mm
Seismic mass m	27 g
Fixed magnet radius	2.38 mm
Fixed magnet height	2.38 mm
Residual magnetism (N40)	1.275 T
Copper wire radius r_w	0.075 mm
Coil outer radius	18.5 mm
Coil inner radius	12.5 mm
Coil height	8.5 mm
Number of turns N	2200

3 Analytical Model

The dynamics of the energy harvester can be reduced to the rigid-body motions of the moving magnet under the gravitational field, the magnetic fields of the end magnets, and the electromagnetic field of the coil. The equilibrium position of the seismic mass under the resulting force is shown in Fig. 1.

When the harvester casing (tube) experiences an acceleration $\ddot{y}(t)$, the suspended mass will oscillate with an acceleration $\ddot{x}(t)$ both are with respect to a fixed frame of reference [3]. The relative displacement of the seismic mass with respect to the tube, can therefore be written as the difference between the absolute displacements of the tube and mass:

$$z(t) = x(t) - y(t). \tag{1}$$

Similarly, the equation of motion of the seismic mass can be written in terms of the relative displacement as:

$$m\ddot{z} + \big(b_m + b_e(z)\big)\dot{z} + F_{ms} + mg = -m\ddot{y}, \tag{2}$$

where F_{ms} is the resultant restoring force applied by the two end magnets to the seismic mass, b_m is a viscous damping coefficient representing mechanical losses, and b_e is an electrical damping coefficient representing the energy dissipated by the Lorentz force imposed by the coil and is expressed as

$$b_e(z) = \frac{\left(\frac{\mathrm{d}\phi}{\mathrm{d}z}\right)^2}{R_c + R_L}, \tag{3}$$

where R_c is the coil resistance, R_L the load resistance, and ϕ the magnetic flux [4].

3.1 Magnetic Restoring Force

Permanent magnets can be modeled as magnetic dipoles [5] located at the magnet center. The magnetic flux density of a dipole with a moment m_p, expressed in cylindrical coordinates, is

$$B_\zeta(\rho,\zeta) = \frac{\mu_\circ}{4\pi}\frac{2\zeta^2-\rho^2}{(\zeta^2+\rho^2)^{5/2}}m_p \tag{4}$$

$$B_\rho(\rho,\zeta) = \frac{3\mu_\circ}{4\pi}\frac{\zeta\rho}{(\zeta^2+\rho^2)^{5/2}}m_p, \tag{5}$$

where (ρ, ζ) are the radial and axial coordinates, respectively. Accordingly, B_ζ and B_ρ represent the flux densities in the axial and radial directions, and μ_o is the permeability of free space. The magnetic field is computed and visualized as shown in Fig. 1. The interaction potential energy between another magnet, with magnetic moment m_f, and this magnet is given by

$$U = \mathbf{B}_f \cdot \mathbf{B}_p = \frac{\mu_\circ}{4\pi}\frac{2\zeta^2-\rho^2}{(\rho^2+\zeta^2)^{5/2}}m_p m_f + \frac{3\mu_\circ}{4\pi}\frac{\zeta\rho}{(\zeta^2+\rho^2)^{5/2}}m_p m_f. \tag{6}$$

Since the end magnets are placed along the tube axis and the moving magnet translates along the same axis, the radial component of the position vector locating the dipole of an end magnet with respect to the moving magnet is negligible throughout motion. Setting the radial component equal to zero ($\rho = 0$) and using Castigliano's theorem, we obtain the axially exerted interaction force between the moving magnet and an end magnet as:

$$F_\zeta = \frac{\partial U}{\partial \zeta} = \frac{3\mu_o m_p m_f}{2\pi\zeta^4}. \tag{7}$$

3.2 Coupling Factor

The field passing through the coil depends on the magnet's relative position with respect to the coil $z(t)$. The induced voltage in a single coil turn is equal to variation in the magnetic flux crossing it:

$$V_i(t) = \frac{d\Phi}{dt}, \tag{8}$$

where

$$\Phi = \mathbf{B}(\rho, z) \cdot \mathbf{A}$$

and A is the coil area crossed by the field. Using the chain rule, we can write for a magnet moving axially with respect to a coil turn located at point $z_\circ$ along the tube axis:

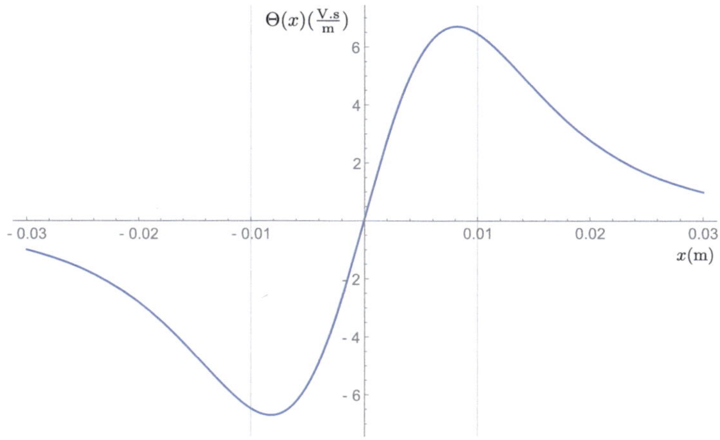

Fig. 2 The coupling factor $\Theta(x)$ as a function of the relative position of the magnet with respect to the center point along the harvester axis

$$V_i(t) = \dot{z}\frac{\partial}{\partial z}(\mathbf{B}\cdot\mathbf{A})\bigg|_{z=z_o}. \tag{9}$$

The voltage across the coil terminals can be written as

$$V(t) = \dot{z}\,\Theta(z), \tag{10}$$

where $\Theta(z)$ is a "coupling factor" representing the flux across the coil [6, 7]:

$$\Theta(z) = 2\pi\sum_{k=1}^{N_x}\sum_{j=1}^{N_r}\int_0^{r_j}\frac{\partial}{\partial z}B_z(\rho,z)\rho d\rho \tag{11}$$

r_j is the radial coordinate of the center of the jth turn located at the kth rung of the coil. Figure 2 shows the coupling factor for the fabricated energy harvester.

4 Finite Element Model

The fabricated energy harvester was also simulated using the multi-physics module in the commercial FEM software COMSOL to obtain the open circuit voltage across the coil terminals. An axisymmetric model, Fig. 3a, of the harvester, including the magnets and coil, and the air box surrounding it was developed. The model was meshed using triangular elements, Fig. 3b. The maximum element edge length was set to 3.7 mm. The complete mesh consisted of 1153 domain elements and 113 boundary elements, corresponding to 2394 degrees-of-freedom. Figure 3c shows

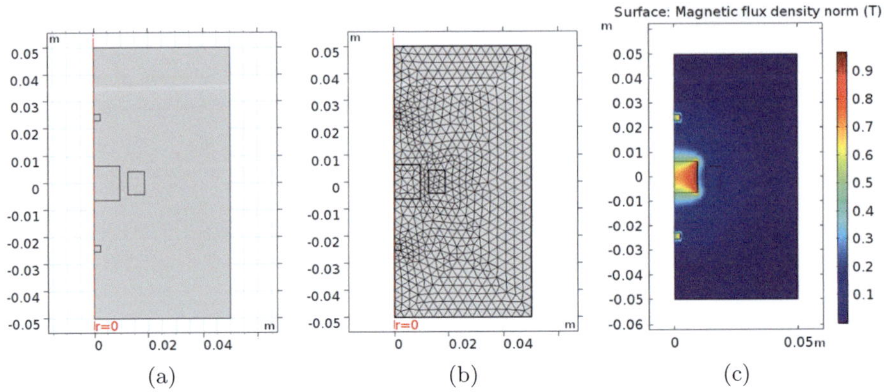

Fig. 3 FEM model. (**a**) Axisymmetric model. (**b**) The mesh. (**c**) Magnetic flux density at $z = 0$

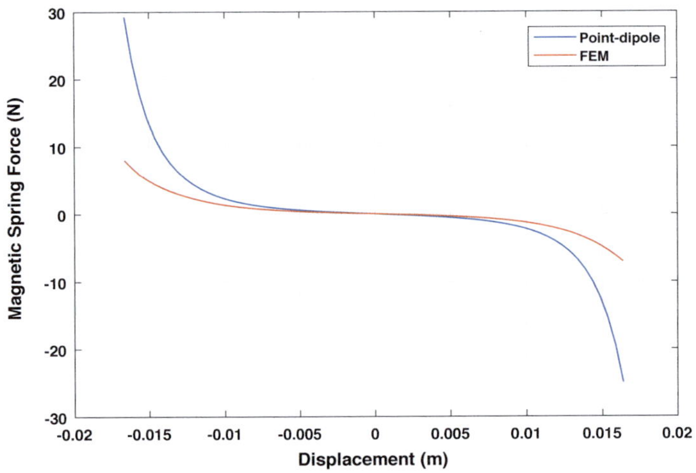

Fig. 4 Comparison of the magnetic restoring force $F_{ms}(z)$ obtained from FEM simulations (red line) and the analytical model (blue line)

the magnitude of the magnetic flux density $B(\rho, z)$ when the centers of the moving magnet and tube axis are coincident.

First, the magnetic restoring force F_{ms} was calculated through a parametric sweep simulation. The position of the moving magnet z was varied along the tube axis, the magnetic field was evaluated for each static position $\mathbf{B}(\rho, z)$, and the net force acting on moving magnet at that position $F_{ms}(z)$ was calculated. Figure 4 shows the resulting FEM magnetic restoring force in red. In addition, the blue line in the figure shows the magnetic restoring force $F_{ms}(z)$ calculated through Eq. (12). This equation is derived from (7) by substituting ζ with the distance between the moving magnet and each fixed magnet. The resulting Eq. (12) is therefore the net magnetic restoring force acting on the moving magnet due to its interaction with both fixed magnets.

$$F_{ms} = -\frac{3\mu_\circ}{2\pi} m_p m_f \,\mathrm{sgn}(z) \left(\frac{1}{(L/2 - z)^4} + \frac{1}{(L/2 + z)^4} \right). \tag{12}$$

The FEM simulations and analytical model are in qualitative agreement. When the moving magnet is located at the center point of the tube axis, the magnetic restoring force vanishes due to symmetry. As it moves towards either end of the tube, the restoring force increases in an antisymmetric fashion to oppose that motion. The two models are also in good quantitative agreement within the limits of tube length.

Next, the solid mechanics model was coupled to the FEM model. The forces acting on the moving magnet, namely weight, mechanical damping, and magnetic restoring force, were defined in the former model and the harvester's base acceleration $\ddot{y}$ was defined in a moving mesh. Finally, the magnet was allowed to move inside the tube governed by the simultaneous simulations of the magnet displacement in the solid mechanics model and the magnetic field in the FEM model and the mesh was updated accordingly. The coupled model was used to calculate the induced voltage due to variations in the magnetic field cutting the coil as the magnet move past it.

5 Results and Discussion

We compare predictions of the seismic mass displacement under a base acceleration amplitude and frequency of 0.1 g m/s^2 and 8 Hz in Fig. 5. The result of FEM simulations appears in red-colored line and those obtained from numerical integration of the analytical model, Eq. (2), appears in blue-colored line. The discrepancy in

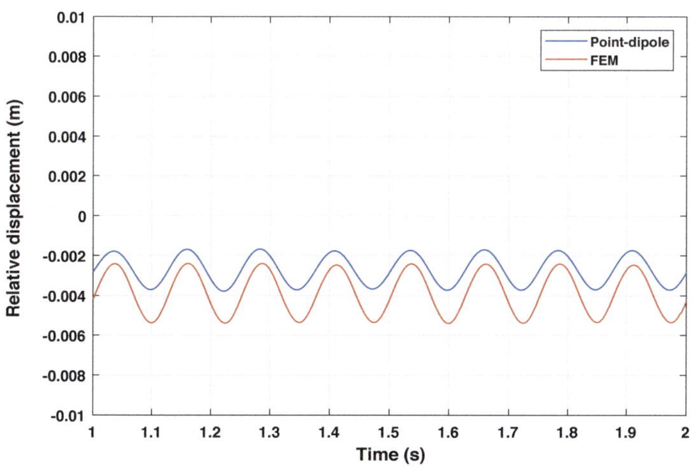

Fig. 5 The relative displacements of the moving magnets from both approaches

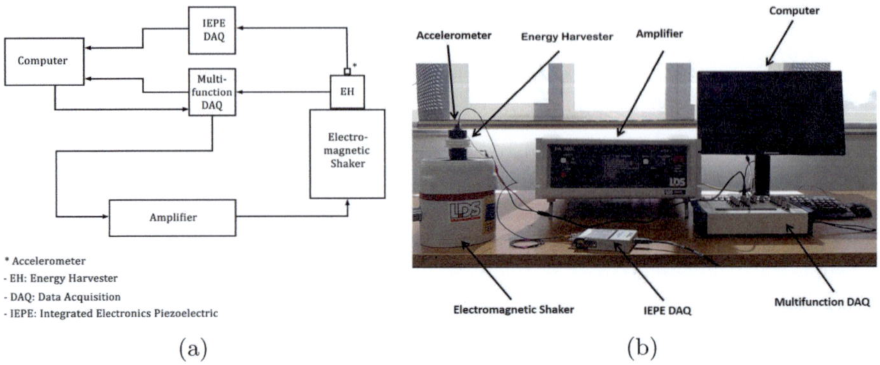

Fig. 6 Experimental setup (**a**) Schematic diagram (**b**) Experimental setup

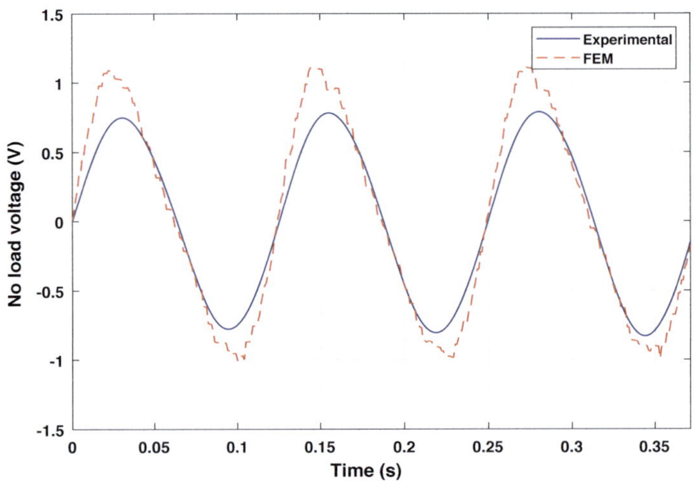

Fig. 7 FEM of no-load voltage compared with experimental results

both the equilibrium position and the amplitude of oscillation is an indicator of the inaccuracy of the dipole-moment model of the permanent magnet.

The fabricated harvester was tested using the experimental setup shown in Fig. 6. The harvester was rigidly attached to an electromagnetic shaker that delivered base accelerations with amplitude and frequency of $0.1\,\mathrm{g\,m/s^2}$ and 8 Hz. The shaker was closed-loop controlled to maintain the base acceleration constant throughout the experiment.

Figure 7 shows a much better agreement between the experimental measurements and the FEM model results regarding the no-load voltage, thus confirming the inadequacy of the dipole-moment model for the permanent magnet.

6 Conclusions

The dipole-moment approach for modeling permanent magnets may be sufficient for the design of the springless energy harvesters, yet based on the comparison with a more detailed studies based on the FEM method. The steady state oscillations suffer from a loss of prediction accuracy. This is due to the point-dipole model being inaccurate for small distances. The experimental measurement compared with both the FEM results and the point-dipole moment confirms the above conclusion.

Acknowledgement This work was supported by The Academy of Scientific Research and Technology, Egypt under grant no. JESOR 2015121

References

1. Berdy, D.F., Valentino, D.J., Peroulis, D.: Kinetic energy harvesting from human walking and running using a magnetic levitation energy harvester. Sens. Actuators A **222**, 262–271 (2015)
2. Yang, B., et al.: A generator with nonlinear spring oscillator to provide vibrations of multi-frequency. Appl. Phys. Lett. **99**(22), 2009–2012 (2011)
3. Synge, J.L., Griffith, B.: Principles of Mechanics, 3rd edn, pp. 130–131. McGraw Hill, New York (1984)
4. Kirolos, A., Moustafa, A., El-Badawy, A., Habib, M., Mohamed, R., Soliman, M., Abdel-Rahman, E.: Design and development of an electromagnetic micropower generator for activity tracking. In: 2018 Twentieth International Middle East Power Systems Conference (MEPCON), pp. 522–527. IEEE, Piscataway (2018)
5. Hahn, K.D., et al.: Eddy current damping of a magnet moving through a pipe. Am. J. Phys. **66**(12), 1066–1076 (1998)
6. Constantinou, P., Mellor, Phil H., Wilcox, Paul, D.: A magnetically sprung generator for energy harvesting applications. IEEE Trans. Mechatron. **17**(3), 415–424 (2012)
7. Dallago, E., Marchesi, M., Venchi, G.: Analytical model of a vibrating electromagnetic harvester considering nonlinear effects. IEEE Trans. Power Electron. **25**(8), 1989–1997 (2010)

Part III
Nonlinear Phenomena in Bio- and Ecosystem Dynamics

Critical Behaviors of Regular Pattern Selection in Neuronal Networks with Chemical Synapses

Peihua Feng, Rong Wang, and Ying Wu

Abstract Regular waves play an important role in cortical information processing in mammalian neocortex. Spatiotemporal patterns change not only among plane waves, spiral, and irregular waves, but also among different types of regular patterns. Neuronal network is composed by two kinds of neurons, respectively, namely neurons coupled by fast excitatory and fast inhibitory synapse. The scenario of pattern transition seems completely different with these two types of synapse. Specifically, phenomenon called critical slowing down exists in pattern transition when neurons are connected by fast excitatory synapses, but there is no labyrinth pattern in this case. However, in the case of fast inhibitory synapse, regular pattern loses its stability and becomes labyrinth pattern. Patterns become more and more chaotic respect to coupling strength between neurons. The results indicate the way neurons couples determines the way in which pattern loses its stability.

Keywords Neuronal network · Chemical synapse · Critical slowing down

1 Introduction

Regular spatiotemporal pattern, as a typical collective behavior of large amount of neurons or other somatic cells, is of great significance to biological systems, like mammalian cortex [1] and cardiac tissue [2], etc. A great number of experimental

P. Feng
School of Aerospace Engineering, Xi'an Jiaotong University, Xi'an, China
e-mail: fphd2017@xjtu.edu.cn

R. Wang
Xi'an University of Science Technology, College of Science, Xi'an, China

Y. Wu (✉)
State Key Laboratory for Strength and Vibration of Mechanical Structures, Shaanxi Engineering Laboratory for Vibration Control of Aerospace Structures, School of Aerospace Engineering, Xi'an Jiaotong University, Xi'an, China
e-mail: wying36@xjtu.edu.cn

W. Lacarbonara et al. (eds.), *New Trends in Nonlinear Dynamics*,
https://doi.org/10.1007/978-3-030-34724-6_17

studies show that the formation, duration, and breakup of regular patterns are related to modulate electrical activities closely [3]. It is believe that many cardiac arrhythmias are usually characterized by spiral waves [4], therefore, the nature of defibrillation is elimination of spiral waves [5]. Spiral waves also exist in rat cortex when the rat is in a sleep-like state [1]. Also, spiral waves are discovered in epileptic seizures, which implies that spiral wave may promote or terminate the process of epilepsy [6]. All these truths indicate that regular wave as a pacemaker is able to organize and modulate cortical population activities both normal and pathological cortical processing [7].

Given the importance of spiral wave formation, several technologies are developed to induce spiral waves. White noise [8] and electric field [9] as external stimulus are able to produce spiral waves when the amplitude of external force is beyond threshold. Brown motion of spiral wave could also be caused by spatiotemporal structured noise [10, 11]. Besides, a defect in the neural network as an initial condition is often used, inducing a spiral wave growing and rotating around it. A line defect is proved to give birth to spiral wave and change its stability [7, 12, 13]. A spiral tip, as a singular point mathematically, is another type of topological defect. Wang adopts a spiral tip to realize the pattern transition in neuronal networks with chemical synapses considered [14]. Therefore, a spiral tip is also used as the initial condition in our study.

Beside the model selection of neuron, the coupling type between neurons mainly determines the dynamical properties of neuronal network. The linear electrical synapses (by gap junction) and the nonlinear chemical synapses are the two cases by which signals transfer among neurons. Although (linear) electrical synapses connect voltage of the two neurons at all times, chemical synapses are more efficient in achieving stochastic coherence of the system [15] and weak signal propagation [16]. Compared to electrical synapses, chemical synapses are more beneficial to give birth to firing transition in time-delayed network [17]. More important, chemical synaptic coupling is more efficient for weak and local signal input, which is more common in biological systems [18]. However, little attention has been paid to spatiotemporal pattern formation in neuronal network when neurons are coupled by chemical synapses.

Fortunately, pattern transition is discovered in neuronal network with chemical coupling. And Wang [14] discusses different mechanisms of pattern selection of two types of chemical synapses, namely fast inhibitory synapses and fast excitatory synapses. Spatiotemporal pattern only alternate in regular patterns in case of fast excitatory synaptic coupling but labyrinth pattern is achieved in the case of fast inhibitory synaptic coupling when the coupling strength is large enough. However, temporal properties of regular pattern formation are missed in previous researches. In this paper, we start from Wang's work and study carefully the critical phenomenon in pattern formation and transition with the same neuronal network model, identical parameter values in Ref. [14].

2 Neuron Model and Neuronal Network

Morris–Lecar model [19] is employed, in which membrane potential V and conducting probability n of potassium channels of the neuron with a position number (i, j) are governed by

$$C_M \frac{dV_{ij}}{dt} = I_{ij}^{\mathrm{ion}} + I_{ij}^{\mathrm{ext}} + I_{ij}^{\mathrm{syn}}, \tag{1}$$

$$\frac{dn_{ij}}{dt} = \lambda(n_\infty(V_{ij}) - n_{ij})/\tau_{n_{ij}}(V_{ij}), (i, j = 1, \ldots, N), \tag{2}$$

where C_M is the membrane capacitance per unit area; λ is the decay rate of n_{ij}; and I_{ij}^{ion}, I_{ij}^{ext}, and I_{ij}^{syn} are the ionic current, external current, and synaptic currents, respectively. I_{ij}^{ext} is an adjustable constant, which can be considered a parameter to control nonlinear behavior like Hopf bifurcation and saddle-node bifurcation [20]. It is assumed that leaks, potassium, and calcium currents contribute to ionic current which is described consequently by

$$I_{ij}^{\mathrm{ion}} = -g_{\mathrm{L}}(V_{ij} - V_{\mathrm{L}}) - g_{\mathrm{K}} n_{ij}(V_{ij} - V_{\mathrm{K}}) - g_{\mathrm{Ca}} m_\infty(V_{ij})(V_{ij} - V_{\mathrm{Ca}}), \tag{3}$$

where g_{L}, g_{K}, and g_{Ca} are leaks, potassium, and calcium conductance, and V_{L}, V_{K}, and V_{Ca} are the corresponding reversal potentials.

The synaptic current I_{ij}^{syn} is coupling term which determines the construction of neuronal network, given by Li et al. [21]

$$I_{ij}^{\mathrm{syn}} = -D(s_{i-1,j} + s_{i+1,j} + s_{i,j-1} + s_{i,j+1})(V_{ij} - V_{\mathrm{syn}}), \tag{4}$$

where D is the channel conductance of the chemical synapse. V_{syn} is the corresponding reversal potential, which is a constant in our simulation. And s_{ij} are synaptic variables for the fraction of open channels. Eq. (4) shows that every neuron couples with its four nearest neurons and the intensity of every coupling is the same. The synaptic variable s_{ij} can be reduced as follows [21]:

$$s_{ij} = -\frac{1}{a_d}\dot{s}_{ij} \qquad V_{ij}^{\mathrm{pre}} < V_Q,$$
$$1 \qquad \text{otherwise.}$$

s_{ij} quickly jumps to 1 when $V_{ij}^{\mathrm{pre}} \geqslant V_Q$ according to Ref. [21]. Different synapses have different neurotransmitters and receptors, whose kinetic properties are described by the values of parameters in Eq. (4). In particular, an excitatory synapse has high reversal potential V_{syn}, while an inhibitory synapse has low (typically sub-threshold) reversal potential. In Ref. [14], the roles of both of two types of synapses are discussed carefully. The fast excitatory and fast inhibitory synapses are termed as AMPA and GABA, respectively, as their neurotransmitters.

But our study just focuses on formation of regular spatiotemporal pattern, therefore we pay more attention on the networks whose neurons are coupled by fast excitatory synapses. Other functions in Eqs. (1)–(3) are defined Ref. [14]. The values of parameters are also displayed in Ref. [14]. The network composes 101×101 neurons. The position number of neurons from lower-left corner to upper-right corner is from $(1, 1)$ to $(101, 101)$. The initial conditions are taken as: $V_{ij} = -33.35$ mV, $n_{ij} = 0.46$ for $i = 41{:}43$, $j = 1{:}50$; $V_{ij} = 0.03$ mV, $n_{ij} = 0.51$ for $i = 44{:}46$, $j = 1{:}50$; $V_{ij} = 33.32$ mV, $n_{ij} = 0.31$ for $i = 47{:}49$, $j = 1{:}50$, and $V_{ij} = -50.33$ mV, $n_{ij} = 0.31$ for the rest of neurons. We apply periodic boundary condition and use Euler forward difference algorithm to integrate the set of differential equations with time steps of 0.01 ms.

3 Regular Pattern Formation and Critical Behaviors

In order to present the entire process of pattern formation in detail, we display several series of snapshots of spatiotemporal patterns at two coupling strength (synaptic conductance D), see Fig. 1a–h for $D = 3.0$ and (A)∼(H) for $D = 3.3$. The final spatiotemporal pattern when the coupling strength equals 3.0 is rectangle spiral wave (RSW). The entire process of formation of it can be divided into two stages roughly. In the first stage, a circular spiral wave (CSW) generates from the end of the rod of non-uniform initial condition, and grows from the center to boundary of the network. At the same time, rectangular spiral wave generates from the boundary to center. These two kinds of the spiral waves spread in the opposite direction and encounter each other at about 3000 ms. The difference of Fig. 1c and (C) is so small that one cannot distinguish one of them from another because spiral waves develop very slowly after generation of face-off situation of the two regular waves. Figure 1d–h shows that CSW shrinks steadily and slowly and vanishes at about 50,000 ms (Fig. 1h), which is termed the second stage of pattern formation. And after that the rectangular spiral wave exists and rotates steadily.

When D equals 3.3, it seems that the first stage is almost the same as the situation when $D = 3.0$. But in second stage, CSW seems much "stronger" and encroaches on the RSW slowly and occupies the whole domain finally.

The process of rectangular pattern formation can also be considered as the diminution of the separatrix of two types of the pattern. Therefore it is equivalent to study the motion of separatrix when one studies the law of spiral wave evolution. The first question is to determine the time instant at which the separatrix sweeps past the appointed neuron. Take the neuron with the position number $(35, 35)$, for example (the red dot in Fig. 2a), its membrane potential history is shown in Fig. 2b. The entire history obviously comprises two stages and an interim stage between them. Previous study shows that amplitude of membrane potential jumps at a critical value of D. The reason is that the neuron firing at different domains of spiral waves obeys periodic motions with different amplitudes and periods.

Fig. 1 Snapshots at a series of instants during coexistence of RSW and CSW, the time instants the snapshots from (**a**) to (**h**) are 1000, 2000, 3000, 10,000, 20,000, 30,000, 40,000, and 50,000 ms as $D = 3.0$. And snapshots from (**a**) to (**h**) are 1000, 2000, 3000, 10,000, 20,000, 30,000, 40,000, and 50,000 ms as $D = 3.3$

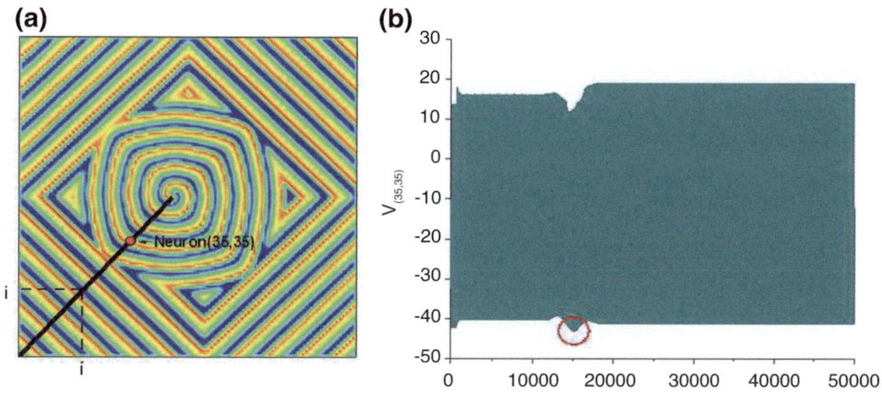

Fig. 2 Studied neuron and its membrane potential history. (**a**) The red dot is the neuron with the position number (35, 35), the snapshot is at $t = 5000$ ms with $D = 3.0$. (**b**) History of membrane potential of neuron (35, 35), the red circle indicates the interim period between the two kinds of oscillation

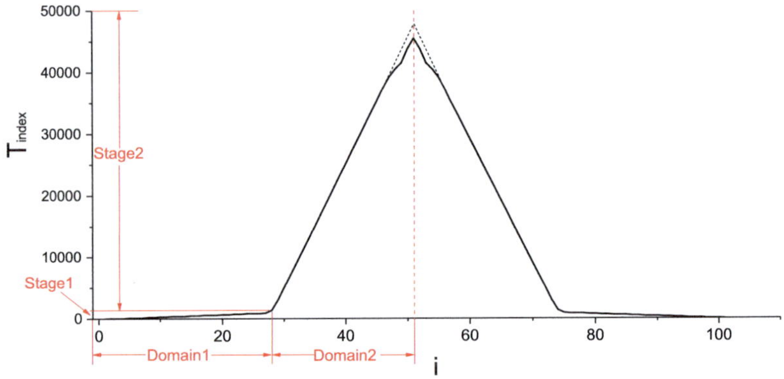

Fig. 3 Time index respect to neuron position number at $D = 3.0$

Between the two periodic motions, there is an interim period during which one periodic motion becomes to another. In this period, the amplitude of oscillation drops so that the membrane potential achieves the minimum (see the red circle in Fig. 2b) through the whole time history. We check all the neurons and find the interim period and minimum value of membrane potential in every one of them. Therefore, the time at which the potential achieves the minimum can be considered as the time index for distinguishing the two kinds of oscillation, and also as the index for the time at which the separatrix passes the studied neuron.

We choose half of diagonal line from the lower-left corner to the center of network, namely neurons $(i, i), i = 1, 2, \ldots, 51$ (see the black rod in Fig. 2a). We plot the time index of neurons along the rod in Fig. 3. In order to make the whole trend clear, we actually plot 101 neurons along the diagonal line from (1, 1) neuron at the lower-left corner to (101, 101) at the upper-right corner. The vertical axis is time at which the separatrix passes the neuron and the horizontal axis is the number of neurons. For example, $i = 1$ indicates the neuron in lower-left corner and $i = 51$ indicates the neuron in the center of the network.

The time index line is axisymmetric about vertical axis $i = 51$. Therefore, we only study left half of the time index line. It includes two parts, the line with small slope and line with much greater slop. The connection point of the two parts of line divides vertical axis into two stages and the horizontal axis into two domains. Consistent with our previous observations, in the first stage, the face-off situation of two types of spiral wave is established, and RSW encroaches the CSW until it vanishes in the second stage. Correspondingly, when the face-off situation has come into being (the connection point), the rectangular spiral wave occupies the domain 1 and circular occupies domain 2. From the line with greater slope, it is discovered that the CSW shrinks steadily for a long time. But when it is getting small, the velocity of RSW drops a little (compared to the dashed line).

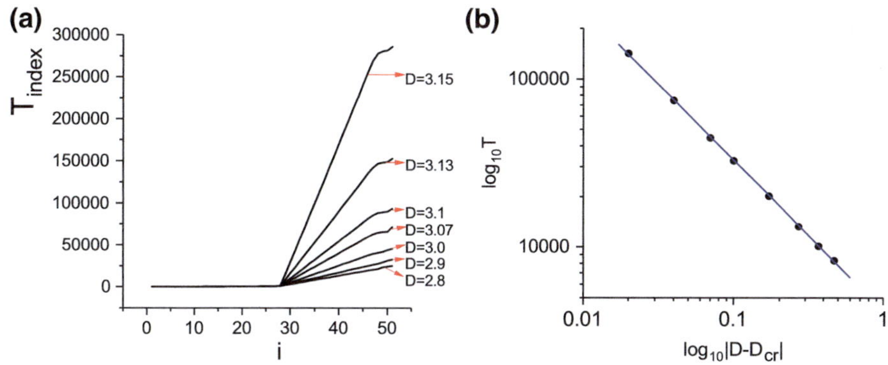

Fig. 4 Time index curves with $D = 2.8, 2.9, 3.0, 3.07, 3.1, 3.13, 3.15$ from bottom to top (**a**), and their distribution in double logarithmic coordinates (**b**)

We plot several time index curves in one figure to reveal the evolution speed of spiral waves. The curves from bottom to top correspond to $D = 2.8, 2.9, 3.0, 3.07, 3.1, 3.13, 3.15$, respectively (Fig. 4a). Apparently with all these values of D, CSW shrinks and the network is occupied by the rectangular wave at last. It is worth noting that the first stages and first domains overlap with each other. The truth implies two facts at least: (a) the time durations the network need to generate face-off situation of two types of regular patterns are the same no matter what value of D takes as long as rectangular spiral waves occupy the entire network, (b) once the face-off situation forms, the areas the two patterns occupy do not change with D varying.

With D getting greater, the whole time duration of rectangular spiral wave formation grows very fast. But there exists a critical value of D_{cr} at which transition of patterns occurs. It is reasonable to assume that when the parameter D gets approach to critical value D_{cr}, the time duration of pattern formation tends to infinity. If power law of time duration and absolute value $|D - D_{cr}|$ holds, the phenomenon called critical slowing down is discovered.

$$T \sim |D - D_{cr}|^{-\lambda}, \tag{5}$$

where $\lambda > 0$ is the critical slowing down index. We plot the time duration and absolute of $|D - D_{cr}|$ in double logarithmic coordinates in Fig. 4b. The blue line is the fitted curve whose slope gives exact value of the critical slowing down index λ. In our simulation λ is 0.92. And the critical value of D_{cr} is about 3.172 by carefully checking. As a matter of fact, the phenomenon of critical slowing down exists in both sides of the critical value of D_{cr}. Similarly, λ of $D > D_{cr}$ is determined as 0.91, which is close to that of $D < D_{cr}$.

4 Conclusions

In previous studies, researchers only focused on the spatial characteristics of patterns during their transition in AMPA network. In our study we investigated the temporal properties of regular pattern formation.

Before one type of spiral wave forms in the network, coexistence of two types of spiral waves called face-off situation occurs. After that CSW spreads to the whole domain of network as synaptic conductance is greater than the critical value D_{cr}, otherwise it shrinks and RSW is left in the entire network as $D < D_{cr}$.

Both below and above the critical value D_{cr}, the time duration of pattern formation with respect to D obeys the power law relation given by Eq. (5), which is a convincing proof of the critical slowing down phenomenon. The critical slowing down index does not change much near D_{cr}. It seems that critical slowing down is ubiquitous during the regular pattern transition.

Acknowledgements The project is supported by National Natural Science Foundation of China under Grant Nos.11472202 and Nos.11772242, China National Postdoctoral fund under Grant Nos.2018M621140, and Overseas Expertise Introduction Project for Discipline Innovation Nos. B18040.

References

1. Huang, X., Xu, W., Liang, J., Takagaki, K., Gao, X., Wu, J.Y.: Neuron **68**(5), 978 (2010)
2. Gray, R.A., Pertsov, A.M., Jalife, J.: Nature **392**(6671), 75 (1998)
3. Wang, C., Ma, J.: Int. J. Mod. Phys. B **32**, 1830003 (2017)
4. Davidenko, J.M., Pertsov, A.V., Salomonsz, R., Baxter, W., Jalife, J.: Nature **355**(6358), 349 (1992)
5. Panfilov, A.V., Müller, S.C., Zykov, V.S., Keener, J.P.: Phys. Rev. E Stat. Phys. Plasmas Fluids Relat. Interdiscip. Topics **61**(4 Pt B), 4644 (2000)
6. Huang, X., Troy, W.C., Yang, Q., Ma, H., Laing, C.R., Schiff, S.J., Wu, J.Y.: J. Neurosci. **24**(44), 9897 (2004)
7. Ma, J., Hu, B., Wang, C., Jin, W.: Nonlinear Dyn. **73**(1–2), 73 (2013)
8. Gu, H.-G., Jia, B., Li, Y.-Y.: Phys. A: Stat. Mech. Appl. **392**(6), 1361 (2013)
9. Wang, C.N., Ma, J., Jin, W.Y., Wu, Y.: Appl. Math. Comput. **218**(8), 4467 (2011)
10. Sendina-Nadal, I.I., Alonso, S., Perez-Munuzuri, V.V., Gomez-Gesteira, M., Perez-Villar, V.V., Ramirez-Piscina, L., Casademunt, J., Sancho, J.M., Sagues, F.: Phys. Rev. Lett. **84**(12), 2734 (2000)
11. Alonso, S., Sagués, F.: Phys. Rev. E Stat. Nonlinear Soft Matter Phys. **63**(4 Pt 2), 046205 (2001)
12. Park, J.S., Lee, K.J.: Phys. Rev. Lett. **88**(22), 224501 (2002)
13. Zhan, M., Kapral, R.: Phys. Rev. E Stat. Nonlinear Soft Matter Phys. **72**(2), 046221 (2005)
14. Wang, R., Li, J., Du, M., Lei, J., Wu, Y.: Commun. Nonlinear Sci. Numer. Simul. **40**, 80 (2016)
15. Balenzuela, P., García-Ojalvo, J.: Phys. Rev. E Stat. Nonlinear Soft Matter Phys. **72**(1), 021901 (2005)
16. Wang, J., Li, X., Feng, D.: Comparisons of Chemical Synapses and Gap Junctions in the Stochastic Dynamics of Coupled Neurons. Springer, Berlin (2007)
17. Xu, B., Gong, Y.B., Wang, B.Y.: Sci. China Chem. **56**(2), 222 (2013)

18. Li, X., Wang, J., Hu, W.: Phys. Rev. E Stat. Nonlinear Soft Matter Phys. **76**(4 Pt 1), 041902 (2007)
19. Morris, C., Lecar, H.: Biophys. J. **35**(1), 193 (1981)
20. Ermentrout, G.B., Terman, D.H.: Mathematical Foundations of Neuroscience. Springer, New York (2010)
21. Li, X., Zhang, J., Small, M.: Chaos Interdiscip. J. Nonlinear Sci. **19**(1), 872 (2009)

Dynamics of a Homeostatically Regulated Neural System with Delayed Connectivity

Eva Kaslik and Raluca Mureşan

Abstract In this work, we explore the effect on the dynamical behavior of the interplay of homeostatic regulation and coupling time delay in a pair of reciprocally coupled Wilson–Cowan networks. The occurrence of rich dynamical behavior is explored both theoretically and by means of numerical simulations.

Keywords Coupled Wilson–Cowan networks · Neural networks · Delayed differential equations · Homeostatic regulation · Coupling delay · Stability · Bifurcations

1 Introduction

Low-dimensional dynamics are known to occur in different regions of the human brain, including theta and gamma oscillations in the hippocampus [3, 4] or low dimensional chaotic dynamics during different sleep phases or epileptic seizures [1]. In spite of low-dimensionality, the emergence of these dynamical behaviors remains unexplained at present time. The general mechanism that might produce these dynamical regimes could be explained by plasticity in the synaptic coupling weights between neurons [9].

Homeostatic plasticity [12, 13] is assumed to be responsible for preventing run-away excitation in the circuit which could produce pathological states such epileptic seizures, but also for countering a catastrophic loss of neuronal activity which may lead to network quiescence. Even though homeostasis has been traditionally considered to have an overall stabilizing effect on neuronal activity [12], recent work revealed that these homeostatic mechanisms may produce complex dynamics, such as neuronal avalanches, in large networks [7].

E. Kaslik · R. Mureşan (✉)
Department of Mathematics and Computer Science, West University of Timisoara, Timisoara, Romania
e-mail: eva.kaslik@e-uvt.ro; raluca.muresan@e-uvt.ro

W. Lacarbonara et al. (eds.), *New Trends in Nonlinear Dynamics*,
https://doi.org/10.1007/978-3-030-34724-6_18

In this work, our aim is to analyze the effect of both homeostatic plasticity and synaptic transmission delays [5] by analyzing a pair of coupled Wilson–Cowan networks [2, 14]. The case without delayed connectivity has been very recently investigated in [9], where three separate situations have been considered: single-nodes, reciprocally coupled dual-nodes, and large coupled networks, with connection strengths derived from functional neuroimaging data [7]. However, it is important to note that transmission delays unavoidably exist in virtually any type of dynamic network, being responsible for the occurrence of rich dynamic behavior and hence, they should be incorporated in the considered mathematical models [8, 10, 11, 15].

2 Mathematical Model

We consider the following symmetric double-node system of Wilson–Cowan networks with delayed coupling:

$$\begin{aligned}
E_1' &= -E_1 + \phi(\alpha E_{2\tau} - W_1 I_1)\\
I_1' &= -I_1 + \phi(\beta E_1)\\
W_1' &= \gamma I_1(E_1 - p)\\
E_2' &= -E_2 + \phi(\alpha E_{1\tau} - W_2 I_2)\\
I_2' &= -I_2 + \phi(\beta E_2)\\
W_2' &= \gamma I_2(E_2 - p),
\end{aligned} \tag{1}$$

where E_i, I_i represent the subpopulations of excitatory and inhibitory neurons in each node, W_i represent the excitatory–inhibitory synaptic weight which are considered to be homeostatically regulated [7], with the homeostatic set point p. For simplicity, the time scales of the excitatory and inhibitory populations are considered equal to 1. The coupling terms α, β as well as the learning rate γ are assumed to be positive, while the self-inhibition weights of each node are assumed to be zero. The function ϕ is assumed to be a sigmoid threshold function (e.g., $\phi(x) = (1 + \exp(-ax))^{-1}$ with $a > 0$). The terms $E_{i\tau}(t) = E_i(t - \tau)$, $i = \overline{1, 2}$ represent the time-delayed connectivity between the two nodes, with $\tau \geq 0$.

It can be easily seen that

$$\Sigma = \{(E_1, I_1, W_1, E_2, I_2, W_2) \; : \; E_1 = E_2, \; I_1 = I_2, \; W_1 = W_2\}$$

is an invariant synchronization subspace to system (1).

3 Local Stability Analysis

The unique equilibrium state $S^\star$ of system (1) has the components:

$$E_1^\star = E_2^\star = p, \quad I_1^\star = I_2^\star = \phi(\beta p), \quad W_1^\star = W_2^\star = \frac{\alpha p - \phi^{-1}(p)}{\phi(\beta p)}.$$

Linearizing the system at the equilibrium point $S^\star$ we obtain the following system of delay differential equations:

$$\dot{x}(t) = Ax(t) + Bx(t-\tau), \tag{2}$$

where

$$x(t) = \left[E_1(t) - E_1^\star,\ I_1(t) - I_1^\star,\ W_1(t) - W_1^\star,\ E_2(t) - E_1^\star,\ I_2(t) - I_1^\star,\ W_2(t) - W_1^\star\right]^{\mathsf{T}}$$

and

$$A = \begin{pmatrix} A_0 & 0 \\ 0 & A_0 \end{pmatrix}, \quad B = \begin{pmatrix} 0 & B_0 \\ B_0 & 0 \end{pmatrix}$$

$$A_0 = \begin{pmatrix} -1 & -W^\star \Phi'(\Phi^{-1}(p)) & -I^\star \Phi'(\Phi^{-1}(p)) \\ \beta \Phi'(\beta p) & -1 & 0 \\ \gamma I^\star & 0 & 0 \end{pmatrix},$$

$$B_0 = \begin{pmatrix} \alpha \Phi'(\Phi^{-1}(p)) & 0 & 0 \\ 0 & 0 & 0 \\ 0 & 0 & 0 \end{pmatrix}.$$

The characteristic equation of system (2) is

$$\det(sI - A - Be^{-\tau s}) = 0. \tag{3}$$

Denoting $J(s) = A + Be^{-\tau s} = \begin{pmatrix} A_0 & B_0 e^{-\tau s} \\ B_0 e^{-\tau s} & A_0 \end{pmatrix}$, based on the special form of this matrix, we deduce that:

$$\begin{aligned} \det(sI - J(s)) &= \begin{vmatrix} sI - A_0 & -B_0 e^{-\tau s} \\ -B_0 e^{-\tau s} & sI - A_0 \end{vmatrix} \\ &= \det(sI - A_0 - B_0 e^{-\tau s}) \det(sI - A_0 + B_0 e^{-\tau s}). \end{aligned}$$

Therefore, s is a root of the characteristic equation (3) if and only if it is the root of one of the following equations:

$$\begin{aligned} \det(sI - A_0 - B_0 e^{-\tau s}) &= 0 \\ \det(sI - A_0 + B_0 e^{-\tau s}) &= 0. \end{aligned} \tag{4}$$

Computing the determinants from (4), the stability and bifurcation analysis of the equilibrium state $S^\star$ given above is reduced to the investigation of the roots of the characteristic equation

$$\pm e^{-\tau s} s(s+1) = c_1 s(s+1)^2 + c_2 s + \gamma c_3 (s+1), \tag{5}$$

where $\delta = \Phi'(\Phi^{-1}(p))$, $c_1 = (\alpha\delta)^{-1}$, $c_2 = W^* \beta \Phi'(\beta p)\alpha^{-1}$ and $c_3 = (I^*)^2 \alpha^{-1}$ are positive constants that do not depend on the parameters γ and τ.

In the following, we first explore sufficient conditions for the asymptotic stability and instability of the equilibrium point $S^\star$ of system (1).

Proposition 1

(a) In the non-delayed case ($\tau = 0$) the equilibrium point $S^\star$ of system (1) *is locally asymptotically stable if and only if the following inequalities hold:*

$$2c_1 > 1 \quad \text{and} \quad \gamma c_3 (1 - c_1) < (2c_1 - 1)(c_1 + c_2 - 1).$$

(b) If $c_1 > 1$, the equilibrium point $S^\star$ of system (1) *is locally asymptotically stable for any $\tau \geq 0$ and $\gamma > 0$.*

(c) If the following inequality is fulfilled:

$$\tau\gamma c_3 + c_1 \cdot \frac{\tau + 1}{\tau} + c_2 \cdot \frac{\tau}{\tau + 1} < e^{-1} \tag{6}$$

then the equilibrium point $S^\star$ of system (1) *is unstable.*

Proof

(a) If $\tau = 0$, the characteristic equation (5) becomes

$$\pm s(s+1) = c_1 s(s+1)^2 + c_2 s + \gamma c_3 (s+1)$$

which is equivalent to

$$c_1 s^3 + (2c_1 \pm 1)s^2 + (c_1 + c_2 + \gamma c_3 \pm 1)s + \gamma c_3 = 0.$$

Taking into account that the constants c_i, $i = \overline{1,3}$ and γ are positive, by the Routh–Hurwitz criterion it follows that all the roots of the above equation have negative real part if and only if

$$2c_1 \pm 1 > 0 \quad \text{and} \quad (2c_1 \pm 1)(c_1 + c_2 + \gamma c_3 \pm 1) > \gamma c_1 c_3.$$

From the positivity of c_1, c_2, c_3, it can be easily seen that if the plus sign is taken into consideration, the above inequalities hold. Hence, the inequalities of the Routh–Hurwitz criterion are fulfilled if and only if

$$2c_1 > 1 \quad \text{and} \quad (2c_1 - 1)(c_1 + c_2 + \gamma c_3 - 1) > \gamma c_1 c_3.$$

The latter inequality can be rewritten as

$$\gamma c_3(1 - c_1) < (2c_1 - 1)(c_1 + c_2 - 1).$$

(b) Let $c_1 > 1$. The characteristic equation (5) can be rewritten as

$$\pm e^{-\tau s} = c_1(s+1) + \frac{c_2}{s+1} + \frac{\gamma c_3}{s}.$$

Assuming that the characteristic equation has a root s in the right half-plane, i.e., $\Re(s) \geq 0$, taking the real part in both sides of the above equation, we deduce:

$$\pm e^{-\tau \Re(s)} \cos(\tau \Im(s)) = c_1(\Re(s) + 1) + \frac{c_2(\Re(s)+1)}{|s+1|^2} + \frac{\gamma c_3 \Re(s)}{|s|^2}.$$

As $c_1 > 1$ and $\Re(s) \geq 1$, it is easy to see that the left-hand side of this equation is subunitary, while the right-hand side is larger than 1. Hence, a contradiction is obtained, leading to statement b.

(c) We rewrite the characteristic equation (5) as

$$s e^{-\tau s} - \gamma c_3 = c_1 s(s+1) + c_2 \frac{s}{s+1}.$$

We denote the left- and right-hand side terms of the above equation as

$$f(s) = s e^{-\tau s} - \gamma c_3 \quad \text{and} \quad g(s) = c_1 s(s+1) + c_2 \frac{s}{s+1},$$

where $f, g : [0, \infty) \to \mathbb{R}$. We observe that the function f has a maximum at $s^\star = \tau^{-1}$, $f(0) = f(\infty) = -\gamma c_3$, and the function g is increasing on $[0, \infty)$, with $g(0) = 0$, $g(\infty) = \infty$.

Therefore, a sufficient condition for the characteristic equation of (1) to have a real positive root is $f(\tau^{-1}) \geq g(\tau^{-1})$, which is equivalent to inequality (6). □

4 Hopf Bifurcation Analysis

Proposition 1 gives us a tool to assess the occurrence of rich dynamical behavior in system (1). If $c_1 > 1$, Proposition 1(b.) shows that oscillatory behavior cannot be expected in a neighborhood of the equilibrium point, regardless of the time delay τ considered in the system. On the other hand, if inequality (6) is fulfilled, the unique equilibrium of system (1) is unstable, and hence, complex dynamic behavior is to be observed. The transition from asymptotic stability to instability involves the occurrence of bifurcation phenomena. The following result characterizes the Hopf bifurcation curves in the (γ, τ) parameter plane.

Proposition 2 *If* $c_1 \in (0, 1)$, *then the characteristic equation* (5) *has a pair of pure imaginary roots if and only if there exists* $k \in \mathbb{Z}_+$ *such that* $(\gamma, \tau) \in \Gamma_k$, *where* Γ_k *is the parametric curve defined by*

$$\Gamma_k : \quad \begin{array}{l} \gamma = \frac{1}{c_3}\left(\frac{2c_1-(-1)^k\cos(\omega)}{h_k(\omega)^2} + (-1)^k\sin(\omega)\frac{1}{h_k(\omega)}\right) \\ \tau = \omega h_k(\omega) \end{array}$$

for $\omega_k \in (k\pi - \arccos(c_1), k\pi + \arccos(c_1))$, *where*

$$h_k(\omega) = \sqrt{\frac{(-1)^k\cos(\omega) - c_1}{c_1 + c_2 - (-1)^k\cos(\omega)}}.$$

Proof The parametric equations of the curves Γ_k are easily deduced by imposing that the characteristic equation (5) has a pair of complex conjugated imaginary roots of the form $\pm i\frac{\omega}{\tau}$. □

Proposition 2 gives the critical values for the parameters (γ, τ) where Hopf bifurcations [6] may occur in a neighborhood of the equilibrium point. The curves Γ_k given by Proposition 2 are plotted in Fig. 1.

5 Numerical Simulations

For the numerical simulations, the following parameter values have been chosen: $\alpha = 1.5$, $\beta = 2.5$; $p = 0.2$ and the sigmoid function is

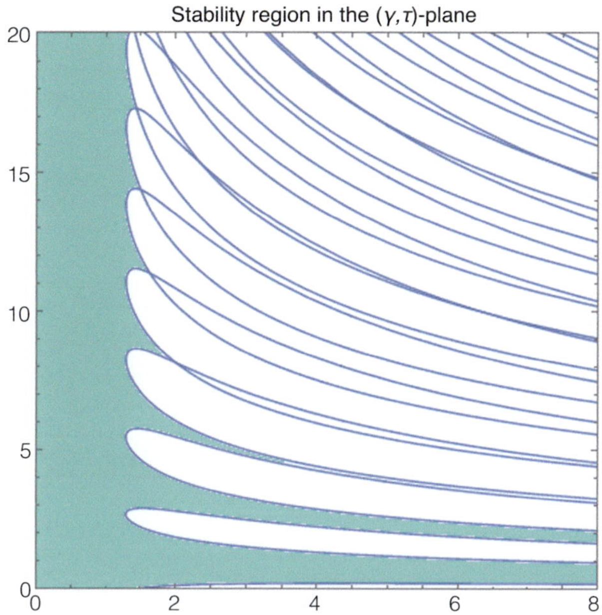

Fig. 1 Stability region (cyan) in the (γ, τ)-plane for the equilibrium of system (1), for $\alpha = 1.5$, $\beta = 2.5$; $p = 0.2$ and $\phi(x) = (1 + \exp(-5x))^{-1}$. The blue curves represent the Γ_k Hopf bifurcation curves provided by Proposition 2

$$\phi(x) = \frac{1}{1 + \exp(-5x)}.$$

The equilibrium point of system (1) is

$$S^{\star} = \{0.2, 0.924142, 0.624643, 0.2, 0.924142, 0.624643\}.$$

The stability region in the (γ, τ) parameter plane of this equilibrium point is represented in Fig. 1. It can be observed that if the learning rate γ is sufficiently small ($\gamma < 1.28529$ computed numerically), the equilibrium state $S^{\star}$ is asymptotically stable, for any value of the coupling delay $\tau \geq 0$.

For a fixed value of the learning rate γ (e.g., $\gamma = 6$ such as in Figs. 2 and 3), as the coupling time delay τ increases, several critical values are encountered which produce a cascade of Hopf bifurcations. In Fig. 2, the dynamics on the invariant subspace Σ has been explored, for different values of the coupling delay τ.

On the other hand, in Fig. 3, the trajectories of system (1) have been represented in the (E_1, E_2)-phase plane, for different valued of the coupling time delay,

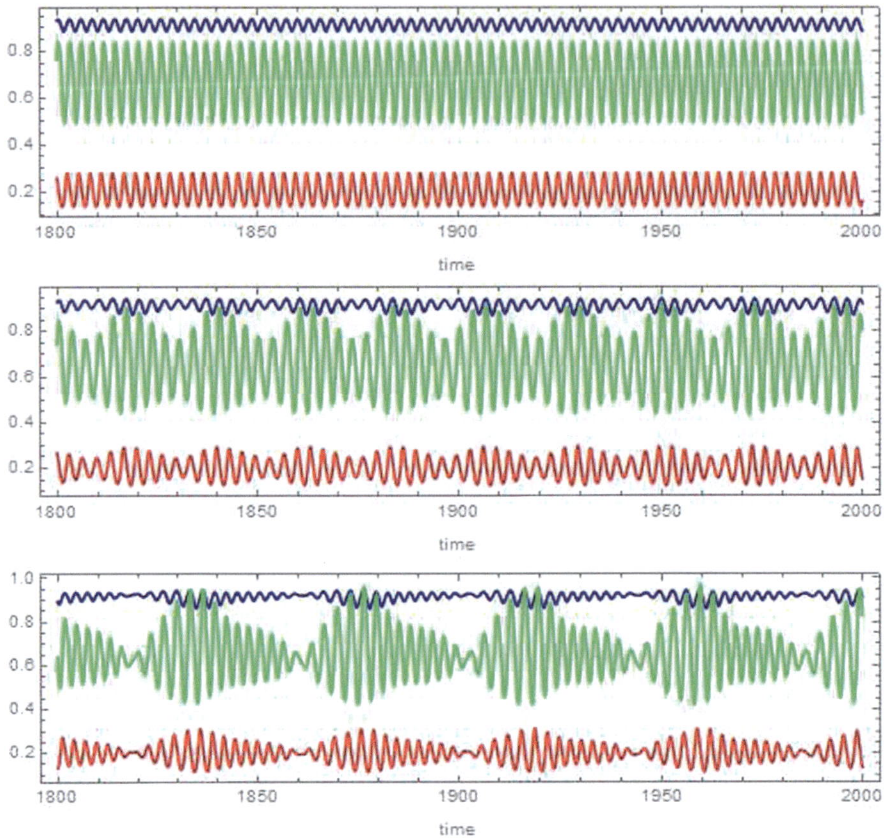

Fig. 2 Evolution of E_i (red), I_i (blue) and W_i (green) in time for $\gamma = 6$ and $\tau = 5.5$, $\tau = 20$ and $\tau = 40$, respectively (top to bottom). An initial condition from the invariant subspace Σ has been chosen

choosing a random initial condition in a neighborhood of the equilibrium point $S^\star$, not belonging to the invariant subspace Σ. Rich dynamical behavior can be observed, ranging from periodic and quasi-periodic orbits to chaotic attractors.

6 Conclusions

The effect on the dynamical behavior of homeostatic regulation versus coupling time delay has been explored for a pair of reciprocally coupled Wilson–Cowan networks. Necessary and sufficient conditions have been established in terms of the system parameters which guarantee the stability or instability of the unique equilibrium

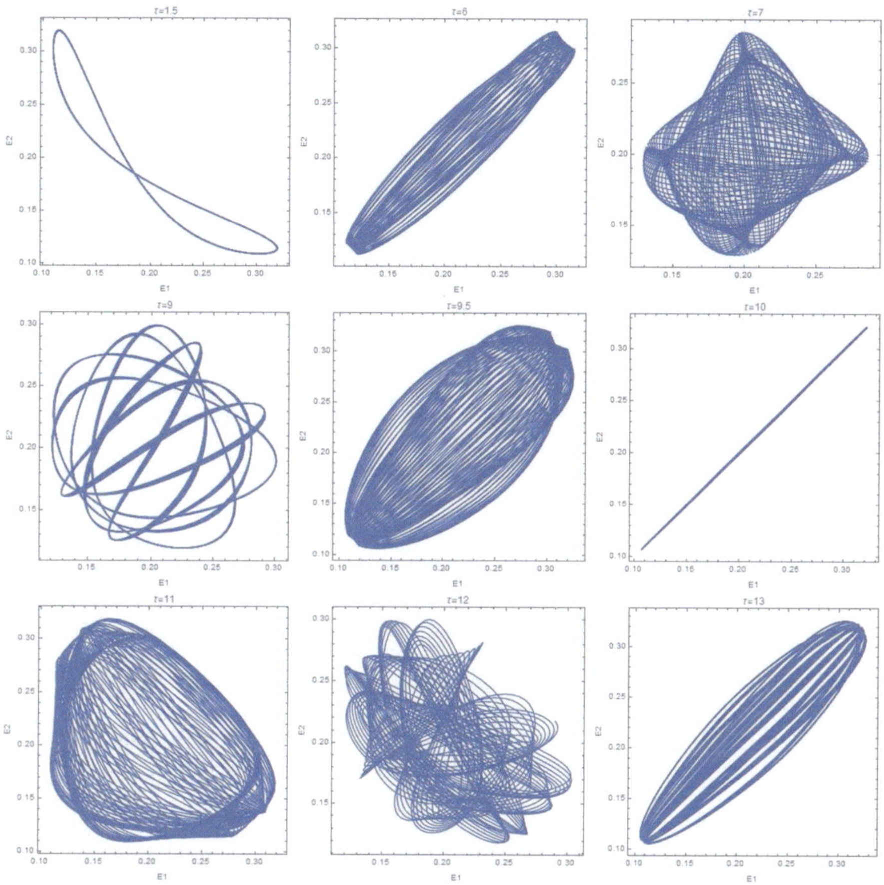

Fig. 3 Trajectories of system (1) in the (E_1, E_2)-phase plane for different values of the coupling time delay τ, with fixed $\alpha = 1.5$, $\beta = 2.5$; $\gamma = 6$, $p = 0.2$ and $\phi(x) = (1 + \exp(-5x))^{-1}$. The initial condition has been chosen randomly from a neighborhood of the equilibrium $S^{\star}$

point. Hopf bifurcation curves have been described analytically in the parameter plane consisting of the learning rate and coupling delay. Numerical simulations reveal rich dynamical behavior in the coupled system.

References

1. Babloyantz, A., Destexhe, A.: Low-dimensional chaos in an instance of epilepsy. Proc. Natl. Acad. Sci. **83**(10), 3513–3517 (1986)
2. Borisyuk, G.N., Borisyuk, R.M., Khibnik, A.I., Roose, D.: Dynamics and bifurcations of two coupled neural oscillators with different connection types. Bull. Math. Biol. **57**(6), 809–840 (1995)

3. Buzsáki, G.: Theta oscillations in the hippocampus. Neuron **33**(3), 325–340 (2002)
4. Buzsáki, G., Wang, X.J.: Mechanisms of gamma oscillations. Annu. Rev. Neurosci. **35**, 203–225 (2012)
5. Deco, G., Jirsa, V., McIntosh, A.R., Sporns, O., Kötter, R.: Key role of coupling, delay, and noise in resting brain fluctuations. Proc. Natl. Acad. Sci. **106**(25), 10,302–10,307 (2009)
6. Hassard, B.D., Hassard, D., Kazarinoff, N.D., Wan, Y.H., Wan, Y.W.: Theory and applications of Hopf bifurcation, vol. 41. Cambridge University Press, Cambridge (1981)
7. Hellyer, P.J., Jachs, B., Clopath, C., Leech, R.: Local inhibitory plasticity tunes macroscopic brain dynamics and allows the emergence of functional brain networks. NeuroImage **124**, 85–95 (2016)
8. Kuang, Y.: Delay differential equations: with applications in population dynamics, vol. 191. Academic Press, London (1993)
9. Nicola, W., Hellyer, P., Campbell, S.A., Clopath, C.: Chaos in homeostatically regulated neural systems. Chaos: Interdiscip. J. Nonlinear Sci. **28**(083104) (2018)
10. Pasillas-Lépine, W.: Delay-induced oscillations in Wilson and Cowan's model: an analysis of the subthalamo-pallidal feedback loop in healthy and parkinsonian subjects. Biol. Cybern. **107**(3), 289–308 (2013)
11. Song, Z., Yang, K., Xu, J., Wei, Y.: Multiple pitchfork bifurcations and multiperiodicity coexistences in a delay-coupled neural oscillator system with inhibitory-to-inhibitory connection. Commun. Nonlinear Sci. Numer. Simul. **29**(1–3), 327–345 (2015)
12. Turrigiano, G.G., Nelson, S.B.: Homeostatic plasticity in the developing nervous system. Nat. Rev. Neurosci. **5**(2), 97 (2004)
13. Vogels, T.P., Sprekeler, H., Zenke, F., Clopath, C., Gerstner, W.: Inhibitory plasticity balances excitation and inhibition in sensory pathways and memory networks. Science **334**(6062), 1569–1573 (2011)
14. Wilson, H.R., Cowan, J.D.: Excitatory and inhibitory interactions in localized populations of model neurons. Biophys. J. **12**(1), 1 (1972)
15. Xie, S., Huang, Z.: Almost periodic solutions for Wilson-Cowan type model with time-varying delays. Discret. Dyn. Nat. Soc. **2013**(683091), 1–7 (2013)

Autapse-Induced Complicated Oscillations of a Ring FHN Neuronal Network with Multiple Delayed Couplings

Xiaochen Mao, Xiangyu Zhou, Tiantian Shi, and Lei Qiao

Abstract This chapter studies the effects of the autaptic connection on the behaviors of ring-coupled FitzHugh–Nagumo (FHN) neurons. Different transmission time delays between neurons and one autaptic time delay on the self-connection are considered. The delay-independent and delay-dependent stability and Hopf bifurcation are analyzed. Illustrative examples are given and interesting phenomena are observed, such as stability switches, different patterns of multi-periodic oscillations, chaotic motions, and coexisting attractors.

Keywords Neuronal networks · Time delays · Autapse · Coexisting attractors

1 Introduction

Over the past few decades, numerous studies have been made on the dynamical mechanisms of neural systems since the pioneering work of the Hodgkin and Huxley [1]. By extracting the excitability of the dynamics in the Hodgkin–Huxley model, FitzHugh and Nagumo presented a two-dimensional system to describe the electrical properties of neural activities [2, 3]. Neural systems can often be described by networks consisting of many neurons and connections. The behaviors arising from the interactions of the neurons are often different from those in isolation [4–6].

Due to the finite time of information transformation occurring in both synapse and dendrite, time delays are unavoidable in neuronal networks. Time delays are also associated with a finite response period to internal signal processing and derived from the closed-loop feedback, such as autapses. It is known that time delay can give rise to a variety of interesting phenomena for a single neuron and neuronal networks [6–11].

X. Mao (✉) · X. Zhou · T. Shi · L. Qiao
Department of Engineering Mechanics, College of Mechanics and Materials, Hohai University, Nanjing, China
e-mail: maochen@hhu.edu.cn

W. Lacarbonara et al. (eds.), *New Trends in Nonlinear Dynamics*,
https://doi.org/10.1007/978-3-030-34724-6_19

In neural systems, innumerable neurons are interconnected via synapses among them. The topological structures of these reciprocal interactions are important and diverse, including catenulate, starlike, annular, latticed, random, small-world, scale-free, and many others [12]. For example, the real cortical connectivity pattern is extremely sparse: most connections are between nearby cells [13]. Ring-shaped networks are common and ubiquitous in many neural structures such as nematodes C. elegans, sea stars, hippocampus, cerebellum, neocortex, etc [13–17].

Biological and anatomical experiments have shown that some neurons connect not only to neighboring cells but also to its own soma or dendrite via a specific loop. This self-connection structure, which is named by autapse, has been found in cerebral cortex, neocortex, cerebellum, rat hippocampus, pyramidal cells, substantia nigra, the visual cortex of cats, etc [8, 18, 19]. For example, in the cerebral cortex, "fast-spiking cells" can generate rich autapses [19]. The autaptic connection could establish a time-delayed feedback mechanism at the cellular level. In addition, it can modulate the behaviors of neurons and regulate the brain function [8, 20–23]. For example, experimental evidence shows that autaptic connections have the capacity to improve the precision of spike times in neocortical fast-spiking interneurons [23].

The purpose of this paper is to study the dynamical properties of a delayed FitzHugh–Nagumo (FHN) neural network with a single autaptic connection. This network can be described by a set of delay differential equations as follows:

$$
\begin{aligned}
&\dot{x}_1(t) = a_1x_1(t) - x_1^3(t) - y_1(t) + c_n f_n\left(x_n\left(t-\tau_n\right)\right) + k\left(x_1\left(t-\sigma\right) - x_1\right),\\
&\dot{y}_1(t) = x_1(t) - b_1y_1(t),\\
&\dot{x}_2(t) = a_2x_2(t) - x_2^3(t) - y_2(t) + c_1 f_1\left(x_1\left(t-\tau_1\right)\right),\\
&\dot{y}_2(t) = x_2(t) - b_2y_2(t),\\
&\cdots\cdots\cdots\cdots\\
&\dot{x}_n(t) = a_nx_n(t) - x_n^3(t) - y_n(t) + c_{n-1} f_{n-1}\left(x_{n-1}\left(t-\tau_{n-1}\right)\right),\\
&\dot{y}_n(t) = x_n(t) - b_ny_n(t)
\end{aligned}
\tag{1}
$$

where x_i denotes the membrane potential of the i-th neuron, y_i represents the slow refractory variables, which model the time dependence of several physical quantities related to electrical conductances of the relevant ion currents across the membrane, a_i and b_i are positive constants, k and σ represent the gain and time delay in the electric autapse of the neuron 1, c_i and τ_i are the coupling strength and time delay, f_i is the activation function of neuron, $i = 1, 2, \ldots, n$. Without loss of generality, these functions are absolutely smooth and assumed to satisfy $f_i(0) = 0$.

The rest of this chapter is organized as follows. In Sect. 2, the local stability of the network equilibrium and the existence of the periodic oscillations are analyzed. Case studies of numerical simulations are given to support the obtained results and abundant activities are shown in Sect. 3. Finally, conclusions are drawn in Sect. 4.

2 Stability Analysis

This section focuses on the stability of the trivial equilibrium position of the system (1). The analysis for other equilibrium positions is similar after a simple variable translation. The characteristic equation of the linearization of Eq. (1) at the trivial equilibrium of the network can be written as follows:

$$\Delta(\lambda,\ \sigma,\ \tau)=\begin{vmatrix}\lambda I-A_1 & 0 & \cdots & -B_n\\ -B_1 & \lambda I-A_2 & \cdots & 0\\ \vdots & \vdots & \ddots & \vdots\\ 0 & 0 & \cdots & \lambda I-A_n\end{vmatrix}=0 \tag{2}$$

where I is a 2×2 identity matrix, $\gamma_i=c_if_i^{'}(0)$, $i=1,2,\ldots,n, j=2,\ldots,n$, and
$A_1=\begin{bmatrix}a_1+ke^{-\lambda\sigma}-k & -1\\ 1 & -b_1\end{bmatrix}$, $A_j=\begin{bmatrix}a_j & -1\\ 1 & -b_j\end{bmatrix}$, $B_i=\gamma_ie^{-\lambda\tau_i}\begin{bmatrix}1 & 0\\ 0 & 0\end{bmatrix}$

After some calculations, one arrives at

$$\begin{aligned}\Delta(\lambda,\ \sigma,\ \tau)&=\prod_{i=1}^{n}|\lambda I-A_i|-\prod_{i=1}^{n}(\lambda+b_i)\gamma_ie^{-\lambda\tau_i}\\ &=\left[\left(\lambda-a_1+k-ke^{-\lambda\sigma}\right)(\lambda+b_1)+1\right]\prod_{j=2}^{n}\left[\left(\lambda-a_j\right)\left(\lambda+b_j\right)+1\right]-\gamma e^{-\lambda\tau}\prod_{i=1}^{n}(\lambda+b_i)\\ &=S(\lambda)-kQ(\lambda)e^{-\lambda\sigma}-\gamma R(\lambda)e^{-\lambda\tau}=0\end{aligned} \tag{3}$$

where $S(\lambda)=P(\lambda)+kQ(\lambda)$, $P(\lambda)=\prod_{i=1}^{n}[(\lambda-a_i)(\lambda+b_i)+1]$, $Q(\lambda)=(\lambda+b_1)\prod_{j=2}^{n}\left[\left(\lambda-a_j\right)\left(\lambda+b_j\right)+1\right]$, $R(\lambda)=\prod_{i=1}^{n}(\lambda+b_i)$, $\tau=\sum_{i=1}^{n}\tau_i$, and $\gamma=\gamma_1\gamma_2\cdots\gamma_n$.To determine the boundary of the stability, the analysis begins with the case when $\sigma=0$. The necessary and sufficient conditions for all roots of $\Delta(\lambda,0,0)=0$ having negative real parts can be determined by the Routh–Hurwitz criteria. As τ varies, let $\lambda=\pm\mathrm{i}\upsilon\ (\upsilon>0)$ be a pair of purely imaginary roots of $\Delta(\lambda,0,\tau)=0$. Then, one has

$$\Delta(\mathrm{i}\upsilon,0,\tau)=P_R(\upsilon)+\mathrm{i}P_I(\upsilon)-\gamma\left[R_R(\upsilon)+\mathrm{i}R_I(\upsilon)\right]\left[\cos(\upsilon\tau)-\mathrm{i}\sin(\upsilon\tau)\right]=0 \tag{4}$$

where $P_R(\upsilon)=\mathrm{Re}\,[P(\mathrm{i}\upsilon)]$, $P_I(\upsilon)=\mathrm{Im}\,[P(\mathrm{i}\upsilon)]$, $R_R(\upsilon)=\mathrm{Re}\,[R(\mathrm{i}\upsilon)]$, and $R_I(\upsilon)=\mathrm{Im}\,[R(\mathrm{i}\upsilon)]$.

By separating the real and imaginary parts of Equation (4) and eliminating the harmonic terms, one arrives at $D(\upsilon)=P_R^2(\upsilon)+P_I^2(\upsilon)-\gamma^2\left[R_R^2(\upsilon)+R_I^2(\upsilon)\right]=0$. If $D(\upsilon)=0$ has a number of positive and simple roots υ_j, then $\Delta(\lambda,0,\tau)=0$ has sets of critical time delays $\tau_{j,k}=(\theta_j+2k\pi)/\upsilon_j$, $j=1,2,\ldots,k=0,1,2,\ldots,$ where $\theta_j\in[0,\ 2\pi)$ and θ_j satisfies

$$\cos\theta_j = \frac{P_R(\upsilon_j)\,R_R(\upsilon_j) + P_I(\upsilon_j)\,R_I(\upsilon_j)}{\gamma\left[{R_R}^2(\upsilon_j) + {R_I}^2(\upsilon_j)\right]},$$
$$\sin\theta_j = \frac{P_R(\upsilon_j)\,R_I(\upsilon_j) - P_I(\upsilon_j)\,R_R(\upsilon_j)}{\gamma\left[{R_R}^2(\upsilon_j) + {R_I}^2(\upsilon_j)\right]} \tag{5}$$

The following analysis turns to the case when the autaptic time delay is taken into account, i.e., $\sigma > 0$. In this case, the autaptic time delay can be regarded as a parameter for fixed value of τ. When the characteristic equation of the system has a pair of purely imaginary roots $\lambda = \pm\,\mathrm{i}\omega\ (\omega > 0)$, one obtains

$$\begin{aligned}\Delta(\mathrm{i}\omega,\ \sigma,\ \tau) = {} & S_R(\omega) + \mathrm{i}S_I(\omega) - k\left[Q_R(\omega) + \mathrm{i}Q_I(\omega)\right]\left[\cos(\omega\sigma) - \mathrm{i}\sin(\omega\sigma)\right] \\ & - \gamma\left[R_R(\omega) + \mathrm{i}R_I(\omega)\right]\left[\cos(\omega\tau) - \mathrm{i}\sin(\omega\tau)\right]\end{aligned} \tag{6}$$

where $S_R(\omega) = \mathrm{Re}\,[S(\mathrm{i}\omega)]$, $S_I(\omega) \doteq \mathrm{Im}\,[S(\mathrm{i}\omega)]$, $Q_R(\omega) = \mathrm{Re}\,[Q(\mathrm{i}\omega)]$, $Q_I(\omega) = \mathrm{Im}\,[Q(\mathrm{i}\omega)]$, $R_R(\omega) = \mathrm{Re}\,[R(\mathrm{i}\omega)]$, and $R_I(\omega) = \mathrm{Im}\,[R(\mathrm{i}\omega)]$. Separating the real and imaginary parts of $\Delta(\mathrm{i}\omega,\ \sigma,\ \tau) = 0$ yields,

$$\begin{aligned}& S_R(\omega) - k\left[Q_R(\omega)\cos(\omega\sigma) + Q_I(\omega)\sin(\omega\sigma)\right] \\ & -\gamma\left[R_R(\omega)\cos(\omega\tau) + R_I(\omega)\sin(\omega\tau)\right] = 0, \\ & S_I(\omega) - k\left[Q_I(\omega)\cos(\omega\sigma) - Q_R(\omega)\sin(\omega\sigma)\right] \\ & -\gamma\left[R_I(\omega)\cos(\omega\tau) - R_R(\omega)\sin(\omega\tau)\right] = 0\end{aligned} \tag{7}$$

Eliminating the autaptic time delay in the above equation gives

$$\begin{aligned}F(\omega) = {} & S_R^2(\omega) + S_I^2(\omega) - 2\gamma S_R(\omega)\left[R_R(\omega)\cos(\omega\tau) + R_I(\omega)\sin(\omega\tau)\right] \\ & -2\gamma S_I(\omega)\left[R_I(\omega)\cos(\omega\tau) - R_R(\omega)\sin(\omega\tau)\right] + \gamma^2 R_R^2(\omega) \\ & +\gamma^2 R_I^2(\omega) - k^2\left[Q_R^2(\omega) + Q_I^2(\omega)\right] = 0\end{aligned} \tag{8}$$

If $F(\omega) = 0$ has any positive and simple roots ω_j, then $\Delta(\lambda,\ \sigma,\ \tau) = 0$ has sets of critical time delays $\sigma_{j,l} = (\varphi_j + 2l\pi)/\omega_j$, $j = 1, 2, \ldots, l = 0, 1, 2, \ldots,$ where $\varphi_j \in [0, 2\pi)$ and φ_j satisfies

$$\begin{aligned}\cos(\varphi_j) = {} & \frac{1}{N}\Big\{S_R(\omega_j)\,Q_R(\omega_j) + S_I(\omega_j)\,Q_I(\omega_j) - \gamma Q_R(\omega_j)\,R_R(\omega_j)\cos(\omega_j\tau) \\ & -\gamma Q_R(\omega_j)\,R_I(\omega_j)\sin(\omega_j\tau) - \gamma Q_I(\omega_j) \\ & \times\left[R_I(\omega_j)\cos(\omega_j\tau) - R_R(\omega_j)\sin(\omega_j\tau)\right]\Big\}, \\ \sin(\varphi_j) = {} & \frac{1}{N}\Big\{S_R(\omega_j)\,Q_I(\omega_j) - S_I(\omega_j)\,Q_R(\omega_j) - \gamma Q_I(\omega_j)\,R_R(\omega_j)\cos(\omega_j\tau) \\ & -\gamma Q_I(\omega_j)\,R_I(\omega_j)\sin(\omega_j\tau) + \gamma Q_R(\omega_j) \\ & \times\left[R_I(\omega_j)\cos(\omega_j\tau) - R_R(\omega_j)\sin(\omega_j\tau)\right]\Big\}\end{aligned} \tag{9}$$

where $N = k|Q(\mathrm{i}\omega_j)|^2$.

By differentiating λ with respect to the autaptic time delay σ in $\Delta(\lambda, \sigma, \tau) = 0$, one has $\mathrm{Re}[\lambda(\sigma)|_{\lambda=\mathrm{i}\omega}]' = [M_R(\mathrm{i}\omega)T_R(\mathrm{i}\omega) + M_I(\mathrm{i}\omega)T_I(\mathrm{i}\omega)]/|T^{\mp}(\mathrm{i}\omega)|^2$, where $M(\mathrm{i}\omega) = -k(\mathrm{i}\omega)Q(\mathrm{i}\omega)e^{-\mathrm{i}\omega\sigma}$, $M_R(\mathrm{i}\omega) = \mathrm{Re}\ M(\mathrm{i}\omega)$, $M_I(\mathrm{i}\omega) = \mathrm{Im}\ M(\mathrm{i}\omega)$, $T(\mathrm{i}\omega) = S'(\mathrm{i}\omega) - \gamma R'(\mathrm{i}\omega)e^{-\mathrm{i}\omega\tau} + \gamma\tau R(\mathrm{i}\omega)e^{-\mathrm{i}\omega\tau} - kQ'(\mathrm{i}\omega)e^{-\mathrm{i}\omega\sigma} + k\sigma Q(\mathrm{i}\omega)e^{-\mathrm{i}\omega\sigma}$, $T_R(\mathrm{i}\omega) = \mathrm{Re}\ T(\mathrm{i}\omega)$, and $T_I(\mathrm{i}\omega) = \mathrm{Im}\ T(\mathrm{i}\omega)$.

After some calculations and from Eq. (9), the numerator of $\mathrm{Re}[\lambda(\sigma)|_{\lambda=\mathrm{i}\omega}]'$ can be written as follows:

$$\begin{aligned}
&\omega S_I(\omega)\{S_I'(\omega) - \gamma[R_I'(\omega)\cos(\omega\tau) - R_R'(\omega)\sin(\omega\tau)] + \gamma\tau R_R(\omega)\cos(\omega\tau) \\
&\quad + \gamma\tau R_I(\omega)\sin(\omega\tau)]\} - \omega\gamma S_I'(\omega)[R_I(\omega)\cos(\omega\tau) - R_R(\omega)\sin(\omega\tau)] \\
&- \omega S_R(\omega)\{-S_R'(\omega) + \gamma[R_R'(\omega)\cos(\omega\tau) + R_I'(\omega)\sin(\omega\tau)] + \gamma\tau R_I(\omega)\cos(\omega\tau) \\
&\quad - \gamma\tau R_R(\omega)\sin(\omega\tau)]\} - \omega\gamma S_R'(\omega)[R_R(\omega)\cos(\omega\tau) + R_I(\omega)\sin(\omega\tau)] \\
&+ \omega\gamma^2[R_R(\omega)R_R'(\omega) + R_I(\omega)R_I'(\omega)] - \omega k^2[Q_R(\omega)Q_R'(\omega) + Q_I(\omega)Q_I'(\omega)] \\
&= 0.5\omega F'(\omega)
\end{aligned}$$

Thus, $\mathrm{sgn}\ \mathrm{Re}\ [\lambda(\sigma)|_{\lambda=\mathrm{i}\omega}]' = \mathrm{sgn}\ F'(\omega)$ holds.

When $F(\omega) = 0$ has only one positive simple root, one can check that the system has a new pair of conjugate roots with positive real parts when the autaptic time delay is crossing a critical value determined by Eq. (9). Thus, if the system free of the autaptic time delay is asymptotically stable, there exists a constant σ_c such that the system remains stable for $\sigma \in [0, \sigma_c)$ and it undergoes a Hopf bifurcation for $\sigma = \sigma_c$. When $F(\omega) = 0$ has at least two simple positive roots, a finite number of stability switches occur as the autaptic time delay increases from zero to the infinity, and the system will become unstable at last [7]. Moreover, if $F(\omega) = 0$ has no positive root, then the system is delay-independent stable or unstable for any given autaptic time delay, depending on whether or not the system free of autaptic time delay is stable.

3 Illustrative Examples

In this section, the activation function is taken as $f_i(x) = \tanh(x)$, which is often used in neuronal networks.

1. $n = 4$, $a_1 = 0.3$, $a_2 = 0.4$, $a_3 = 0.2$, $a_4 = 0.5$, $b_1 = 1.5$, $b_2 = 0.9$, $b_3 = 0.6$, $b_4 = 1.2$, $c_1 = 0.5$, $c_2 = 0.4$, $c_3 = 0.3$, $c_4 = 1$, $\tau_1 = 0.2$, $\tau_2 = 0.3$, $\tau_3 = 0.1$, $\tau_4 = 0.4$, and $k = -0.2$. In this case, the network free of the autaptic time delay is locally stable for $\tau = 1 \in [0, \tau_{1,0})$, where $\tau_{1,0} = 1.36$ based on Eq. (5). After some calculations and from Equation (9), the network always adds a new pair of conjugate roots with positive real parts for each crossing at $\sigma_{1,l} = 0.78$, 6.97, ... and $\sigma_{3,l} = 4.66$, 30.43, ..., but reduces a pair of conjugate roots with positive real parts for each crossing at $\sigma_{2,l} = 3.72$, 10.29, ... Thus, the network is locally asymptotically stable when $\sigma \in [0, \sigma_{1,0}) \cup (\sigma_{2,0}, \sigma_{3,0})$ and unstable for $\sigma \in (\sigma_{1,0}, \sigma_{2,0}) \cup (\sigma_{3,0}, +\infty)$.

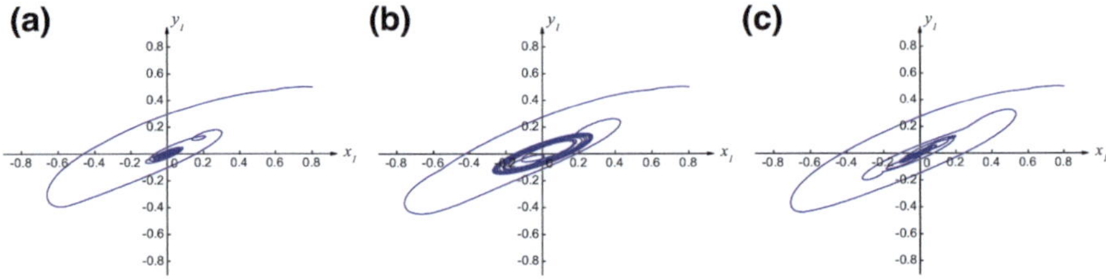

Fig. 1 The phase trajectories on $x_1 - y_1$ plane of four coupled neurons. (**a**) $\sigma = 0.5$; (**b**) $\sigma = 2$; (**c**) $\sigma = 4$

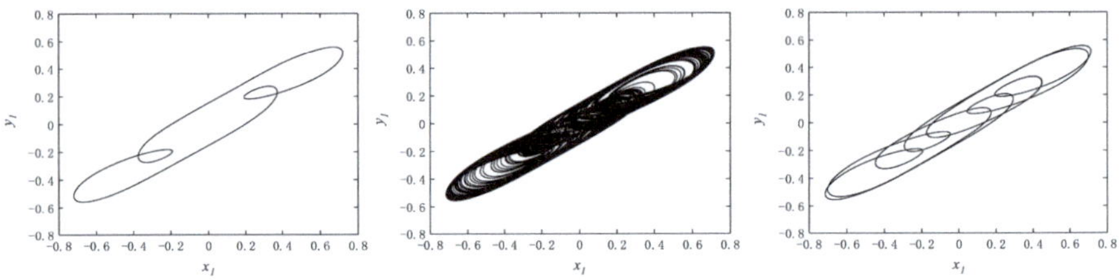

Fig. 2 The phase trajectories on $x_1 - y_1$ plane of six coupled neurons. (**a**) $k = -0.2$; (**b**) $k = -0.1$; (**c**) $k = -0.02$

Figure 1a illustrates that the phase trajectories converge to the stable rest state when $\sigma = 0.5$. Figure 1b gives a stable limit cycle arising from Hopf bifurcation when $\sigma = 2$. Figure 1c shows that the trivial rest point becomes stable again when $\sigma = 4$. The responses of the network, as shown in Fig. 1, well coincide with the theoretical assertions.

2. $n = 6$, $a_i = 1.1$, $b_1 = 1.2$, $b_2 = 0.6$, $b_3 = 0.5$, $b_4 = 0.9$, $b_5 = 0.6$, $b_6 = 0.8$, $c_i = 0.2$, $\tau_i = 0.5$, and $i = 1, 2, \ldots, 6$. To discuss the effects of the gain of the autapse on the dynamics of the network, the autaptic time delay is chosen as $\sigma = 0.1$. Figure 2a gives a period-3 orbit for $k = -0.2$. When $k = -0.1$, as displayed in Fig. 2b, the period-3 oscillation loses its stability and a chaotic attractor comes into being. Figure 2c illustrates a period-7 oscillation when $k = -0.02$.

   Figure 3a gives a complicated chaotic attractor when $k = 0.1$. Figure 3b illustrates that the chaotic attractor separates into two strange attractors when $k = 0.15$. When $k = 0.19$, as shown in Fig. 3c, the chaotic attractors disappear and two period-8 oscillations come into being. With a slight increase of the gain of the autapse, two coexisting period-4 orbits arise for $k = 0.2$, as shown in Fig. 3d. Figure 3e gives that the period-4 orbits lose their stability and two period-2 solutions occur when $k = 0.3$. When $k = 0.5$, two separated period-1 solutions appear, as shown in Fig. 3f. As shown in Fig. 3, the red and blue lines represent the responses of the system under different initial conditions. When the gain of the autapse varies, the system exhibits the process of periodic-doubling bifurcation.

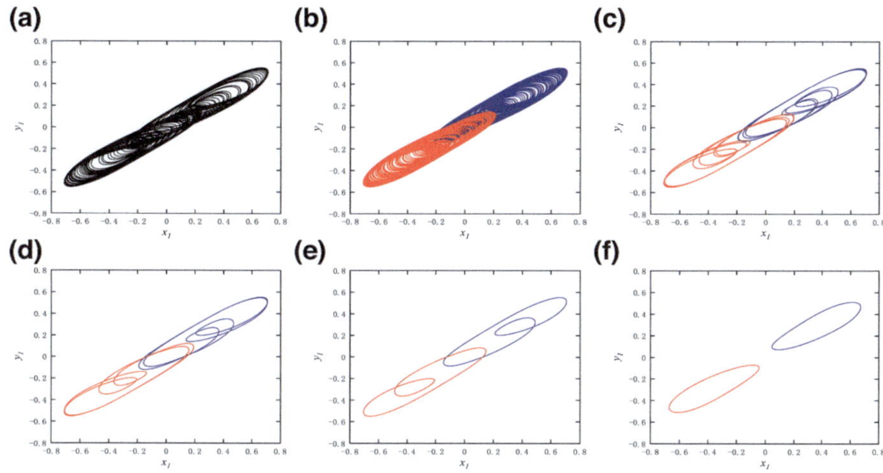

Fig. 3 The phase trajectories on $x_1 - y_1$ plane of six coupled neurons. (**a**) $k = 0.1$; (**b**) $k = 0.15$; (**c**) $k = 0.19$; (**d**) $k = 0.2$; (**e**) $k = 0.3$; (**f**) $k = 0.5$

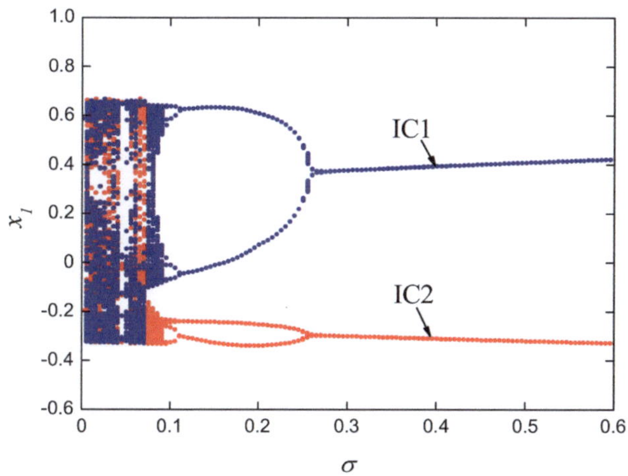

Fig. 4 The bifurcation diagram on the Poincaré section of six coupled neurons for $k = 0.2$

The following study turn to the case when the autaptic time delay varies and the gain of the autapse is fixed as $k = 0.2$. Figure 4 shows the detailed bifurcation diagram as the function of the autaptic time delay under different conditions IC1 and IC2, where the Poincaré section is defined by $\Sigma = \{(\sigma, x_1) : (x_2 = 0, \ \dot{x}_2 > 0)\}$. By changing the value of the autaptic time delay, the system can exhibit chaotic phenomena and different coexisting multi-periodic oscillations. As shown in Fig. 4, the system undergoes the periodic-doubling bifurcation that leads to chaos.

4 Conclusions

In this paper, the behaviors of a delay-coupled ring FitzHugh–Nagumo network with one autaptic connection are studied. The stability switches and periodic responses of the network are discussed. Case studies of numerical simulations are given and various phenomena are observed, such as multi-periodic oscillations and chaos. The detailed bifurcation diagram as the function of the autaptic time delay on Poincaré section is given. It is shown that the autaptic connection plays important roles in the dynamics of the network and can give rise to abundant behaviors, such as quiescent state, periodic and chaotic firing patterns, and multistability.

Acknowledgments This work was supported by the National Natural Science Foundation of China under Grant Nos. 11872169 and 11472097 and Fundamental Research Funds for the Central Universities under Grant No. 2018B17514.

References

1. Hodgkin, A.L., Huxley, A.F.: A quantitative description of membrane current and its application to conduction and excitation in nerve. J. Physiol. **117**(4), 500–544 (1952)
2. Fitzhugh, R.: Impulses and physiological states in theoretical models of nerve membrane. Biophys. J. **1**(6), 445–466 (1961)
3. Nagumo, J., Arimoto, S., Yoshizawa, S.: An active pulse transmission line simulating nerve axon. Proc. IRE. **50**(10), 2061–2070 (1962)
4. Buric, N., Todorovic, D.: Dynamics of FitzHugh-Nagumo excitable systems with delayed coupling. Phys. Rev. E. **67**(6), 066222 (2003)
5. Ueta, T., Miyazaki, H., Kousaka, T., Kawakami, H.: Bifurcation and chaos in coupled BVP oscillators. Int. J. Bifurcat. Chaos. **14**(4), 1305–1324 (2004)
6. Wang, Q., Zheng, Y., Ma, J.: Cooperative dynamics in neuronal networks. Chaos. Soliton. Fract. **56**(SI), 19–27 (2013)
7. Hu, H.Y., Wang, Z.H.: Dynamics of Controlled Mechanical Systems with Delayed Feedback. Springer-Verlag, Heidelberg (2002)
8. Ma, J., Tang, J.: A review for dynamics in neuron and neuronal network. Nonlinear Dyn. **89**(3), 1569–1578 (2017)
9. Krupa, M., Touboul, J.D.: Complex oscillations in the delayed FitzHugh-Nagumo equation. J. Nonlinear Sci. **26**(1), 43–81 (2016)
10. Mao, X.C.: Complicated dynamics of a ring of nonidentical FitzHugh-Nagumo neurons with delayed couplings. Nonlinear Dyn. **87**(4), 2395–2406 (2017)
11. Song, Y.L., Xu, J.: Inphase and antiphase synchronization in a delay-coupled system with applications to a delay-coupled FitzHugh-Nagumo system. IEEE Trans. Neural Netw. Learn. Syst. **23**(10), 1659–1670 (2012)
12. Strogatz, S.H.: Exploring complex networks. Nature. **410**(6825), 268–276 (2001)
13. Kaslik, E., Balint, S.: Complex and chaotic dynamics in a discrete-time-delayed Hopfield neural network with ring architecture. Neural Netw. **22**(10), 1411–1418 (2009)
14. Khokhlova, T.N., Kipnis, M.M.: The breaking of a delayed ring neural network contributes to stability: the rule and exceptions. Neural Netw. **48**, 148–152 (2013)
15. Guo, S.J., Huang, L.H.: Hopf bifurcating periodic orbits in a ring of neurons with delays. Phys. D. **183**, 19–44 (2003)

16. Horikawa, Y.: Stabilization of metastable dynamical rotating waves in a ring of unidirectionally coupled nonmonotonic neurons. Neurocomputing **225**, 148–156 (2017)
17. Xu, X.: Complicated dynamics of a ring neural network with time delays. J. Phys. A. **41**(3), 035102 (2008)
18. Tamas, G., Buhl, E.H., Somogyi, P.: Massive autaptic self-innervation of GABAergic neurons in cat visual cortex. J. Neurosci. **17**(16), 6352–6364 (1997)
19. Van der Loos, H., Glaser, E.: Autapses in neocortex cerebri: synapses between a pyramidal cell's axon and its own dendrites. Brain Res. **48**, 355–360 (1972)
20. Flight, M.H.: Neuromodulation exerting self-control for persistence. Nat. Rev. Neurosci. **10**(5), 316–316 (2009)
21. Uzun, R., Yilmaz, E., Ozer, M.: Effects of autapse and ion channel block on the collective firing activity of Newman-Watts small-world neuronal networks. Phys. A. **486**, 386–396 (2017)
22. Yue, Y.W., Liu, L.W., Liu, Y.J., Chen, Y., Chen, Y.L., Yu, L.C.: Dynamical response, information transition and energy dependence in a neuron model driven by autapse. Nonlinear Dyn. **90**(4), 2893–2902 (2017)
23. Bacci, A., Huguenard, J.R.: Enhancement of spike-timing precision by autaptic transmission in neocortical inhibitory interneurons. Neuron. **49**(1), 119–130 (2006)

A Time-Delay Nonlinear Model of Dopamine-Modulated Prefrontal-Limbic Interactions in Schizophrenia

Eva Kaslik, Mihaela Neamţu, and Anca Rădulescu

Abstract We present a nonlinear mathematical model of dopamine-modulated prefrontal-limbic interactions in schizophrenia, including discrete time delays. An extensive stability and bifurcation analysis is undertaken in a neighborhood of the positive equilibrium of the system. The results reveal the importance of time delays in modulating dopamine reactivity.

Keywords Schizophrenia · Mathematical model · Time delay · Stability · Bifurcations

1 Introduction

Schizophrenia is an incurable neuropsychiatric illness, with dramatic personal and social implications [2, 4, 6, 11]. Its diagnosis and treatment are currently based on clinical symptoms rather than on the neurophysiological basis (which remains unknown). The "stress-diathesis model" remains a popular hypothesis that attributes stress vulnerability in schizophrenia to a pre-existing impairment in hippocampal and prefrontal inhibitory control of the limbic arousal response. The

E. Kaslik
Department of Mathematics and Computer Science, West University of Timisoara, Timisoara, Romania
e-mail: eva.kaslik@e-uvt.ro

M. Neamţu (✉)
West University of Timişoara, Timişoara, Romania

Academy of Romanian Scientists, Bucharest, Romania
e-mail: mihaela.neamtu@e-uvt.ro

A. Rădulescu
SUNY New Paltz, New Paltz, NY, USA
e-mail: radulesa@newpaltz.edu

W. Lacarbonara et al. (eds.), *New Trends in Nonlinear Dynamics*,
https://doi.org/10.1007/978-3-030-34724-6_20

subsequent exacerbated fear reaction raises cortisol levels, with toxic effects on the hippocampus, further deepening the pre-existing impairment of the inhibitory unit. This is a possible trigger for the main neurodegenerative cycle in schizophrenia. While empirical work cannot fully explain the complex mechanics of the prefrontal-limbic system, a mathematical model can approach system dysregulation with analytical techniques, quantifying the nonlinear components of a self-interacting network. It can be used to test hypotheses that bridge connectivity with functional dynamics and subsequently with behavior observed empirically.

It has been proposed that schizophrenia symptoms constitute an end-stage of a cyclic and neurodegenerative process [1], in which a hereditary predisposition reduces the individual psychological threshold toward stimuli to the point where even minor daily stresses will directly trigger psychotic experiences. However, the etiology of this systemic degeneration has been challenging any simple explanation, and current antipsychotic medications are likely treating the outward symptoms rather than their cause.

2 Modeling Methods

One of our earlier studies on empirical fMRI time series from human subjects suggested that key dynamic differences between patients with schizophrenia and healthy controls can be captured in the existence and geometry of oscillations in a two-dimensional subspace of prefrontal-limbic regions [13]. This inspired us to consider, in our subsequent modeling work [12], a two-dimensional prefrontal-amygdala system, and understand analytically how the coupled dynamics can play the major role that had been demonstrated empirically in regulation of emotional arousal. In the current paper, we refine the model with a focus on the dopamine regulatory aspect, which in previous work was represented mathematically by non-linear terms. New literature shows that dopamine-modulated mechanisms, unlike those mediated by other neurotransmitters, operate based on a system of actual biophysical delays. It has been suggested that the three different timescales across which dopamine operates [14, 15] (fast, intermediate and low) may underlie the broadness of dopamine's effects on executive, cognitive, and motivational function (disrupted in schizophrenia). In particular, at the lowest timescale, "dopamine exerts an almost tonic influence on postsynaptic structures." Deficits in this delayed/tonic dopamine release have been shown to affect post-synaptic function (which cannot be otherwise explained by reductions in phasic dopamine changes) and may further lead to the deficits in movement, attention, and cognition—characteristic of pathologies like Parkinson's disease or schizophrenia.

A realistic model of brain function which encompasses the regulatory effect of dopamine must therefore take into consideration delays, which may have been crucial, if subtle effects that go beyond the nonlinearities are included and discussed in our original framework. The use of nonlinear delayed equations as a distinct and important approach in modeling schizophrenia-like neural patterns has been recently investigated in [15]. In this paper, we improve our previous work to address dopamine delay mechanisms, by introducing a delayed neural response in the target regions of dopamine-mediated pathways.

Our model represents the time activations of the amygdala, the hippocampus, and the prefrontal cortex as three distinct variables a, p, and h, while a fourth variable δ stands for the activation of the dopamine system, controlled via the nucleus accumbens and the ventral tegmental area. The system of equations reads

$$\begin{aligned}
\dot{a} &= -\mu_1 a - k_1 p - \gamma_1 h + I, \\
\dot{p} &= k_2 a - \mu_2 f(p, \delta_\tau) + \frac{\gamma_2}{a_1 C(a) + 1} h, \\
\dot{h} &= k_3 f(p, \delta_\tau^2) - a_2 C(a), \\
\dot{\delta} &= -\xi_1 f(a, \delta) + \xi_2 f(p, \delta) + \xi_3 f(h, \delta)
\end{aligned} \tag{1}$$

Here, the function f is of the form $f(x, \delta) = xg(\delta)$, where x can be a, p, or h, and the function g is increasing, such that $g(0) = 1$. The term $\delta_\tau(t) = \delta(t - \tau)$ represents the delay term, where the parameter τ is the delay in the dopamine action. The linear coefficients are positive system parameters, representing the strengths of the connections between the respective brain areas. As in prior works [3, 10, 12], the dependence of cortisol on arousal levels (measured as amygdala activation) is represented as $C(a) = \frac{\alpha e^a}{e^a + 1}$.

In prior work [12], we investigated the dependence of the system's temporal dynamics on a larger set of physiological parameters, representing connectivity strengths between the same key brain areas, but also including vulnerability to stress-induced cortisol, dopamine regulation, and autoimmunity.

3 Results

We performed stability analyses studying the system sensitivity to parameter perturbations, and we computed the bifurcations. We obtained analytical conditions for the existence of a positive stable equilibrium, and for this equilibrium to undergo a supercritical Hopf transition into stable oscillations. Hopf bifurcations are illustrated in the presence and the absence of delays, with respect to different parameters.

3.1 Nonlinear Model

For the system without delays, we focused on locating Hopf bifurcation curves in the parameter plane defined by μ_1 (level of anxiety) and a_2 (vulnerability to stress cortisol). Our results suggest that varying a_2 for fixed μ_1 can readily

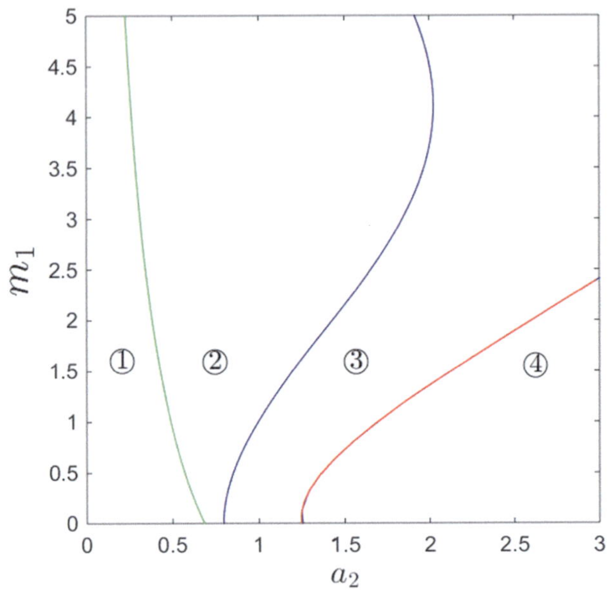

Fig. 1 *Transitions in between dynamic regimes in the* (a_2, μ_1) *parameter plane*. The Saddle Node (green), Hopf (blue), and Fold (red) curves delimit the plane into the parameter regions (1)–(4), with different asymptotic behaviors, as explained in the text

push the system through qualitative changes in asymptotic dynamics (see Fig. 1), while changing μ_1 and keeping a_2 fixed, is more likely to introduce more subtle quantitative/kinetic changes in the convergence to the equilibrium, or in the duty cycle. A small sensitivity to stress cortisol in the system is necessary to stabilize the system to the equilibrium characteristic to a healthy functional (region 2). When this sensitivity is increased past a "vulnerability" threshold, the system crosses the Hopf curve and enters oscillations (region 3), exhibiting out of phase swings in the amygdala arousal reaction to the stimulus I, and in the prefrontal activation, attempting to (unsuccessfully) provide appropriate inhibition. When a_2 is increased past a "pathological" value, the system loses the oscillatory stability, and enters unstable oscillations, with escaping trajectories (region 4).

3.2 Delay Model

The equilibrium states of the model (1) are the solutions of the following system:

$$\begin{aligned}
&\mu_1 a + k_1 p + \gamma_1 h = I,\\
&k_2 a - \mu_2 p g(\delta) + C_2(a) h = 0,\\
&k_3 p g(\delta^2) - a_2 C(a) = 0,\\
&\xi_1 a = \xi_2 p + \xi_3 h
\end{aligned} \tag{2}$$

with $C_2(a) = \dfrac{\gamma_2}{a_1 C(a) + 1}$.

In the delayed case, the linearization of system (1) at an equilibrium state $E = (a^\star, p^\star, h^\star, \delta^\star)$ has the form:

$$\dot{x}(t) = Ax(t) + Bx(t - \tau),$$

where $x(t) = (a(t) - a^\star, p(t) - p^\star, h(t) - h^\star, \delta(t) - \delta^\star)^T$ and

$$A = \begin{pmatrix} -\mu_1 & -k_1 & -\gamma_1 & 0 \\ k_2 + C_2'(a^\star)h^\star & -\mu_2 g(\delta^\star) & C_2(a^\star) & 0 \\ -a_2 C'(a^\star) & k_3 g((\delta^\star)^2) & 0 & 0 \\ -\xi_1 g(\delta^\star) & \xi_2 g(\delta^\star) & \xi_3 g(\delta^\star) & 0 \end{pmatrix} \text{ and } B = \begin{pmatrix} 0\,0\,0 & 0 \\ 0\,0\,0 & -\mu_2 p^\star g'(\delta^\star) \\ 0\,0\,0 & 2k_3 p^\star \delta^\star g'(\delta^\star) \\ 0\,0\,0 & 0 \end{pmatrix}$$

Therefore, the characteristic equation is of the form:

$$\Delta(\lambda, \tau) := \lambda P_3(\lambda) - P_2(\lambda) e^{-\tau\lambda} = 0 \tag{3}$$

where

$$P_3(\lambda) = \lambda^3 + r_2\lambda^2 + r_1\lambda + r_0 = \lambda^{-1} \det(\lambda I_4 - A),$$
$$P_2(\lambda) = s_2\lambda^2 + s_1\lambda + s_0$$

with r_i and s_i expressed in terms of the elements of the matrices A and B.

Proposition 1 (Local Asymptotic Stability in the Non-delayed Case) *In the non-delayed case, if the following inequalities are satisfied:*

$$\begin{gathered} r_2 > 0, \quad r_1 > s_2, \quad r_0 > s_1, \quad s_0 < 0, \\ r_2(r_1 - s_2)(r_0 - s_1) > (r_0 - s_1)^2 - s_0 r_2^2, \end{gathered} \tag{4}$$

then the equilibrium point E of system (1) is locally asymptotically stable.

Following [5, 7–9], we obtain:

Theorem 1 *Assume that inequalities (4) are satisfied and consider*

$$\tau_0^+ = \frac{1}{\omega_0} \arccos\left(\frac{1}{\omega_0} \cdot \Im\left(\frac{P_2(i\omega_0)}{P_3(i\omega_0)}\right)\right) \tag{5}$$

where $\omega_0 > 0$ is the smallest positive solution of the equation $|P_2(i\omega)| = \omega|P_3(i\omega)|$.

The equilibrium point E is asymptotically stable for $\tau \in [0, \tau_0^+)$. At $\tau = \tau_0^+$, system (1) undergoes a Hopf bifurcation at the equilibrium point E.

Proof The characteristic equation (3) has a pair of complex conjugate roots $z = \pm i\omega$ (with $\omega > 0$) on the imaginary axis if and only if

$$i\omega P_3(i\omega) = P_2(i\omega)e^{-i\omega\tau}. \tag{6}$$

Taking the absolute value in (6) we obtain $|P_2(i\omega)| = \omega|P_3(i\omega)|$. The roots of this equation are the solutions of the equation $R(\omega) = 1$ where

$$R(\omega) = \left(\frac{|P_2(i\omega)|}{\omega|P_3(i\omega)|}\right)^2.$$

The continuous function R satisfies $R(0) = \infty$ and $R(\infty) = 0$; therefore, there exists at least one $\omega_0 > 0$ such that $R(\omega_0) = 1$. If ω_0 denotes the smallest such solution, it can easily be seen that $R'(\omega_0) < 0$, i.e.,

$$R'(\omega_0) = -\frac{2}{\omega_0}\left(1 + \omega_0 \Im\left[\frac{P_2'(i\omega_0)}{P_2(i\omega_0)} - \frac{P_3'(i\omega_0)}{P_3(i\omega_0)}\right]\right) < 0.$$

From Eq. (6), we obtain the critical value τ_0^+ given by (5). Based on Proposition 1, the equilibrium point E is asymptotically stable when $\tau = 0$, and therefore, due to the continuous dependence of the roots of the characteristic equation on the parameter τ, we have that E is asymptotically stable for any $\tau \in (0, \tau_0^+)$.

Let $\lambda(\tau)$ denote the root of the characteristic equation (3) satisfying $\lambda(\tau_0^+) = i\omega_0$. Therefore

$$\lambda'(\tau_0^+) = -\frac{\partial\Delta/\partial\tau}{\partial\Delta/\partial\lambda}\bigg|_{\tau=\tau_0^+} = -\frac{\lambda P_2(\lambda)e^{-\tau\lambda}}{P_3(\lambda) + \lambda P_3'(\lambda) - P_2'(\lambda)e^{-\tau\lambda} + \tau P_2(\lambda)e^{-\tau\lambda}}\bigg|_{\tau=\tau_0^+}$$

and hence, a straightforward computation leads to

$$\text{sign}\left(\Re\left[\lambda'(\tau_0^+)\right]\right) = \text{sign}\left(1 + \omega_0\Im\left[\frac{P_2'(i\omega_0)}{P_2(i\omega_0)} - \frac{P_3'(i\omega_0)}{P_3(i\omega_0)}\right]\right) = \text{sign}\left(-R'(\omega_0)\right) = 1.$$

This nondegeneracy condition for the Hopf bifurcation shows that the equilibrium point E is asymptotically stable if $\tau \in [0, \tau_0^+)$ and at $\tau = \tau_0^+$, system (1) undergoes a Hopf bifurcation at the equilibrium point E. □

We conclude that the parameter dependence observed in the nonlinear system is further modulated by the degree of delay, in the sense that: (1) the system will be prompted to cross from a regime of stable equilibrium into a stable oscillation one at lower levels of stress vulnerability a_2 and/or anxiety μ_1 for slower dopamine reactivity τ, and will be more "resilient" for higher dopamine reactivity; (2) for given stress vulnerability and anxiety, the lack of appropriate dopamine reactivity (too large τ) may in and of itself push the system into oscillations.

4 Numerical Simulations

For the numerical simulations, we have chosen the following parameter values:

$$\mu_1 = 3, \quad \mu_2 = 1, \quad k_1 = 2, \quad k_2 = 1, \quad \xi = 1,$$

$$\gamma_2 = 1, \quad a_1 = 2, \quad a_2 = 1, \quad \alpha = 0.8, \quad I = 0.83.$$

For these values, we find the positive equilibrium state of system (1):

$$E = (a^\star, p^\star, h^\star, \delta^\star) = (0.2075, 0.146798, 0.0607023, 0.015552).$$

The set of inequalities (4) are satisfied and the equilibrium E is therefore asymptotically stable when there is no delay in system (1), i.e., when $\tau = 0$. Based on Theorem 1, we compute $\omega_0 = 0.209336$ and we obtain the critical value of the time delay for the occurrence of a Hopf bifurcation: $\tau_0^+ = 0.08416$. In Figs. 2 and 3, the trajectories of system (1) are shown for two different values of the time delay: $\tau = 0$ and $\tau = 0.1$ (after the Hopf bifurcation). The appearance of a stable limit cycle is observed numerically, suggesting a supercritical Hopf bifurcation. A theoretical investigation of the criticality of the Hopf bifurcation and the stability of the resulting limit cycle will be provided in a future paper.

5 Conclusions

Dopamine reactivity is a crucially determinant factor of prefrontal-limbic systemic behavior, and subsequently of emotional regulation. The timing factor involved in dopamine-regulated pathways seems to have in particular a strong effect on the regulation efficiency. This effect could only be captured by a theoretical model

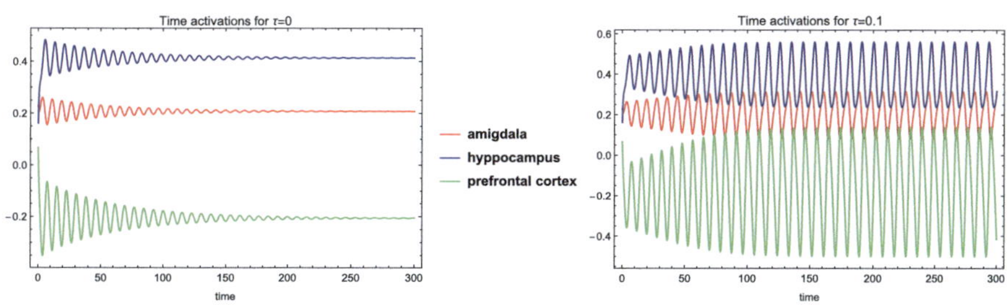

Fig. 2 Trajectories of system (1) when $\tau = 0$ (left) and $\tau = 0.1$ (right). When $\tau = 0$ (shown on the left), the solution of (1) converges to the asymptotically stable equilibrium state E. In the second case, $\tau = 0.1$ (shown on the right), the solution of (1) converges to the stable limit cycle, occurring due to the Hopf bifurcation which takes place at $\tau = \tau_0^+ = 0.08416$

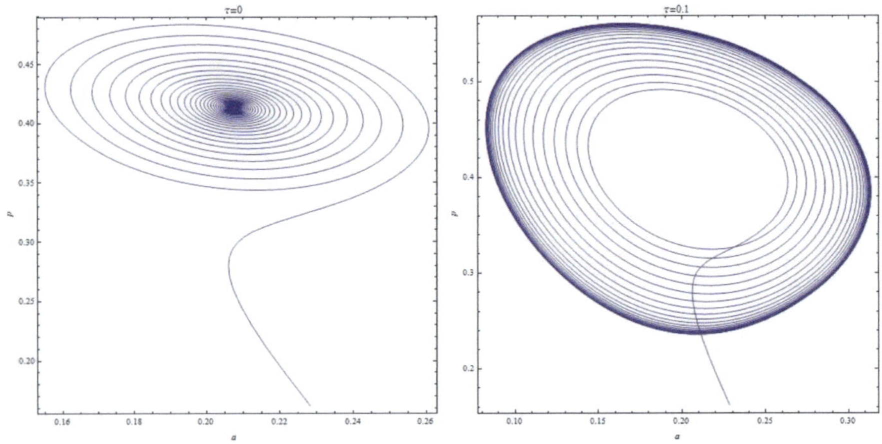

Fig. 3 Evolution of the trajectories of system (1) in the phase-plane (a, p) when $\tau = 0$ (left) and $\tau = 0.1$ (right). When $\tau = 0$ (shown on the left), the solution of (1) converges to the asymptotically stable equilibrium state E. In the second case, $\tau = 0.1$ (shown on the right), the solution of (1) converges to the stable limit cycle, occurring due to the Hopf bifurcation which takes place at $\tau = \tau_0^+ = 0.08416$

incorporating dopamine reactivity as a time delay, and was invisible in a classical nonlinear model of prefrontal-limbic interactions.

For the considered nonlinear mathematical model of dopamine-modulated prefrontal-limbic interactions in schizophrenia with time delay, we performed a thorough local asymptotic stability and bifurcation analysis. The critical value of the time delay corresponding to a Hopf bifurcation in a neighborhood of the equilibrium point has been determined theoretically. Numerical simulations have been presented to substantiate the theoretical results, which show that the resulting limit cycle due to the Hopf bifurcation is asymptotically stable. The theoretical analysis into the stability of this limit cycle will be explored in a future work. Moreover, the effect of different types of distributed time delays on the system dynamics will also be investigated.

References

1. Ashe, P.C., Berry, M.D., Boulton, A.A.: Schizophrenia, a neurodegenerative disorder with neurodevelopmental antecedents. Prog. Neuro-Psychopharmacol. Biol. Psychiatry **25**(4), 691–707 (2001)
2. Corrigan, P.W., Watson, A.C., Barr, L.: The self-stigma of mental illness: implications for self-esteem and self-efficacy. J. Soc. Clin. Psychol. **25**(8), 875–884 (2006)
3. Dickerson, S.S., Kemeny, M.E.: Acute stressors and cortisol responses: a theoretical integration and synthesis of laboratory research. Psychol. Bull. **130**(3), 355 (2004)

4. Goswami, S., Singh, G., Mattoo, S., Basu, D.: Courses of substance use and schizophrenia in the dual-diagnosis patients: is there a relationship? Indian J. Med. Sci. **57**(8), 338 (2003)
5. Hassard, B.D., Hassard, D., Kazarinoff, N.D., Wan, Y.H., Wan, Y.W.: Theory and applications of Hopf bifurcation, vol. 41. Cambridge University Press, Cambridge (1981)
6. Heilä, H., Heikkinen, M.E., Isometsä, E.T., Henriksson, M.M., Marttunen, M.J., Lönnqvist, J.K.: Life events and completed suicide in schizophrenia: a comparison of suicide victims with and without schizophrenia. Schizophr. Bull. **25**(3), 519 (1999)
7. Iooss, G., Joseph, D.D.: Elementary Stability and Bifurcation Theory. Springer, Berlin (2012)
8. Kuang, Y.: Delay Differential Equations: With Applications in Population Dynamics, vol. 191. Academic Press, London (1993)
9. Kuznetsov, Y.A.: Elements of Applied Bifurcation Theory, vol. 112. Springer, Berlin (2013)
10. LeDoux, J.: The emotional brain, fear, and the amygdala. Cell. Mol. Neurobiol. **23**(4–5), 727–738 (2003)
11. Olfson, M., Mechanic, D., Hansell, S., Boyer, C.A., Walkup, J.: Prediction of homelessness within three months of discharge among inpatients with schizophrenia. Psychiatr. Serv. **50**(5), 667–673 (1999)
12. Rădulescu, A.: Schizophrenia—a parameters' game? J. Theor. Biol. **254**(1), 89–98 (2008)
13. Rădulescu, A.R., Mujica-Parodi, L.R.: A principal component network analysis of prefrontal-limbic functional magnetic resonance imaging time series in schizophrenia patients and healthy controls. Psychiatry Res. Neuroimaging **174**(3), 184–194 (2009)
14. Schultz, W.: Dopamine reward prediction-error signalling: a two-component response. Nat. Rev. Neurol. **17**(3), 183 (2016)
15. Zendehrouh, S., Bakouie, F., Gharibzadeh, S.: Modeling schizophrenic-like neuronal patterns using nonlinear delayed differential equations. Comput. Biol. Med. **39**(11), 1058–1062 (2009)

Wilson–Cowan Neuronal Interaction Models with Distributed Delays

Eva Kaslik, Emanuel-Attila Kokovics, and Anca Rădulescu

Abstract A generalization of the well-known Wilson–Cowan model of excitatory and inhibitory interactions in localized neuronal populations is presented, by taking into consideration distributed time delays. A stability and bifurcation analysis is undertaken for the generalized model, with respect to two characteristic parameters of the system. The stability region in the characteristic parameter plane is determined and a comparison is given for several types of delay kernels. It is shown that if a weak Gamma delay kernel is considered, as in the original Wilson–Cowan model without time-coarse graining, the resulting stability domain is unbounded, while in the case of a discrete time delay, the stability domain is bounded. This fact reveals an essential difference between the two scenarios, reflecting the importance of a careful choice of delay kernels in the mathematical model. Numerical simulations are presented to substantiate the theoretical results. Important differences are also highlighted by comparing the generalized model with the original Wilson–Cowan model without time delays.

Keywords Wilson–Cowan model · Distributed delays · Stability · Bifurcations · Chaos

E. Kaslik
Department of Mathematics and Computer Science, West University of Timisoara, Timisoara, Romania
e-mail: eva.kaslik@e-uvt.ro

E.-A. Kokovics (✉)
West University of Timişoara, Timişoara, Romania
e-mail: emanuel.kokovics92@e-uvt.ro; emanuel.kokovics@e-uvt.ro

A. Rădulescu
SUNY New Paltz, New Paltz, NY, USA

W. Lacarbonara et al. (eds.), *New Trends in Nonlinear Dynamics*,
https://doi.org/10.1007/978-3-030-34724-6_21

1 Introduction

The original mathematical model describing excitatory and inhibitory interactions in localized neuronal populations has been derived in 1972 by Wilson and Cowan [11]. In this model, denoting by $E(t)$ and $I(t)$ the proportions of excitatory and inhibitory cells firing per unit of time, at the time instant t, it has been assumed that $E(t+\tau)$ and $I(t+\tau')$ are equal to the proportion of cells which are sensitive (i.e., not refractory) and which also receive at least threshold excitation at the moment of time t. Therefore, as a first step, the following system of integral equations has been obtained:

$$\begin{aligned} E(t+\tau) &= \left(1 - \int_{t-r}^{t} E(s)ds\right) \cdot \mathcal{S}_e \left[\int_{-\infty}^{t} h(t-s)\,(c_1 E(s) - c_2 I(s) + P_e(s))\,ds\right], \\ I(t+\tau') &= \left(1 - \int_{t-r'}^{t} I(s)ds\right) \cdot \mathcal{S}_i \left[\int_{-\infty}^{t} h(t-s)\,(c_3 E(s) - c_4 I(s) + P_i(s))\,ds\right] \end{aligned} \tag{1}$$

In this system, the first factors in the right-hand side represent the proportion of sensitive excitatory/inhibitory cells, where r is the absolute refractory period (msc), the functions $\mathcal{S}_e$, $\mathcal{S}_i$ are sigmoid threshold functions, their arguments denoting the mean field level of excitation/inhibition generated in an excitatory/inhibitory cell at time t (assuming that individual cells sum their inputs and that the effect of the stimulation decays exponentially with a time course $h(t)$). Moreover, $c_i > 0$ are connectivity coefficients representing the average number of excitatory/inhibitory synapses per cell and P_e, P_i denote external inputs.

Applying time-coarse graining, the well-known Wilson–Cowan model [11] has been obtained and analyzed, consisting of a system of ordinary differential equations without time delays. Generalizations of this model including discrete time delays have been investigated in several papers, often considering refractory periods r, r' being equal to zero. Based on the integral terms appearing in the original model (1) as arguments of the threshold functions, the following model with distributed delays will be analyzed in this paper:

$$\begin{aligned} \dot{u}(t) &= -u(t) + f\left[\theta_u + \int_{-\infty}^{t} h(t-s)\,(au(s) + bv(s))\,ds\right], \\ \dot{v}(t) &= -v(t) + f\left[\theta_v + \int_{-\infty}^{t} h(t-s)\,(cu(s) + dv(s))\,ds\right] \end{aligned} \tag{2}$$

where $u(t)$ and $v(t)$ represent the synaptic activities of the two neuronal populations, a, b, c, d are connection weights and θ_u, θ_v are background drives. The activation function f is considered to be increasing and of class C^1 on the real line.

In system (2), the delay kernel $h : [0, \infty) \to [0, \infty)$ is a probability density function representing the probability that a particular time delay occurs. It is assumed to be bounded, piecewise continuous and satisfy

$$\int_0^\infty h(s)ds = 1, \quad \text{with the average time delay} \quad \tau = \int_0^\infty sh(s)ds < \infty.$$

The particular case of discrete time delays (Dirac kernels) has been discussed in [4]. However, there are other important classes of delay kernels often used in the literature, such as Gamma kernels or uniform distribution kernels. It is worth mentioning that in the original Wilson–Cowan model [11], a weak Gamma kernel $h(t) = \tau^{-1} \exp(-t/\tau)$ has been used, so this case should be the original reference point. Analyzing mathematical models with particular classes of delay kernels (e.g., weak Gamma kernel or strong Gamma kernel $h(t) = 4\tau^{-2} t \exp(-2t/\tau)$) may shed a light on how distributed delays affect the dynamics differently from discrete delays. However, in the modeling of real world phenomena, one usually does not have access to the exact distribution, and approaches using general kernels may be more appropriate [1–3, 5–7, 9, 10, 12].

Initial conditions associated with system (2) are of the form:

$$u(s) = \varphi(s), \quad v(s) = \psi(s), \quad \forall s \in (-\infty, 0],$$

where φ, ψ are bounded real-valued continuous functions defined on $(-\infty, 0]$.

2 Main Stability and Bifurcation Results

The equilibrium states of system (2) are the solutions of the following algebraic system:

$$\begin{aligned} u &= f(\theta_u + au + bv), \\ v &= f(\theta_v + cu + dv) \end{aligned} \tag{3}$$

The linearized system at an equilibrium state $(u^\star, v^\star)$ is

$$\begin{aligned} \dot{u} &= -u + \phi_1 \int_{-\infty}^{t} h(t-s)\,(au(s) + bv(s))\,ds, \\ \dot{v} &= -v + \phi_2 \int_{-\infty}^{t} h(t-s)\,(cu(s) + dv(s))\,ds \end{aligned} \tag{4}$$

where $\phi_1 = \phi_1(u^\star, v^\star) = f'(\theta_u + au^\star + bv^\star) > 0$ and $\phi_2 = \phi_2(u^\star, v^\star) = f'(\theta_v + cu^\star + dv^\star) > 0$.

Applying the Laplace transform to the linearized system (4), we obtain

$$\begin{aligned} zU(z) - u(0) &= -U(z) + \phi_1 H(z)\,(aU(z) + bV(z)), \\ zV(z) - v(0) &= -V(z) + \phi_2 H(z)\,(cU(z) + dV(z)) \end{aligned} \tag{5}$$

where $U(z)$ and $V(z)$ represent the Laplace transforms of the state variables u and v, respectively, while $H(z)$ is the Laplace transform of the delay kernel h.

System (5) is equivalent to

$$\begin{pmatrix} z + 1 - a\phi_1 H(z) & -b\phi_1 H(z) \\ -c\phi_2 H(z) & z + 1 - d\phi_2 H(z) \end{pmatrix} \begin{pmatrix} U(z) \\ V(z) \end{pmatrix} = \begin{pmatrix} u(0) \\ v(0) \end{pmatrix} \tag{6}$$

and hence, the characteristic equation associated to the equilibrium state $(u^\star, v^\star)$ is

$$\Delta(z) = (z+1)^2 - \alpha H(z)(z+1) + \beta H^2(z) = 0, \tag{7}$$

where

$$\begin{aligned} \alpha &= a\phi_1(u^\star, v^\star) + d\phi_2(u^\star, v^\star) = af'(\theta_u + au^\star + bv^\star) + df'(\theta_v + cu^\star + dv^\star); \\ \beta &= (ad - bc)\phi_1(u^\star, v^\star)\phi_2(u^\star, v^\star) = (ad - bc) f'(\theta_u + au^\star + bv^\star) \\ &\quad \times f'(\theta_v + cu^\star + dv^\star). \end{aligned}$$

The following delay-independent stability and instability results are easily obtained, based on the properties of the Laplace transform and the particularities of the characteristic equation (7):

Theorem 1 (Delay-Independent Stability and Instability)

1. *In the non-delayed case, the equilibrium state $(u^\star, v^\star)$ of system (2) is locally asymptotically stable if and only if*

$$\alpha < \min\{2, \beta + 1\}. \tag{8}$$

2. *If the following inequality holds*

$$|\alpha| + |\beta| < 1, \tag{9}$$

then the equilibrium state $(u^\star, v^\star)$ of system (2) is locally asymptotically stable for any delay kernel $h(t)$.

3. *If the following inequality holds*

$$\beta < \alpha - 1, \tag{10}$$

then the equilibrium state $(u^\star, v^\star)$ of system (2) is unstable for any delay kernel $h(t)$.

It is important to note that the characteristic equation (7) has a root $z = 0$ if and only if $\beta = \alpha - 1$. To investigate for which combinations of parameters (α, β) the characteristic equation will have complex conjugated roots of the form $\pm i\omega$, we further assume that $\hat{H}(z) = H(z/\tau)$ does not depend on the average time delay τ. In fact, for the most important classes of delay kernels we have

(a) Dirac kernel: $\hat{H}(z) = e^{-z}$;

(b) p-Gamma kernel: $\hat{H}(z) = \left(\frac{p}{p+z}\right)^p$;

(c) Uniform kernel: $\hat{H}(z) = e^{-z} \cdot \frac{\sinh(\rho z)}{\rho z}$.

We will further define $\hat{H}(i\omega) = \rho(\omega)e^{-i\theta(\omega)}$. The following equation will play an important role in the bifurcation analysis to follow:

$$\tau \sin\theta(\omega) + \omega \cos\theta(\omega) = 0. \tag{11}$$

A careful and lengthy theoretical investigation [8] leads to the following theorems which characterize the stability region $S(\alpha, \beta)$ of the equilibrium $(u^\star, v^\star)$ from the (α, β)-plane (see Fig. 1).

Theorem 2 *Assuming that Eq. (11) has at least one positive real root, let us denote:*

$$\omega_\tau = \min\{\omega > 0 \;:\; \tau \sin\theta(\omega) + \omega\cos\theta(\omega) = 0\} \text{ and } \mu_\tau = (\rho(\omega_\tau)\cos\theta(\omega_\tau))^{-1}.$$

The boundary of the stability region $S(\alpha, \beta)$ of the equilibrium state $(u^\star, v^\star)$ of system (2) *is given by the union of the line segments and curve given below:*

$$(l_0):\quad \beta = \alpha - 1, \qquad \alpha \in [1 + \mu_\tau, 2];$$

$$(l_\tau):\quad \beta = \mu_\tau(\alpha - \mu_\tau), \qquad \alpha \in [2\mu_\tau, 1 + \mu_\tau];$$

$$(\gamma_\tau):\quad \begin{array}{l} \alpha = \dfrac{2}{\rho(\omega)}\left[\cos\theta(\omega) - \dfrac{\omega}{\tau}\sin\theta(\omega)\right] \\ \beta = \dfrac{1}{\rho^2(\omega)}\left(1 + \dfrac{\omega^2}{\tau^2}\right) \end{array}, \qquad \omega \in (0, \omega_\tau).$$

At the boundary of the stability domain $S(\alpha, \beta)$, the following bifurcation phenomena take place in a neighborhood of the equilibrium $(u^\star, v^\star)$ of system (2)*:*

(a) Saddle-node bifurcations take place along the open line segment (l_0);
(b) Hopf bifurcations take place along the open line segment (l_τ) and curve (γ_τ);
(c) Bogdanov–Takens bifurcation at $(\alpha, \beta) = (2, 1)$;
(d) Double-Hopf bifurcation at $(\alpha, \beta) = \left(2\mu_\tau, \mu_\tau^2\right)$;
(e) Zero-Hopf bifurcation $(\alpha, \beta) = (1 + \mu_\tau, \mu_\tau)$.

Theorem 3 *If Eq. (11) does not admit any positive real root, the boundary of the stability region $S(\alpha, \beta)$ of the equilibrium state $(u^\star, v^\star)$ of system (2) is given by the union of the half-line and curve given below:*

$$(l_0): \quad \beta = \alpha - 1 \qquad , \ \alpha \in (-\infty, 2];$$

$$(\gamma_\tau): \quad \begin{cases} \alpha = \dfrac{2}{\rho(\omega)}\left[\cos\theta(\omega) - \dfrac{\omega}{\tau}\sin\theta(\omega)\right] \\ \beta = \dfrac{1}{\rho^2(\omega)}\left(1 + \dfrac{\omega^2}{\tau^2}\right) \end{cases} \qquad , \ \omega > 0.$$

At the boundary of the stability domain $S(\alpha, \beta)$, the following bifurcation phenomena take place in a neighborhood of the equilibrium $(u^\star, v^\star)$ of system (2):

(a) Saddle-node bifurcations take place along the open half-line (l_0);
(b) Hopf bifurcations take place along the curve (γ_τ);
(c) Bogdanov–Takens bifurcation at $(\alpha, \beta) = (2, 1)$.

In Fig. 1, the stability domains given by the previous theorems are represented for four different delay kernels with the same average time delay $\tau = 1$. In each subfigure, the blue rhombus represents the delay-independent part of the stability domain given by Theorem 1. It is important to note that compared to discrete time delays, the stability domains in the case of Gamma delay kernels are much larger. Moreover, in the case of a weak Gamma kernel (as it was the one included in the original Wilson–Cowan model [11], and therefore it produces the behavior of the model in its pure form, before the coarse-grain approximation), the stability region is unbounded, as in Theorem 3.

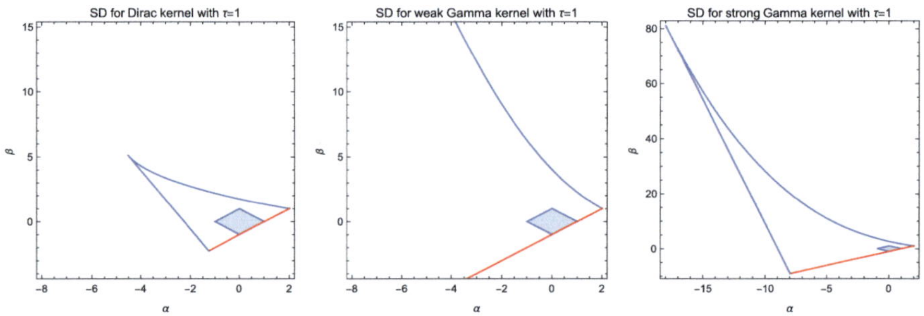

Fig. 1 Stability domain $S(\alpha, \beta)$ for different types of delay kernels, with a fixed average time delay $\tau = 1$. The stability domains are obtained based on Theorem 2 (left and right) and Theorem 3 (middle). The blue shaded region represents a delay-kernel-invariant subset of $S(\alpha, \beta)$. Along the blue curves and line segments Hopf bifurcations take place, while the red line corresponds to saddle-node bifurcations

3 Numerical Simulations

The sigmoid activation function is chosen as in [4]: $f(x) = (1 + \exp(-\delta x))^{-1}$. For all numerical simulations, the following values of the system parameters are chosen: $\theta_u = 0.1$, $\theta_u = 0.2$, $a = d = -6$, $b = c = 3$, and $\delta = 40$. The following equilibrium is computed: $(u^\star, v^\star) = (0.0660694, 0.076733)$, with the characteristic parameters: $\alpha = -31.8118$ and $\beta = 188.846$. Based on Theorem 2, the critical value of the average time delay $\tau^\star$ responsible for the occurrence of a Hopf bifurcation which causes the loss of asymptotic stability of the equilibrium $(u^\star, v^\star)$ is determined in the case of a Dirac kernel $\tau_0^\star = 0.0674893$ and a strong Gamma kernel $\tau_2^\star = 0.202917$. The Hopf bifurcations are supercritical, causing the appearance of stable limit cycles, as it can be seen in Fig. 2.

On the other hand, in the case of a weak Gamma kernel, from Theorem 3 it follows by numerical computations that for the specific values of α and β given above, the equilibrium $(u^\star, v^\star)$ is asymptotically stable, for any $\tau > 0$. Therefore, no oscillations or bursting behavior is expected to occur in a neighborhood of the equilibrium if a weak Gamma kernel is considered in the mathematical model. This reflects an important difference between the different types of behavior that can be observed for different types of delay kernels. The weak Gamma kernel has a particular importance as it has been included in the original Wilson–Cowan model before applying time-coarse graining.

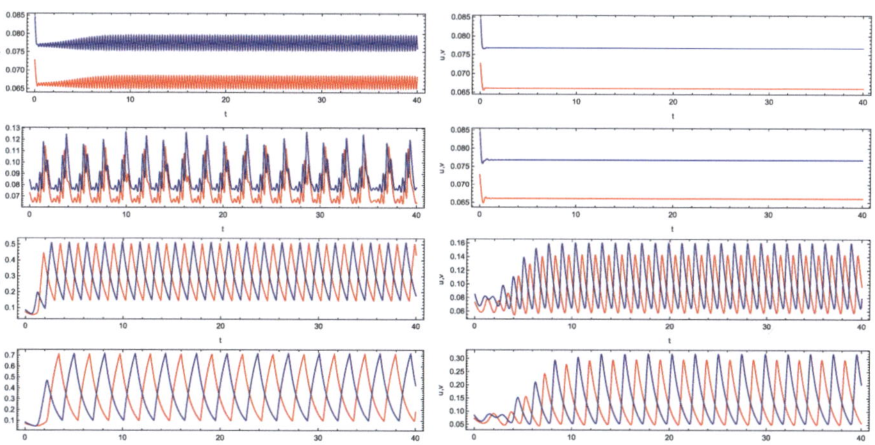

Fig. 2 Evolution of state variables $u(t)$ and $v(t)$ of system (2) with discrete time delay (left) and strong Gamma kernel (right) for valued of the average time delay $\tau \in \{0.07, 0.1, 0.5, 1\}$ (top to bottom). The values of the parameters are fixed: $\theta_u = 0.1$, $\theta_u = 0.2$, $a = d = -6$, $b = c = 3$, and $\delta = 40$. The same initial condition has been chosen in a neighborhood of the equilibrium

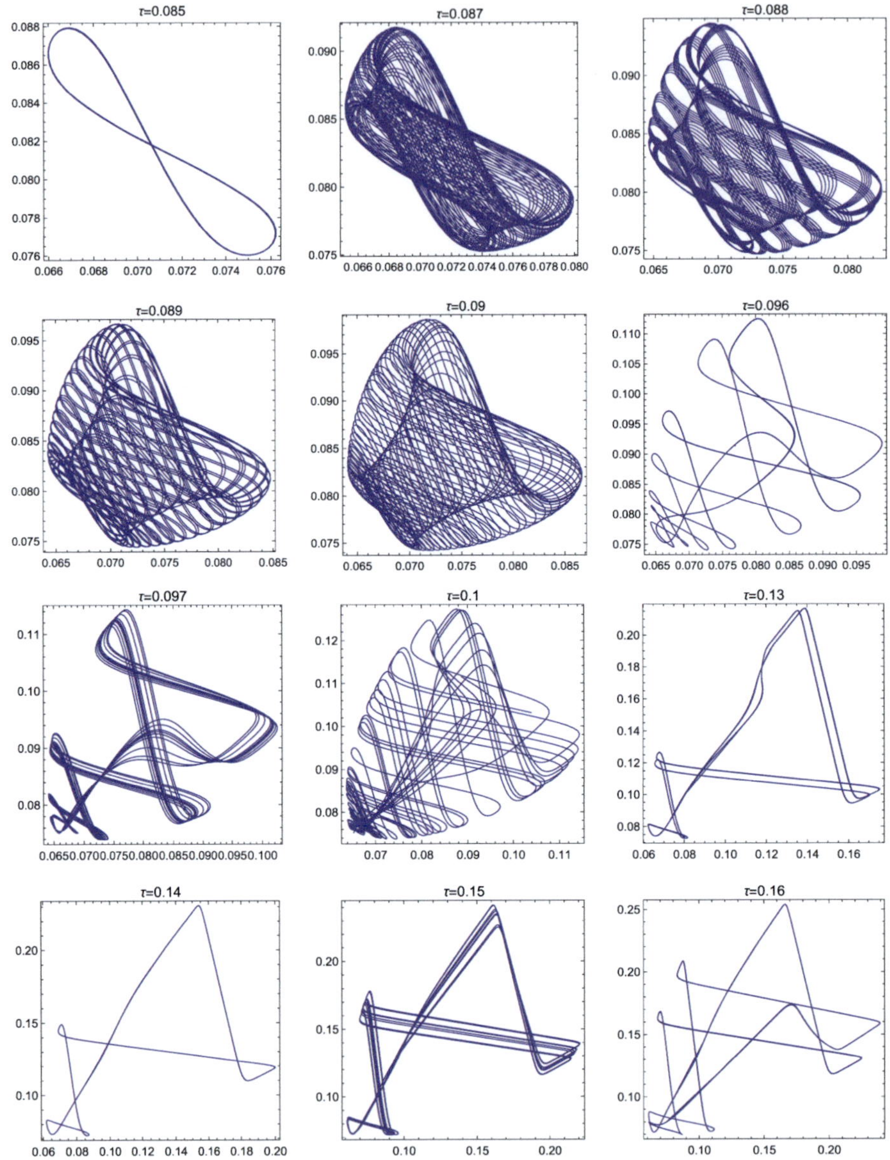

Fig. 3 Periodic, quasi-periodic, and chaotic orbits shown in the (u, v)-phase-plane for the Wilson–Cowan model with discrete time delay, obtained for different values of τ

Numerical simulations also reveal complex bursting and quasi-periodic behavior in the Wilson–Cowan model with a discrete time delay (see Fig. 3), suggesting a series of limit cycle bifurcations. Interestingly, these phenomena could not be observed for strong Gamma kernels with the same system parameters.

4 Conclusions

A local stability and bifurcation analysis has been presented for a generalization of the Wilson–Cowan model of excitatory and inhibitory interactions in localized neuronal populations, incorporating general distributed delays. Essential differences have been pointed out for different scenarios involving diverse delay kernels, emphasizing the importance of a careful choice of delay kernels in the mathematical model.

References

1. Adimy, M., Crauste, F., Ruan, S.: Stability and Hopf bifurcation in a mathematical model of pluripotent stem cell dynamics. Nonlinear Anal. Real World Appl. **6**(4), 651–670 (2005)
2. Bernard, S., Bélair, J., Mackey, M.C.: Sufficient conditions for stability of linear differential equations with distributed delay. Discrete Continuous Dyn. Syst. B **1**(2), 233–256 (2001)
3. Campbell, S., Jessop, R.: Approximating the stability region for a differential equation with a distributed delay. Math. Model. Nat. Phenom. **4**(2), 1–27 (2009)
4. Coombes, S., Laing, C.: Delays in activity-based neural networks. Philos. Trans. R. Soc. Lond. Ser. A **367**(1891), 1117–1129 (2009)
5. Diekmann, O., Gyllenberg, M.: Equations with infinite delay: blending the abstract and the concrete. J. Differ. Equ. **252**(2), 819–851 (2012)
6. Faria, T., Oliveira, J.J.: Local and global stability for Lotka–Volterra systems with distributed delays and instantaneous negative feedbacks. J. Differ. Equ. **244**(5), 1049–1079 (2008)
7. Jessop, R., Campbell, S.A.: Approximating the stability region of a neural network with a general distribution of delays. Neural Netw. **23**(10), 1187–1201 (2010)
8. Kaslik, E., Kokovics, E.A., Radulescu, A.: Stability and bifurcations of two-dimensional systems with distributed delay and applications to a Wilson–Cowan model (2019). https://arxiv.org/abs/1904.12108
9. Özbay, H., Bonnet, C., Clairambault, J.: Stability analysis of systems with distributed delays and application to hematopoietic cell maturation dynamics. In: IEEE Conference on Decision and Control, CDC, pp. 2050–2055 (2008)
10. Ruan, S., Wolkowicz, G.S.: Bifurcation analysis of a chemostat model with a distributed delay. J. Math. Anal. Appl. **204**(3), 786–812 (1996)
11. Wilson, H.R., Cowan, J.D.: Excitatory and inhibitory interactions in localized populations of model neurons. Biophys. J. **12**(1), 1–24 (1972)
12. Yuan, Y., Bélair, J.: Stability and Hopf bifurcation analysis for functional differential equation with distributed delay. SIAM J. Appl. Dyn. Syst. **10**(2), 551–581 (2011)

Nonlinear Hydrodynamics and Numerical Analysis for a Series of Catastrophic Floods/Debris (2011–2017): The Tectonic Wave Processes Possible Impact on Surface Water and Groundwater Flows

Tatiana Trifonova, Sergei Arakelian, Dmitri Trifonov, Sergei Abrakhin, Vyacheslav Koneshov, Alexei Nikolaev, and Mileta Arakelian

Abstract The chapter suggests an original approach to explain and predict the process of a flood and/or mudflow (debris) formation and spreading out over the river beds in mountain conditions taking into account the nonlinear dynamics. The phenomena are under the flash increase of water masses involved (being strongly above the precipitations intensity measured) due to the groundwater possible impact. For the case a collection of different scenarios in conceptual nonlinear dynamic models is under our brief consideration. The 3D crack-Net in the frame of unified rivershed basin in a mountain massif is a natural transportation system (varied by some dynamic stress factors) for a groundwater due to hydrostatic/hydrodynamic pressure redistribution by different reasons (e.g., earthquakes as well). The process has a nonlinear wave character with obvious signs of self-organization and, especially, for surface discharge of water from underground aquifers. It can be described within the soliton model of nonlinear hydrodynamics for propagation of the surface water mass. The approach can result in more reasonable forecast and early warning for the natural water hazard/disaster taking into account the groundwater nonlinear flow contribution as a dominant factor under some conditions to the land surface water.

T. Trifonova (✉)
Lomonosov Moscow State University, Moscow, Russia

S. Arakelian · D. Trifonov · S. Abrakhin
Vladimir State University, Vladimir, Russia
e-mail: arak@vlsu.ru

V. Koneshov · A. Nikolaev
Schmidt Institute of Physics of the Earth, RAS, Moscow, Russia

M. Arakelian
Yerevan State University, Yerevan, Armenia

W. Lacarbonara et al. (eds.), *New Trends in Nonlinear Dynamics*,
https://doi.org/10.1007/978-3-030-34724-6_22

Keywords Catastrophic floods · Groundwater · Crack-Net · Seismicity · 3D-river basin

1 Introduction

Motivation of this chapter is established on the problem that the knowledge level for catastrophic mudflow/debris and floods has still many problems, and there is no reasonable forecast for the process development to prevent the catastrophic consequences [1]. A forward problem here entails solving the 3D-governing liquid flow equations for head in both time and space over a solution domain (i.e., a river basin or an arbitrary-defined field site) with given model input parameters, boundary, and initial conditions [2, 3].

In frame of concept of flow through heterogeneous geologic media the model is under Poiseuille's law for flow through a bundle of cylindrical capillary tubes of radius R being an analog of porous rock core. In the case a discharge parameter $\sim R^4$ and results in dramatic increase of the pressure parameters map in the geological system, especially for a groundwater dislocation (cf. [4]).

For that we have three types of the flow regimes in terms of specific water discharge vs. hydraulic gradient, i.e., linear laminar, nonlinear laminar, and turbulent. In the last case, we have a nonlinear dynamics flow regime when streamlines crossing each other takes place and the flow exhibits chaotic behavior [3, 5]. Under this scenario the analysis should be carried out numerically [6]. Our simplest and most conventional model for the water balance estimation for disastrous floods shows the greater (up to 75%) water mass discharge observed during the events than it could be expected from the rainfall process only in the area under study [7]. The fact gives us the founding to take into account the groundwater possible contribution to the events. This is especially true for long standing water on the land surface during the events. In this aspect the principal item of the present chapter is to recognize the impact of different phenomena taking place for that (cf. [8, 9]), in particular of tectonic state.

The following processes in the Earth crust are considered: from microseismic and local seismic shocks of various magnitudes to global movement of the tectonic plates [10]. The Earth's crust (and/or its particular parts) isn't considered as stable and immutable today, i.e., in this case we can't speak about stable behavior of groundwater regimes (cf. [11, 12]). The collection of different scenarios in conceptual nonlinear dynamic models with saturated and/or unsaturated zones in this aspect is presented, e.g., in [5, 9].

The key point of the proposed approach is the identification of the conditions of earthquake influence on the river basins functioning (cf. [13, 14]). The possible role of the preceding seismic activity for some certain disastrous floods of 2013–2017 is analyzed by us qualitatively. Usually, when the water balance of the river basin is estimated, the soil permeability and its percolation component are considered as the key characteristics [15]. On the one hand, it is the small-scale percolation, defined by the properties of specific soils, prevailing in the considered river basin. But also it is the large-scale percolation, defined by the fracturing of the Earth crust. Without going into details of the forming of the river basin geological structure we highlight fracturing of the Earth crust which makes the intense interaction of surface water and groundwater possible (cf. [6, 16]). The infiltration component of soil (and deeper rock layers) permeability is significantly characterized by slower vertical and lateral movement of water masses but in some cases can be turbulent [17]. It is obvious that present great and universal advantages of nonlinear dynamics should be applied to these triggered processes.

The organization of the main material in this chapter is built according to the following logic. First, we consider models in the framework of nonlinear hydrodynamics with different types of soliton solutions for surface water flows using the relevant key parameters and available data. Then, the possible reasons for the development of such processes in actually registered catastrophic floods and mudflows due to the influence of groundwater are discussed. Finally, an analysis of the forecast conditions is carried out under which a sudden release of groundwater to the surface of the Earth may occur due to seismic factors in the specific areas where these water events occurred.

2 Physical Basis and Reasonable Models

We focus now our consideration on a mathematical modeling of the catastrophic water flow exit on the land surface based on the conception of nonlinear hydrodynamics of the wave processes development with forming of solitons (within the classes of solutions for the Korteweg-deVries (KdV)-equation) [2, 4]. In principle, different regimes can occur, and should be under analysis in respect to the detailed states of the water systems.

Let's consider two general scenarios for the processes (cf. [18, 19]).

Scenario 1. Mathematical modeling of the flood/mudflow process formation as a standard self-organization in soliton-like system.

Scenario 2. Generalized KdV-equation approach: the flood solitary propagation process by groundwater discharge (both continuous and flash) as a multiple soliton process vs. depth of the water channel.

This general approach and practical verification of the results were applied to some water catastrophic events including the conditions for a solitary destructive wave propagation, e.g., for the Krimsk-city fast event (Krasnodar region, Russia, July 6–7, 2012) which occurred over the land surface under the trigger mechanism (cf. [20]).

The results of such analysis for the Krimsk-city event are shown in Fig. 1 for a manybody problem when there is a collision/convergence of the two solitons and their further spread/distribution over the surface. From the point of view of the theory of solitons, this instability and such decay occur starting with a certain critical value of the soliton amplitude (Fig. 1a(1) shows the beginning of this decay), i.e., we are talking about the threshold effect. The number of born satellites is proportional to the amplitude of the original soliton, and their partial propagation speeds are proportional to the amplitude of each satellite. At the same time, according to the proposed model for debris flow, the solitons should pass through each other without changing their amplitude and shape (Fig. 1a(4)) after their divergence. As a result, during the overlapping of the satellites, the profile of the ensemble/train of solitons changes (Fig. 1a(2,3)), but eventually they are ordered (self-organized) in space, which determines, the final character of the mudflow profile (Fig. 1a(4)). It is important that the mudflow process (at the stage corresponding to Fig. 1a(4)) can significantly enhance its one-time destructive impact on the objects on the earth's surface (localization of a large mass/energy of flow in one place). The qualitative

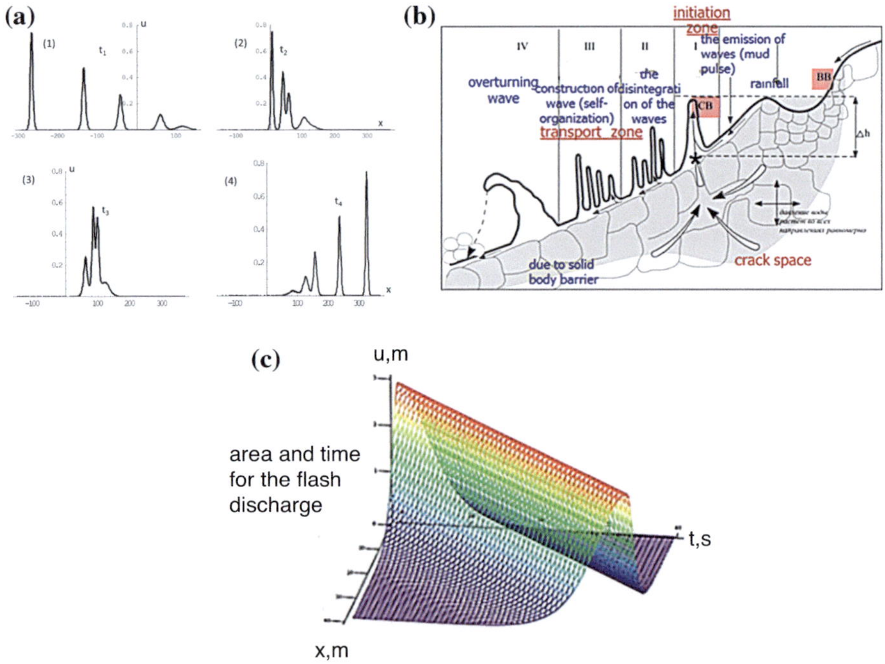

Fig. 1 (**a**) Calculation profile of a propagating soliton wave (in arbitrary units) in various/consecutive moments of time: $t_1 < t_2 < t_3 < t_4$ for the some control parameters: $\sigma = \Delta_0 (u_0/\gamma_2)^{1/2} = 9$, $u_0 = 1.71$, $t = 0.84$; both the space characteristics (horizontally—x) and the amplitude characteristics (vertically—u) of the soliton are given in arbitrary units for a manybody regime propagation. (**b**) Principal scheme for the stage development and propagation of a mudflow wave: BB—collecting funnel; CB—mudflow wave; *—mudflow gate; Δh—hydrostatic thrust/pressure head; —surface water with drainage process contribution; —groundwater deep movement. (**c**) The 3D-soliton image mode by computer simulation for the water-flow propagation in the river channel (from the very beginning ($t = 0$ s) the groundwater flash discharge of amplitude U (in meters, m) comes to the land surface)

analysis shows that a mudflow process can be schematically represented by a four stage development and propagation for a mudflow soliton (see Fig. 1b): (I) there occurs the main mudflow discharge (initial soliton width Δ_0); (II) the process falls into separate soliton satellites; (III) the stage of self-organization for these satellites according to the values of their amplitudes in the process of propagation; (IV) the soliton is breaking, i.e., turning over (great nonlinearity) γ_1—contribution of groundwater), and/or latter the decay (great dissipation/dispersion—γ_2) process takes place. In Fig. 1c, we have shown a fragment of a nonlinear dynamic process of propagation of the soliton by the 3D-soliton mode computer simulation image in the case.

We carried out, as well, some numerical analysis for process of the running water flow by modeling of a sudden breakthrough of dam for an artificial water reservoir in the Central Russian plain near the Vladimir city, i.e., a local Sodyshka river (Fig. 2). We showed that the event duration is not more than several days ($\sim$100 h). Thus, a long standing water on the land surface during each catastrophic event requires a separate analysis in respect of a specific water discharge source, e.g., being groundwater under some conditions.

We carried out also some correlation analysis for the subject. The most accurate correlation (the Pierson coefficient K) can be estimated for two principal parameters: river-discharge during the flood and artesian water level in wells in some localized river-basin areas (cf. [7, 12]). We had $K \gtrsim (-0.97)$ but with some optimal day shift $\sim$10–20 days (for distance $\sim$200 km), e.g., for the Mississippi-river catastrophic flood (April-June, 2011; maximal level—on May). Thus, we can talk about anticorrelation events, i.e., that means that increase/decrease of river discharge is due to decrease/increase of the artesian water level. As to other correlations—between the precipitation level, correspondingly and both the river discharge and the groundwater level during the event—the maximal value of K was less: $K \lesssim +0.7$. Thus, in fact, the groundwater probably plays a dominant role for the case.

3 The Principles of the Earthquake Influence on the River Basin Functioning

The nonlinear dynamic processes, first, of the regional tectonics dynamic map and, second, of three-dimension (3D) nonlinear/shock wave mobility for rock masses and also mechanisms of structural-material reworking of the consolidated Earth's crust (under certain fundamental regularities of the consolidated crust nonlinear geodynamics) [10] are out of the subject of this chapter. With regard to the relationship of seismic and groundwater, then on this occasion there is an extensive literature (see e.g., [11, 12, 21]). Usually, these issues are addressed with the help of diagrams and drawings (cf. [10]).

There are at least two obvious mechanisms explaining how the earthquakes occurred can influence the river basin functioning (cf. [18, 21]). On the one

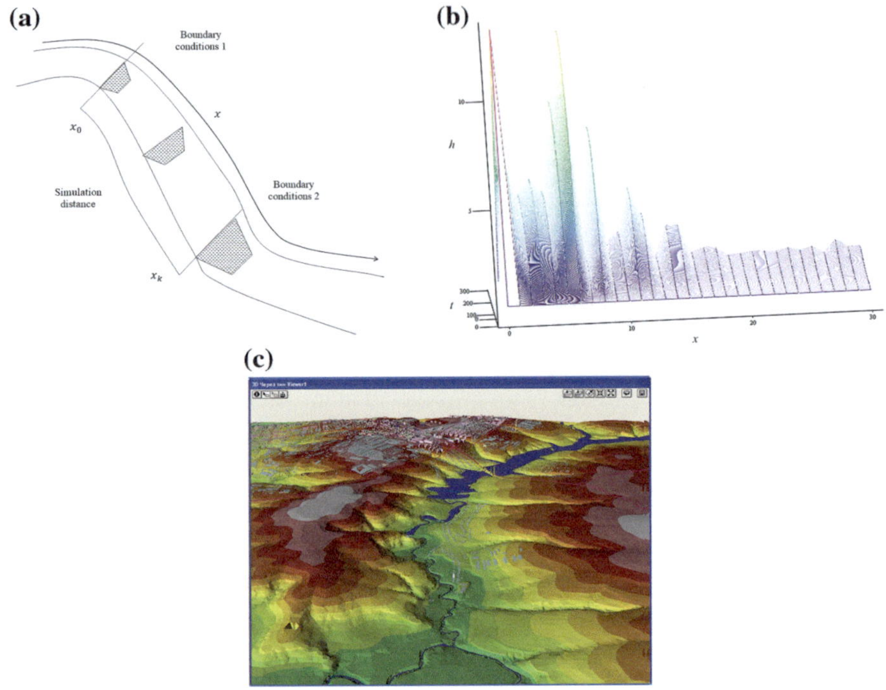

Fig. 2 (**a**) The sections considered for the river bed. The propagation of the water flow has been analyzed by a system of Saint-Venant equations combining the continuity equation and the convection-diffusion equation [15]: $\frac{dh}{dt} + \frac{1}{b}\frac{dQ}{dx} = 0$; $\frac{dQ}{dt} - \frac{1}{2}\,\frac{1\,nh^{\frac{10}{3}}\,d^2Q}{2n^2Q\,dx^2} = 0$, here $Q = vS$, and for initial condition $Q|_{t=t_0} = Q(x)$; $h|_{t=t_0} = h(x)$; for the border conditions in both initial section x_0 and section x_k of moving flow $Q|_{x=x_0} = Q(t)$; $h|_{x=x_0} = h(t)$; $Q|_{x=x_k} = Q_k(t)$; $h|_{x=x_k} = h_k(t)$, where Q—water discharge in the specified section; h—depth of the stream; b—width of the base of the channel; x—distance; t—time; n—roughness coefficient; v—flow velocity, S—cross-section of the living flow. (**b**) Wave height graph vs. x coordinate and time t. (**c**) The flooding area on the Sodyshka River after 15 min of simulation. A computational experiment was conducted on the movement of water flow with the real parameters of the object: $b = 400$ m, $x = 10$ km, $n = 0.041$. The complete simulation time is 3 h, and the analyzed distance is 6 km. Initial conditions: $(x)|_{t=t_0} = 2\ \frac{\text{m}^3}{\text{s}}$; $h(x)|_{t=t_0} = 0.5$ m and the boundary conditions: $Q(t)|_{x=x_0} = 500\frac{\text{m}^3}{\text{s}}$; $h(t)|_{x=x_0} = 13.3$ m. The final result (the flood is over): $h_k\left(t\right)|_{x=x_k} = 0.5$ m, $t_k = 10{,}800$ s ~ 3 h

hand, some parts of the transport-net may suddenly change during the topology restructuring. On the other hand, when the cracks are blocked the dramatic growth of pressure in other parts of the net occurs (cf. [22]).The basis of both is the impact of the earthquakes seismic waves which can propagate to huge distances in the Earth crust (up to several thousand km—cf. [10, 22]). We are now interested in their influence on the topology of the existing groundwater transport ways (cf. [6, 23]).

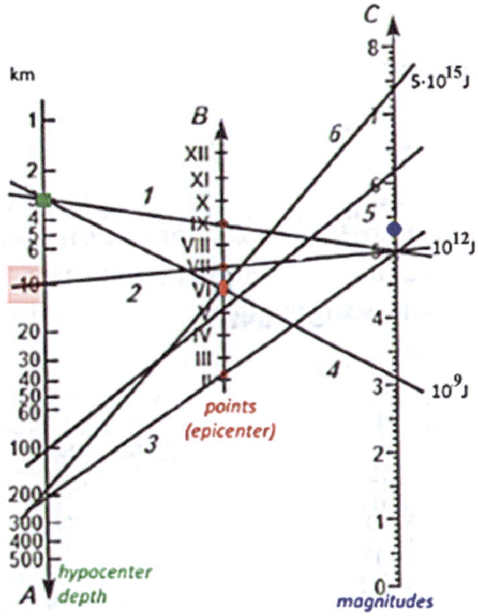

Fig. 3 The statistical data on intensity (points—axis *B* (up to IX), and magnitudes—axis *C* (up to 7 – 5 × 10^{15} J)) for a number of earthquakes occurred taking into account the localization of depth of the earthquake source—hypocenters/focuses (up to 200 km)—axis *A*, from the Earth's surface. Relative numbers of lines 1–6 mark the specific events that occurred. The data are taken from the analysis of different references—see text (cf. [10]). It is obvious that the most dangerous event is classified by the parameter marked by line 2 (i.e., large force of the earthquake both in depth and on surface)

It is important that after the breakout (which has opened a new channel) the groundwater can flow continuously for a long time defining the long water staying on the surface (until the local groundwater resource is exhausted).

The scale difference between groundwater response on the earthquake in both time and distance is sufficiently complicated and depends on many factors being not clear still: may vary in time—from immediate reaction to hours and weeks, and in distance—up to thousands kilometers even [11, 12].

We showed [20] that a more substantial impact on the groundwater exit on the land surface occurs for the earthquake hypocenter depth ~10 km when the magnitude value is about 5.0 (~10^{12} J) which may be associated with VII points in earthquake epicenter on the land surface (see Fig. 3).

4 Preliminary Recommendations for the Identification of the Earthquakes Influence on Disastrous Floods in the 3D-River Basin

In framework of our conception we assume that marked by lines earthquakes in Fig. 3 could influence on the emergence and development of subsequent disastrous floods. Such change of the river basin functioning regime in the part of groundwater dynamics results in its interaction with the surface water. However, the territories, where the floods occur, are located in different areas of the wave path regarding the epicenters of preceding earthquakes (Fig. 4). But the hypocenters localization is

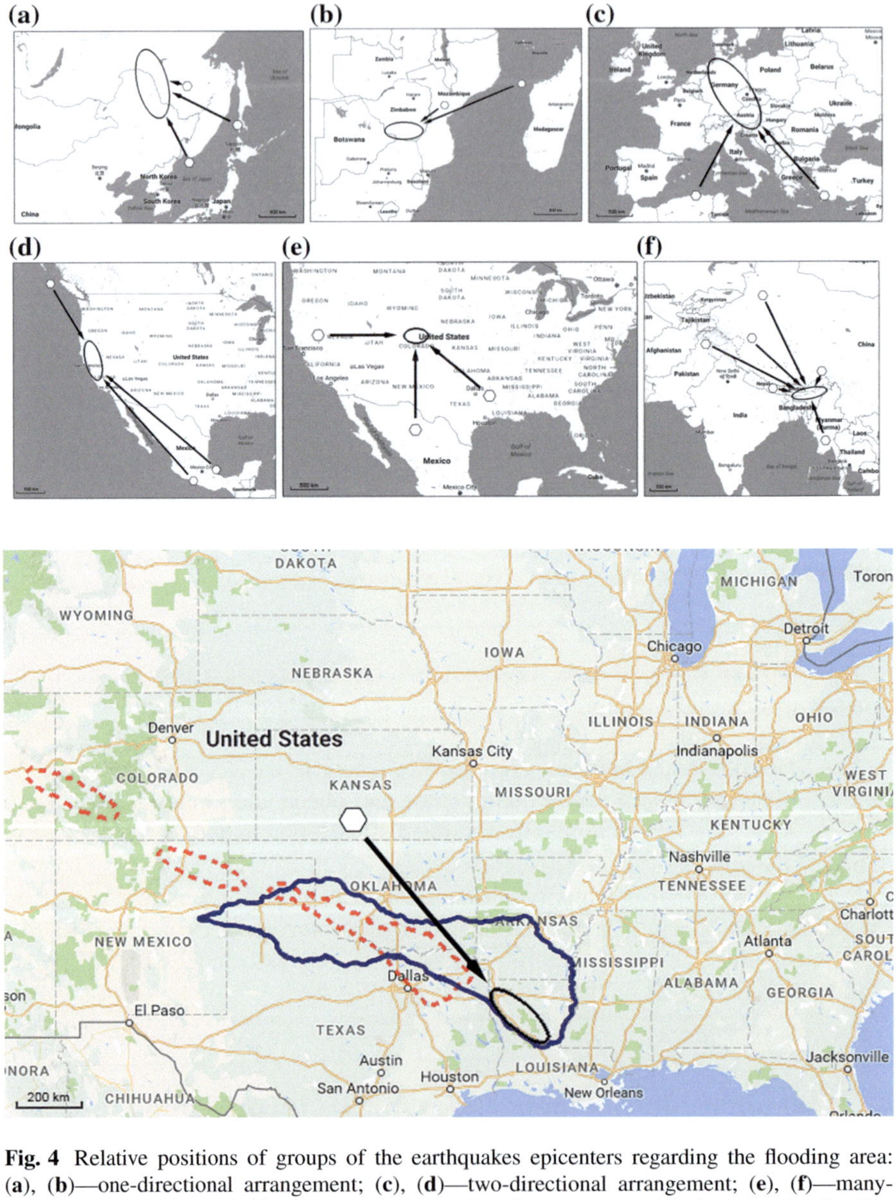

Fig. 4 Relative positions of groups of the earthquakes epicenters regarding the flooding area: (**a**), (**b**)—one-directional arrangement; (**c**), (**d**)—two-directional arrangement; (**e**), (**f**)—many-directional arrangement (white hexagons—the earthquakes epicenters; black ovals—the flooding areas); (**g**) the Red River basin and the Southern Oklahoma Aulacogen: white hexagon—the earthquake epicenter; black oval—the flooding area. Area with solid blue border—the Red River basin; areas with dashed red border—the Southern Oklahoma Aulacogen. The areas that are considered obvious on the maps

more principal for the problem in accordance with Fig. 3. We will nevertheless talk about epicenters as they are usually fixed in practice. We consider in particular three cases for the flood events in the case.

The first case is the floods in Amur region, Russia (2013), and in Matabeleland South, Zimbabwe (2016–2017) (Fig. 4a, b): the epicenters were located almost in one direction from the flooding area. Second case is the floods in Western Europe (2013) and in California, USA (2017) (Fig. 4c, d): two groups of epicenters equidistant from the flooding area are clearly distinguished. Finally, the third case is the floods in Colorado, USA (2013), and Assam, India (2015) (Fig. 4e, f): three and more groups of epicenters located in different directions from the flooding area can be distinguished.

In the second and the third cases, we clearly observed a strong kind of seismic wave interference: the influence of waves going from different directions is maximal in the area equidistant from the groups of epicenters. Herewith, the topology of the groundwater transport-net is the subject of significant restructuring, and the probability of the disastrous flood emergence grows dramatically (cf. [24]). In addition, we should pay our attention to the geological structure due to maximal amplitudes of seismic waves at the boundary between homogeneous media, e.g., for Red River basin where the disaster occurred (Louisiana, USA, May 2015)—see Fig. 4g.

5 Conclusions

The development of adequate approaches and construction of algorithms for the analysis of the reasons of disastrous floods emergence, and finally their forecast need consideration of numerous factors and new methodological foundations. Because the catastrophic floods (soliton like events) are associated with extremal conditions in interaction of the different nature wave processes, a nonlinear dynamics approach is a good candidate for that. One of the main problems here is the formalization of the key factors of such nonlinear interaction including the role of precipitation, seismic activity and/or local geological structure. For the practical verification of the proposed approach our activity in the future should be focused on the simulation modeling in the context of methods of stochastic nonlinear dynamic processes by manipulation of the key uncertainty parameters for the events (precipitation, soil and rock composition, etc).

Acknowledgements The chapter was prepared within the framework of the state task of VlSU № 16.1123.2017/PCh and was supported by the RFBR (Grant № 16-41-330032 p_a), also was partially supported by the Ministry of Science and Higher Education of Russia – Agreement No. 075-15-2019-1837.

References

1. Jakob, M., Hungr, O.: Debris-Flow Hazards and Related Phenomena, p. 745. Springer Science, Berlin (2005)
2. Alwyn, S.: Nonlinear Science Emergence and Dynamics of Coherent Structures, p. 480. Oxford University Press, Oxford (2003)
3. Gelhar, L.W., Axness, C.L.: Three-dimensional stochastic analysis of macro-dispersion in aquifers. Water Resour. Res. **19**(1), 161–180 (1983)
4. Trifonova, T.A., Akimov, V.A., Abrakhin, S.I., Arakelian, S.M., Prokoshev, V.G.: Basic Principles of Modeling and Forecasting of Extreme Natural and Man-Made Disasters, p. 436. Russian Emercom (2014)
5. Rudenko, O.V., Hedberg, C.M.: Nonlinear dynamics of grains in a liquid-saturated soil. Nonlinear Dyn. **35**(2), 187–200 (2004)
6. Vuković, M., Soro, A.: Groundwater Dynamics, p. 544. Water Resources Publications (1997)
7. Trifonova, T., Trifonov, D., Arakelian, S.: The 2015 disastrous floods in Assam, India, and Louisiana, USA: water balance estimation. Hydrology. **3**(4), 41 (2016)
8. Ostrovsky, L.A., Jonson, P.A.: Dynamic nonlinear elasticity in geomaterials. La Rivista del Nuovo Cimento. **24**(7), 1–47 (2001)
9. Ni, C.-F., Yeh, T.-C.J.: Stochastic inversion of pneumatic cross-hole tests and barometric pressure fluctuations in heterogeneous unsaturated formations. Adv. Water Resour. **31**(12), 1708–1718 (2008)
10. Leonov, M.G.: Tectonics of the Consolidated Crust. Transactions of the Geological Institute of RAS, vol. 575, p. 457. Nauka, Moscow (2008)
11. Wang, T., Yongtai, C.: Monitoring and research on the behavior of underground water and its relation with earthquakes in China. Acta Seismol. Sin. **3**(1), 103–114 (1990)
12. Wakita, H.: Water wells as possible indicators of tectonic strain. Science. **189**, 553–555 (1975)
13. Smyth, J.R., Jacobsen, S.D.: Nominally anhydrous minerals and Earth's deep water cycle. In: Earth Deep Water Cycle. Geophysical Monograph Series, vol. 168, (2006)
14. Hydrologic response in well. https://va.water.usgs.gov/earthquakes/
15. Streeter, V.L., Wylie, E.B.: Fluid Mechanics. McGraw-Hill, Auckland (1981)
16. Trifonova, T.A.: Mountain river channel: energy model of development. Rep. Russ. Acad. Sci. **3**, 334–338 (1994)
17. Eyink, G.L., Thomson, D.J.: Free decay of turbulence and breakdown of self-similarity. Phys. Fluids. **12**, 477–479 (2000)
18. Koneshov, V., Arakelian, S., Trifonova, T., et al.: Numerical analysis for a series of recent catastrophic floods (2011-2017): new verifications of the groundwater and tectonic processes possible impact. In: The Ninth International Conference on Environmental Science and Technology, Houston, Texas, USA (2018)
19. Trifonova, T.A., Trifonov, D.V., Arakelian, S.M.: Catastrophic floods—possible contribution of groundwater due to flash reconstruction of the rock mass 3D-cracknet under seismic factors. Mod. Appl. Sci. **9**, 76–86 (2015)
20. Koneshov, V.N., Trifonova, T.A., Trifonov, D.V., Arakelian, S.M.: Dynamics of groundwater basin functioning—a possible manifestation of the catastrophic floods. Russ. J. Geol. Geophys. South Russ. **2**, 50–66 (2017)
21. Trifonova, T.A., Trifonov, D.V.: Influence of seismic activity on the groundwater state and the emergence of disastrous floods. Dyn. Difficult Syst. **9**, 1 (2015)
22. Trifonova, T., Arakelian, S., Arakelian, M.: Surface and subterranean water interaction in catastrophic flood and mudflow for a river mountain basin: basic principles for risk assessment. In: IGU 2013 Kyoto Regional Conference. Japan (2013)
23. Seed, R.B., Cetin, K.O., et al.: Recent Advances in Soil Liquefaction Engineering a Unified and Consistent Framework. College of Engineering University of, Berkeley, CA (2001)
24. Sheriff, R.: Encyclopedic Dictionary of Applied Geophysics, 2nd edn. Cambridge University Press, Cambridge (2002)

Refined Weighted-Permutation Entropy: A Complexity Measure for Human Gait and Physiologic Signals with Outliers and Noise

Huan Zhao, Jian Yu, Junyi Cao, and Wei-Hsin Liao

Abstract The complexity quantification of human gait and physiologic signals has received considerable interest for wearable healthcare. Permutation entropy is one of the most prevalent algorithms for measuring the complexity of time series, but it fails to account for amplitude information with outliers and noise in such time series. Though weighted-permutation entropy aims to incorporate amplitude information by counterweighing the motif with its variance, it mixes noise and outliers with abrupt changes, which have a negative effect on the analysis result. To overcome this problem, this chapter proposes a refined weighted-permutation entropy by assigning fewer weights to outliers and more weights to regular spiky patterns according to the normal distribution function. The refined weighted-permutation entropy is used to analyze simulated synthetic signals with noise and outliers. The comparative analyses with the permutation entropy and weighted-permutation entropy are also performed. Moreover, human gait and ECG experimental data are analyzed by the proposed method. The results demonstrate its better robustness and stability than traditional methods in distinguishing different states of human gait and physiologic signals.

Keywords Permutation entropy · Human gait · Physiologic signals

H. Zhao · J. Cao (✉)
Xi'an Jiaotong University, Xi'an, China
e-mail: caojy@mail.xjtu.edu.cn

J. Yu
China Ship Development and Design Center, Wuhan, China

W.-H. Liao
The Chinese University of Hong Kong, Hong Kong, China

W. Lacarbonara et al. (eds.), *New Trends in Nonlinear Dynamics*,
https://doi.org/10.1007/978-3-030-34724-6_23

1 Introduction

Human gait and physiologic signals had been widely investigated because they can be used for wearable healthcare monitoring. Due to complex multiple coupling phenomenon in the physiological systems, much effort has been devoted to developing complexity measure methods. For the sake of dynamical evaluating models generated from human gait [1, 2] and physiologic time series [3–5], the permutation entropy (PE) method was proposed by Bandt [6]. PE provides the advantage of increasing immunity to complex time series mixed with large artifacts and noise. Moreover, it is suitable for noisy, chaotic, regular, or real-world time series. There are a large number of successful applications such as mechanical fault diagnosis, electroencephalography (EEG), and electrocardiography (ECG) [7, 8].

However, PE may be incapable of distinguishing different patterns of the same motif. In order to address this issue, Fadlallah [9] incorporated amplitude information through assigning weights to design the weighted-permutation entropy (WPE) to achieve a complexity measure for time series. By assigning more weights to the regular spiky pattern, WPE can sensitively track the abrupt changes in a complicated signal regardless of noise effects. This method and its modified algorithms had been successfully employed in many fields. Deng et al. [10] suggested WPE can become a useful tool to detect brain dysfunction in Alzheimer's disease (AD) and other neural diseases. Their comparison results between WPE and PE showed that WPE was more capable of distinguishing the AD patients from the normal controls.

Although WPE can reduce the noise effects and track abrupt changes of amplitude information in the signal, there are many outliers processed as useful amplitude information. In real experimental conditions, environmental, human, and some other adverse factors may lead to unfavorable outliers. Moreover, some outliers or ambiguous data may have a great negative impact on the analysis results. Considering amplitude information and ordinal patterns of time series, a refined weighted-permutation entropy (RWPE) is proposed in this chapter. The normal distribution function is introduced to incorporate significant amplitude information from the signal when computing the weights for each motif. Furthermore, the Gaussian modulation sinusoidal signal as well as the sine signal with white noise and pulses are respectively analyzed by PE, WPE, and RWPE. Results show that WPE has a better performance in noise immunity and tracking abrupt changes of amplitude. Finally, the RWPE calculation results of human gait and ECG signals demonstrate its superiority in the complexity quantification and characteristics extraction of real-world time series.

2 Refined Weighted-Permutation Entropy

For a given time series $\{x(i)\}_{i=1}^{N}$, its time-delay embedding expression $\mathbf{X}(i)$ can be reconstructed as follows,

$$\mathbf{X}(i) = \{x(i), x\left(i+\tau\right), \cdots, x\left[i+(m-1)\,\tau\right]\} \tag{1}$$

where $i = 1,2,\cdots, N-(m-1)\tau$, τ is the time delay and m is the embedding dimension. Then the reconstruction vectors $\mathbf{X}(i)$ are arranged in ascending order,

$$\{x\,(i + (t_1 - 1)\,\tau) \le x\,(i + (t_2 - 1)\,\tau) \le \cdots \le x\,(i + (t_m - 1)\,\tau)\} \qquad (2)$$

where t_m is the time index of the element in the reconstruction vector. When $x(i + (t_{k1} - 1)\tau) = x(i + (t_{k2} - 1)\tau)$, the value of index t_m defines the sequence which can be summed as $x(i + (t_{k1} - 1)\tau) \le x(i + (t_{k2} - 1)\tau)$, $t_{k1} < t_{k2}$. The sequence is represented by motifs $\left\{\pi_j^{m,\tau}\right\}_{j=1}^{m!}$. Obviously, there are m different symbols in $\{t_1, t_2, \cdots, t_m\}$ and thus $m!$ possible motifs. The probability distribution of each motif is calculated and defined as:

$$p_w\left(\pi_j^{m,\tau}\right) = \frac{\sum_{i\le n} \mathbf{1}_{u:\text{type}(u)=\pi_j}\left(\mathbf{X}(i)\right) w_i}{\sum_{i\le n} \mathbf{1}_{u:\text{type}(u)\in\boldsymbol{\Pi}}\left(\mathbf{X}(i)\right) w_i} \qquad (3)$$

where $\mathbf{1}_\mathbf{A}(u)$ denotes the indicator function of set $\mathbf{A}$, if $u \in \mathbf{A}$, $\mathbf{1}_\mathbf{A}(u)=1$ and if $u \notin \mathbf{A}$, $\mathbf{1}_\mathbf{A}(u)=0$, type(.) denotes the map from pattern space to symbol space, Π denotes motifs $\left\{\pi_j^{m,\tau}\right\}_{j=1}^{m!}$, w_i is the weight value calculated by the variance or energy of each reconstruction vector. The normal distribution function is applied as:

$$w_i = \frac{1}{\sqrt{2\pi}\,\sigma} e^{-\frac{(x-\mu)^2}{2\sigma^2}} \quad (\sigma > 0) \qquad (4)$$

where σ is the standard deviation of time series, x is the standard deviation of reconstruction vectors $\mathbf{X}(i)$, μ is not the mean value but $\mu = k\sigma$ (k>0, k<0 is insignificant as k is the parameter to adjust the distance between x and σ.).

It can be seen from Eq. (4) that weight increases when x is close to μ. Therefore, some motifs containing useful amplitude information can be assigned more weights due to their standard deviation are close to the center of the normal distribution function. On the contrary, motifs away from the central of standard deviations can be assigned fewer weights. For example, due to fluctuations like noise and motifs with higher standard deviations resulting from large outliers, motifs that have small standard deviation can be weighted fewer toward the final value of RWPE. Consequently, the proposed method assigns weights to each motif based on the proportion of normal distribution function. Finally, the normalized RWPE is computed as:

$$\text{RWPE}\,(m, \tau) = H\,(m, \tau)\,/\ln\,(m!) \qquad (5)$$

where $H\,(m, \tau) = -\sum_{j:\pi_j^{m,\tau}\in\boldsymbol{\Pi}} p\left(\pi_j^{m,\tau}\right) \ln p\left(\pi_j^{m,\tau}\right)$. Generally, RWPE($m$, τ) vary from 0 to 1 for any parameters, bigger RWPE(m, τ) implies that the time series is more complex and irregular while smaller RWPE(m, τ) indicates that the time series is more regular and periodic.

In this method, k is the parameter to adjust the location of normal distribution function center, k can be adjusted based on data types to assign appropriate weights toward possible motifs. For example, if the pulses contain useful information, k will be increased to track the pulses, otherwise, k will be decreased. This method could not only weaken the effects of outliers and noise by assigning smaller weights but also help to track abrupt changes clearly by assigning larger weights.

3 Validation by Synthetic Data

3.1 *The Gaussian Modulation Sinusoidal Signal with White Noise*

In this part, the Gaussian-modulated sinusoidal signal with a frequency of 10 kHz and a pulse (or outlier) signal with 1 kHz and amplitude attenuation rate of 0.9 are added together to simulate a real signal. The designed signal and with white noise are analyzed by PE, WPE, and RWPE methods as shown in Fig. 1.

PE is unable to detect the pulses signal, but WPE can track the pulses sensitively when there is no noise. Unfortunately, WPE fails to detect the Gaussian-modulated

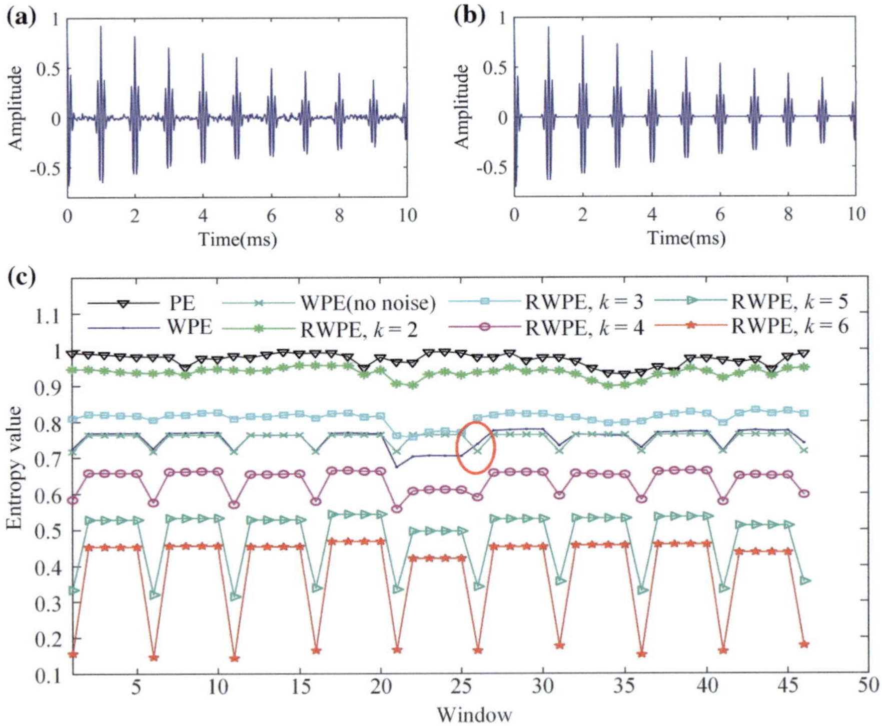

Fig. 1 (**a**) The Gaussian-modulated sinusoidal signal with noise; (**b**) the Gaussian-modulated sinusoidal signal without noise; (**c**) entropy calculation of PE, WPE, and RWPE with windows of 50 samples slid by 10 samples. The embedding dimension $m = 3$ and time delay $\tau = 1$

sinusoidal signal with white noise when the value of the window is 26, as marked in the red oval. It is obvious that RWPE can correctly sense periodical pulses when k increased (especially $k = 5,6$), which shows the excellent immunity to noise. Moreover, though PE and RWPE have a similar trend of entropy fluctuations, RWPE has better anti-noise capability due to its smaller entropy.

3.2 The Sine Signal Without Noise and Impulses

To analyze the data without noise and impulse with PE, WPE, RWPE, the sine signal as well as it with noise and outliers are analyzed respectively, as shown in Fig. 2.

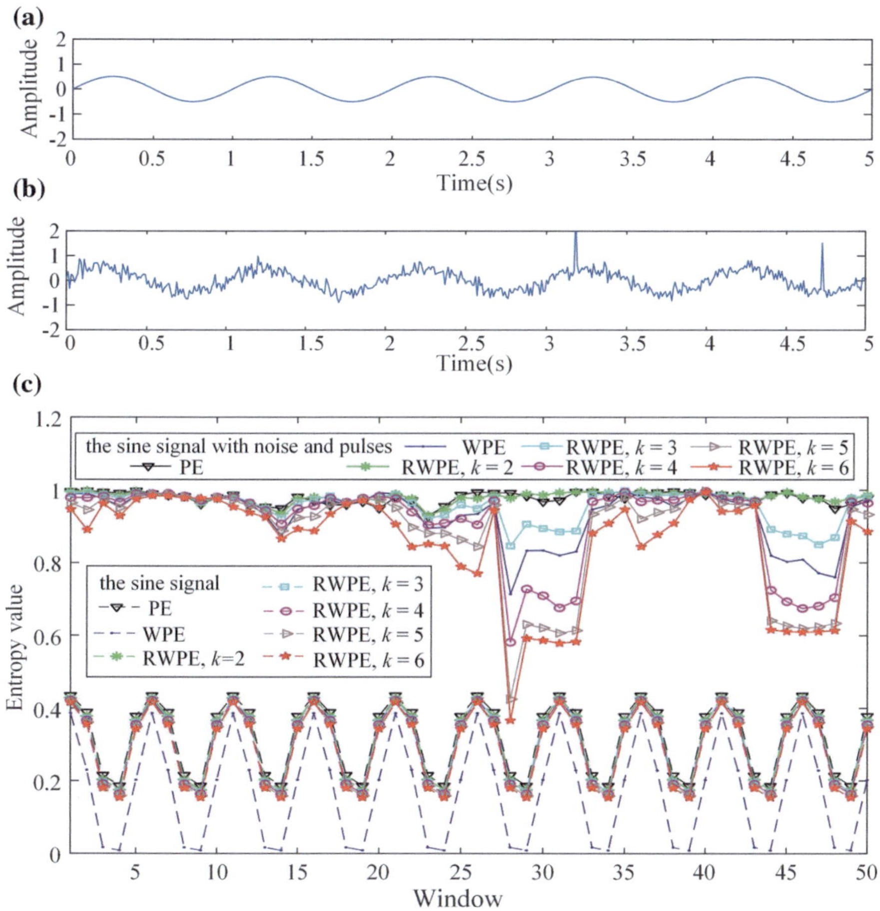

Fig. 2 (**a**) The sine signal without noise and pulses; (**b**) the sine signal with noise and pulses; (**c**) entropy calculation of PE, WPE, and RWPE

Figure 2c shows distinct drops in the value of WPE corresponding to the impulsive regions. Contrary to the WPE method, there is no obvious change in PE at the same regions. As for RWPE, when k is larger ($k = 4, 5, 6$), it tracks more remarkable drops of complexity than WPE in impulse regions. When k becomes smaller ($k = 2, 3$), the value of RWPE is slightly smaller than that of PE and WPE method which reflects good immunity to outliers and noise. Furthermore, for signals without noise and pulses, analyses of PE, WPE, and RWPE show the similar trend and characteristics to the original signal.

4 Analysis of Real Experiment Signals

4.1 Analysis of Human Gait Signals

To recognize whether the gait is healthy or pathological, human gait time-series from PhysioNet database [11] are analyzed by RWPE. The records in this database are from healthy control subjects as well as patients with Parkinson's disease (PD), amyotrophic lateral sclerosis (ALS), and Huntington's disease (HD).

As shown in Fig. 3, RWPE can recognize the health and disease efficiently when $k \geq 3.5$ though there are some outliers exit in the gait signals. As for the characterization of variability, the Mann–Whitney U test is applied, the p-value of PE, WPE, and RWPE ($k = 4$) are calculated. The results are shown in Table 1, in which p-value ($p < 0.05$) shows the comparison of statistical significance of PE, WPE, and RWPE ($k = 4$) for the four states, "Yes" and "No" mean the statistical difference are obvious and unobvious, respectively.

It is seen that PE fails to express the various complexities of those gait states because of $p > 0.05$. For WPE, it only distinguishes the control and HD. However, RWPE shows a better variety of complexity than PE and WPE. Therefore, it is

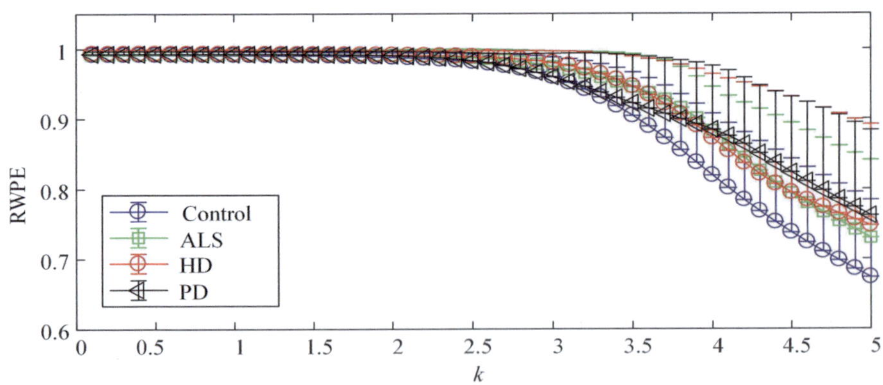

Fig. 3 RWPE analysis of four different human gait

Table 1 Mann–Whitney *U* test of human gait

	p-value (Mann–Whitney *U* test)		
Methods	Control vs. ALS	Control vs. HD	Control vs. PD
PE	0.6501 (No)	0.2114 (No)	0.4945 (No)
WPE	0.6193 (No)	0.0149 (Yes)	0.5453 (No)
RWPE ($k = 4$)	0.0448 (Yes)	0.0215 (Yes)	0.0267 (Yes)

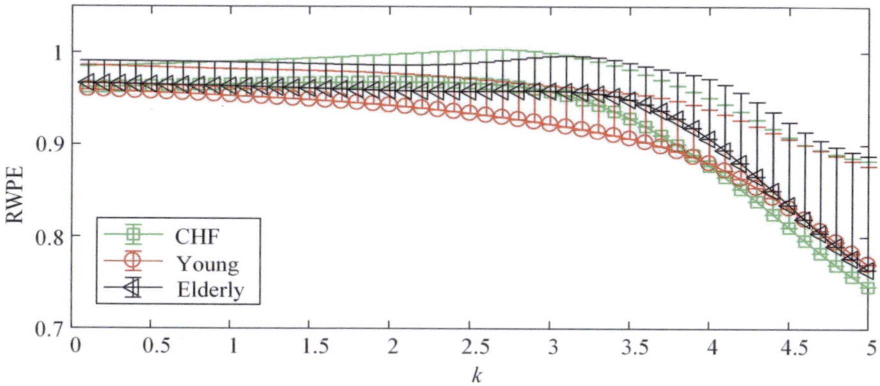

Fig. 4 RWPE analysis of three physiological states

concluded that RWPE has a good immunity to the effect of outliers and noise, which presents better performance in the characterization of signal complexity.

4.2 Analysis of Human ECG Signals

ECG time series from the Fantasia database and the BIDMC congestive heart failure (CHF) database [12, 13] are also analyzed by the proposed method. There are some ambiguous outliers in the ECG signals and the corresponding calculation results by RWPE are shown in Fig. 4.

The analysis results of Mann–Whitney *U* test among the three groups are shown in Table 2. They indicate that PE and WPE could not distinguish all three groups ($p > 0.05$). However, RWPE has effective discrimination ability between young and elderly, young and CHF when $k = 3$. Therefore, RWPE can reduce the effect of outliers and noise by selecting the appropriate value of k, which has better robustness and stability than PE and WPE.

Table 2 Mann–Whitney U test of ECG

Methods	p-values (Mann–Whitney U test)		
	Young vs. Elderly	Young vs. CHF	Elderly vs. CHF
PE	0.2888 (No)	0.2680 (No)	0.9354 (No)
WPE	0.1493 (No)	0.6098 (No)	0.3640 (No)
RWPE ($k = 3$)	0.0051 (Yes)	0.0105 (Yes)	0.9084 (No)

5 Conclusions

The refined weighted-permutation entropy is proposed for incorporating amplitude information of nonlinear signals through weighted entropy based on normal distribution function. For improving the robustness and stability, the proposed method weights the motifs by selecting appropriate values of k to increase immunity to outliers and noise effects. In addition to synthetic signals, real physiological and human gait signals with outliers (or spiky features) and noise are selected to perform experimental verifications of the RWPE. The results show that RWPE is capable of weakening the effects of outliers and noise by assigning fewer weights to them when compared with WPE and PE. Furthermore, the complexity measure of human gait and ECG signals can be characterized distinctly for distinguishing different pathophysiological states. It will be helpful for human wearable devices through the monitoring of physiologic signals and personal healthcare in the future.

Acknowledgment This work was in part supported by the Key Project of Trico-Robot plan of NSFC (Grant No. 91748208).

References

1. Costa, M., Peng, C.K., Goldberger, A.L., Hausdorff, J.M.: Multiscale entropy analysis of human gait dynamics. Phys. A. **330**, 53–60 (2003)
2. Yu, J., Cao, J., Liao, W.-H., Chen, Y., Lin, J., Liu, R.: Multivariate multiscale symbolic entropy analysis of human gait signals. Entropy. **19**, 557 (2017)
3. Li, J., Yan, J., Liu, X., Ouyang, G.: Using permutation entropy to measure the changes in EEG signals during absence seizures. Entropy. **16**, 3049–3061 (2014)
4. Liu, T., Yao, W., Wu, M., Shi, Z., Wang, J., Ning, X.: Multiscale permutation entropy analysis of electrocardiogram. Phys. A. **471**, 492–498 (2017)
5. Staniek, M., Lehnertz, K.: Symbolic transfer entropy. Phys. Rev. Lett. **100**, 158101 (2008)
6. Bandt, C., Pompe, B.: Permutation entropy: a natural complexity measure for time series. Phys. Rev. Lett. **88**, 174102 (2002)
7. Li, Y., Xu, M., Wei, Y., Huang, W.: A new rolling bearing fault diagnosis method based on multiscale permutation entropy and improved support vector machine based binary tree. Measurement. **77**, 80–94 (2016)
8. Morabito, F.C., Labate, D., Foresta, F.L., Bramanti, A., Morabito, G., Palamara, I.: Multivariate multi-scale permutation entropy for complexity analysis of Alzheimer's disease EEG. Entropy. **14**, 1186–1202 (2012)

9. Fadlallah, B., Chen, B., Keil, A., Principe, J.: Weighted-permutation entropy: a complexity measure for time series incorporating amplitude information. Phys. Rev. E. **87**, 022911 (2013)
10. Deng, B., Liang, L., Li, S., Wang, R., Yu, H., Wang, J., Wei, X.: Complexity extraction of electroencephalograms in Alzheimer's disease with weighted-permutation entropy. Chaos. **25**, 043105 (2015)
11. Hausdorff, J.M., Mitchell, S.L., Firtion, R., Peng, C.K., Cudkowicz, M.E., Wei, J.Y., Goldberger, A.L.: Altered fractal dynamics of gait: reduced stride-interval correlations with aging and Huntington's disease. Appl. Physiol. **82**, 262–269 (1997)
12. Iyengar, N., Peng, C.-K., Morin, R., Goldberger, A.L., Lipsitz, L.A.: Age-related alterations in the fractal scaling of cardiac interbeat interval dynamics. Am. J. Physiol. Regul. Integr. Comp. Physiol. **271**, 1078–1084 (1996)
13. Baim, D.S., Colucci, W.S., Monrad, E.S., Smith, H.S., Wright, R.F., Lanoue, A., Gauthier, D.F., Ransil, B.J., Grossman, W., Braunwald, E.: Survival of patients with severe congestive heart failure treated with oral milrinone. J. Am. Coll. Cardiol. **7**, 661–670 (1986)

Simultaneous Multi-Parametric Analysis of Bone Cell Population Model

Julijana Simonović

Abstract Using the bone cell population mathematical model of the system of coupled ordinary differential equations (ODEs) with power-law nonlinearities, it is possible to properly interpret and analyze bone cell communication dynamics. The system of bone cellular communication is complex and not yet properly described and revealed. The structural analysis has been used here for stability analyses of the problem, as like as for analyses of system sensibility to small parameters changes. The usage of the multi-parametric synchronous analysis presented in this chapter is the advantage of Mathematica ODE solver that provides the functional interpretation of important parameters of dynamics. The models explored in numerous numerical (in silico) experiments also provide the more realistic approaches to interpreting the development of interventions for patients with bone trauma and diseases, but also for those who want to prioritize the healthy and strong skeleton. This research is a very practical and clear example of nonlinear theory application for bone cell signalling processes modeling and interpretation.

Keywords Bone cell population model · Cell signalling · Nonlinear multi-parametric analysis

1 Introduction

Mathematical modeling can be developed to characterize complex bone phenomena, from intercellular signalling to cell division, proliferation, migration, and even mutation; from bone remodeling to healing, from tumor growth to cancer metastasis, from osteoarthritis to osteoporosis treatment. Mathematical interpretation and models of this system become even more complex as the higher resolution screening and high-throughput omics research discover the new players involved

J. Simonović (✉)
Biomedical Engineering Department, School of Engineering, Cardiff University, Wels, UK
e-mail: simonovicjulijana@gmail.com; julijana.simonovic@masfak.ni.ac.rs

W. Lacarbonara et al. (eds.), *New Trends in Nonlinear Dynamics*,
https://doi.org/10.1007/978-3-030-34724-6_24

in the process, what puts demands on theories, methods, and assumptions used and the efficiency of numerical methods employed for solving as well as managing of parameters and data. Mathematical modeling is the tool for synthesis—the process of combining separate elements in order to form a coherent whole—the tool of inductive reasoning, allowing reconstruction of the function of bone as an organ based on the acquired knowledge of elementary processes. Additionally, it is well known from the phenomenological analogies and mathematical mapping that we can use the same mathematical framework and formulations to describe phenomena in the disparate sciences [1, 2]. Good example is nonlinear hysteresis phenomena that can be described with simple phase lag between input and output but also with Duffing differential equation amplitude jumps [3] and many more models in different fields from ferromagnetism to biology [4].

The complexity of bone models depends not only on the number of parameters describing different biochemistry and multi-physics influences but also on different time scales, rate of the periodicity of involved processes and numerous delay in transmission of the signal. These issues are challenging even for experienced applied mathematicians and are difficult to be adequately communicated to a biologist who has to accept and endorse the predictive vigor of mathematical models. Thus, it is important to have a friendly model capable of understanding from the point of view to both mechanistic and biologist approaches. Moreover, another unresolved issue remains linked with ambiguities of mathematical results from in silico experiments and their discrepancy with in vivo/vitro experimental results in the modeling of bone tissue. The future advancement of bone biology research will strongly rely on how well experimental and theoretical groups are able to communicate and collaborate with each other. A unique bridge between biological models and experimental validation is more realistic mathematical model together with possibility of numerous in silico experiments. Once established credible mathematical model helps us in decision-making while analyzing systems perturbation, performing sensitivity analysis likewise to develop hypotheses to guide and design new biological experiments.

Generalized form of predator–prey (S-System type) mathematical model, described in next section, is used to analyze the dynamic interaction between several bone cell lineages that are involved in the process of targeted bone remodeling. Functional model, introduced in Sect. 3 aims to present interaction of the cellular functions and activities. This kind of model is unique and different from System Biology Graphical Notation BioModels of bone remodeling and in such a way closer and friendly for communication to mathematicians and mechanics. In Sect. 4, a hands-on and applicable multi-parametric analysis of presented system of equations for specific ranges of parameter's values is presented and discussed together with a few steps of nonlinear structural analysis of system stability. Concluding remarks contain also discussion about future directions and the latest stochastic analysis findings that address S-system type models of bone remodeling.

2 Functional Model

In the process of mathematical modeling of the particular biological processes, it is necessary to interpret and ascertain the meaning of all parameters involved in order to explain all the mutual correlations and sites of possible internal miscommunication and conflicts between all the relevant actors and their functions. In the process of bone cellular signaling at least three bone cell lineages are involved and it is not clear enough and distinguished which of them are responsible for certain functions. Most of them do their function at the same level of importance and simultaneously so that is impossible to designate certain relevant parameter to just one cell lineage, in neither time nor location. Many overlapping and instantaneous processes merge with each other and as the bottom-goal determines the quality and quantity of the new formed bone tissue. Proposing the structural models of biological processes in the form of flow models that has been already published [5, 6] is useful in terms of understanding nominated names of bone cells lineage and their relationship with cooperative binding effects of the large number of significant proteins and molecules, ligands, receptors, and decoys. Nevertheless, it is more important what function has certain cell lineage than how it is designated. No matter how we named these cell lineages they will still serve for their different functions: processes of bone resorbing and forming, receiving and sending the signals, proliferating, differentiating, preserving viability, ageing or dying, which are more important for the sake of balanced remodeling bone tissue process. Defining the functions is also necessary for the purpose to form an appropriate physical model for preparing mathematical models. If some function is properly determined and defined with all aspects of inputs (I), outcomes (O), preconditions (P), resources (R), controls (C), and time (T), noted as a corners of hexagons on the Fig. 1, then it is a straightforward way to find its mathematical function. For instance, using the FRAM (Functional Resonance Analysis Method) [7] it is convenient to associate appropriate aspects having recognized the function to define controllers, preconditions, resources, imputes, and outputs. The time is treated as separate aspect even though it can be incorporated within other aspects. Using this method it is easy to address upstream, downstream, and actual function and their interconnections and mutual influences, even more it is easy to register and assign which one of the functions can have possible conflicts with large number of outputs, controllers, and connections with other functions and present the sources of the internal resonance or dynamical absorption.

Figure 1 shows simple form of the phenomenological model of bone cell signaling where the important function of bone mass qualitative (via initial and total number of involved cells) and quantitative (via relative rate of formation or resorption, parameters k_1 and k_2, respectively) changes. It depends on two functions: "to form" which corresponds to OB lineage (with palette of green hues) and "to resorb" corresponding to OC lineage (with palette of purple hues). These two functions change number of cells from initial number to the total available number of cells. Both of them are described with first two ODEs in the system (2). From

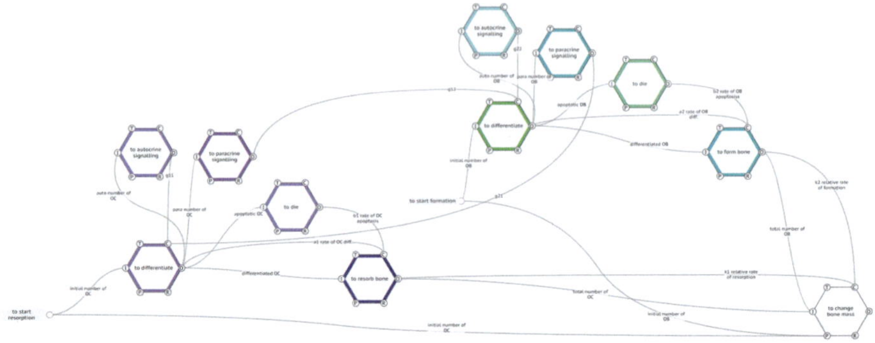

Fig. 1 Functional model of the OC-OB communication processes that influence bone mass changes on the level of bone multicellular unit representing the activities of the one cycle of the targeted bone remodeling. The model corresponds to the system of Equations (2), (3), and (4)

Fig. 1, it is noticeable that these functions have form of fractal similarity of which everyone increase with the number of differentiated cells (positive term in Equations (2) and (3)) and decrease with the number of died cells (negative terms in Equations (2) and (3)). Up and down self-regulation of the cell lineage is presented with autocrine signaling function (determined by $\gamma_{11} = g_{cc}$ and $\gamma_{22} = g_{bb}$ parameters) and mutual influence is assigned as paracrine signaling function (determined by $\gamma_{12} = g_{cb}$ and $\gamma_{21} = g_{bc}$ parameters). Both influence differentiation stages of cells activities. Although there exists far more complicated and complex descriptions of bone cell signaling, this model is used as a good representative and elegant form to describe connection with conceptual biological models and mathematical system of Equations (2)-(4) generated for in silico experiments of this chapter.

3 Mathematical Model

The system of bone cellular communication, which involves at least three main cellular lineages: forming-osteoblasts (OBs), resorbing-osteoclasts (OCs), and orchestrating-osteocytes (OCYs) cells lineage, with their self (autocrine signalling) and mutual (paracrine signalling) interactions and their interactions with the environment, is complex and elucidate a number of parameters that detail the psychological mechanism of bone tissue adaptation processes. One cycle of the targeted bone remodeling consists of the bone resorbing activity of the OCs and the bone forming activity of the OBs, which are both driven by the signals transduced via OCYs from the external loading. The importance of this process goes with the fact that after every ten years of adult life the whole skeleton is regained and renewed due to this process. The following are the general form of models used by several authors [5, 6, 8–11] representing the power of analytical approaches:

$$\begin{aligned} \tfrac{du_i}{dt} &= \alpha_i f_i\,(u_i) + k \cdot f_i\,(u_i, \mu(t)) - \beta_i u_i, \quad i = 1, 2, 3 \\ \tfrac{dz}{dt} &= -k_1 v_1 + k_2 v_2 \text{ for } v_j = \begin{Bmatrix} u_j - \overline{u_j}, \ if \ u_j > \overline{u_j} \\ 0, \ if \ u_j \le \overline{u_j} \end{Bmatrix} \quad j = 1, 2 \end{aligned} \tag{1}$$

where u_i are the densities of OCs, OBs, and OCYs for i = 1, 2, 3, respectively, and $f_i(u_i)$ are the functions giving the growth rates which include the interaction between cell populations by the biochemical regulators in the form of power law approximation: $f_1\,(u_1, u_2) = u_1^{\gamma_{11}} u_2^{\gamma_{21}}$ and $f_2\,(u_1, u_2) = u_1^{\gamma_{12}} u_2^{\gamma_{22}}$, where γ_{ij} for i, j = 1, 2 are defined by their autocrine and paracrine regulation. The last, so called bone mass equation describes the activity of bone resorption and formation where z is total bone mass, k_i represents the normalized activities of bone resorption and formation, $\overline{\mu}_i$ represents the steady states for the OCs and OBs, and k is a positive proportionality constant measured in cells day^{-1}. The term $\mu(t)$ functions as a regulator of the bone-remodeling process and includes external signaling transduced via osteocyte activities that stimulate the production of OC and OB. This input function $\mu(t)$ can model regulators production from osteocytes as well as its regulation by the sclerostin inhibitor [6, 11]. An explicit functional form for $\mu(t)$ linked to osteocytes activity is still to be established and justified by biological experiments.

3.1 *In Silico Experiment and Discussion*

For the calculation, compliance and discussion in this chapter, the simple form of system (1) is in use:

$$\frac{du_1}{dt} = \alpha_1 u_1^{\gamma_{11}} u_2^{\gamma_{12}} - \beta_1 u_1 \tag{2}$$

$$\frac{du_2}{dt} = \alpha_2 u_1^{\gamma_{21}} u_2^{\gamma_{22}} - \beta_2 u_2 \tag{3}$$

$$\frac{dz}{dt} = -k_1 \max\,[0, u_1 - \overline{u_1}] + k_2 \max\,[0, u_2 - \overline{u_2}] \tag{4}$$

The unique nontrivial steady states of $\overline{\mu}_i$ could be calculated and in general depends on relations of eight different parameters involved in this simple example. The behavior of the solution of Equations (2) and (3) can be explored by sign of the real part of the eigen values of Jacobian matric $J\,(\overline{u}_1, \overline{u}_2)$ of the steady state solutions. The nature of the Jacobian matrices' eigen value depends on the following functions:

$$\begin{aligned} \text{trace} \quad & J\,(\overline{u}_1, \overline{u}_2) = \psi = \beta_1\,(\gamma_{11} - 1) + \beta_2\,(\gamma_{22} - 1), \\ \det \quad & J\,(\overline{u}_1, \overline{u}_2) = \Lambda = \beta_1\beta_2\,[(\gamma_{22} - 1)\,(\gamma_{11} - 1) - \gamma_{12}\gamma_{21}] = -\beta_1\beta_2\Pi \\ \text{and} \quad & \Delta = tr\,J^2 - 4\det J = [\beta_1\,(\gamma_{11} - 1) - \beta_2\,(\gamma_{22} - 1)]^2 + 4\beta_1\beta_2\gamma_{12}\gamma_{21}. \end{aligned}$$

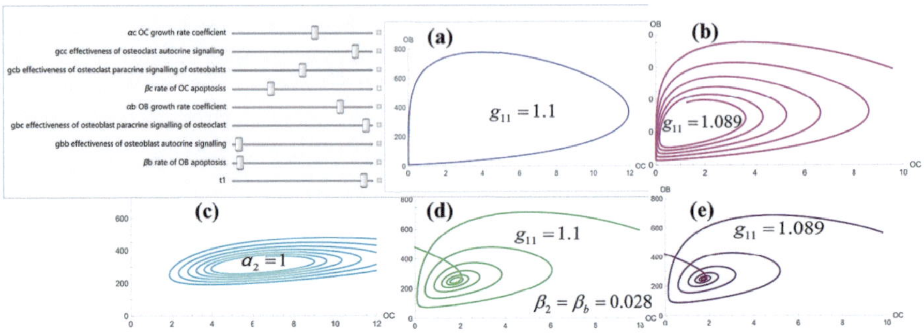

Fig. 2 Phase portraits of OC-OB interaction dynamics for different γ_{11} parameter (effectiveness of osteoclast autocrine signaling) values: (**a**) $\gamma_{11} = 1.1$, (**b**) $\gamma_{11} = 1.098$ and (**c**) $\alpha_2 = 1$, (**d**) $\gamma_{11} = 1.1$ and $\beta_2 = \beta_b = 0.028$ and (**e**) $\gamma_{11} = 1.098$ and $\beta_2 = \beta_b = 0.028$

The solution of systems (2) and (3) exhibit a limit cycle as ψ passes through 0 performing self-sustained oscillation of number of OC and OB, see Fig. 2a. Also, if $\psi > 0$ and $\Lambda > 0 (\Pi < 0)$ solution yields unstable oscillations diverging away from the nontrivial steady state solutions $\overline{\mu}_i$ what defines existence of unstable source (repellers), see Fig. 2c and if $\psi < 0$ and $\Lambda > 0 (\Pi < 0)$ solution yields damped oscillations converging to the $\overline{\mu}_i$ what defines existence of stable attractors (sinks), Fig. 2b. Likewise if $\Delta < 0$ unstable saddle points can be find. The bifurcation diagram for these parameters could be designed based on this analysis but more applicable solution for biologist is a tool where they can simultaneously change values of the parameters and track the shape of the phase diagram. In such a way find the combination of parameters value that satisfy desirable dynamics of solution. The multi-parametric analysis here is proposed where simply by moving sliders for parameter‘s value we can follow the changes in the dynamics of the OC and OB, and find out the same inferences as it comes from structural analysis (see Fig. 2).

The visualization of different system dynamics obtained by small changes of only one parameter (effectiveness of osteoclast autocrine signaling $\gamma_{11} = g_{cc}$) are presented at Fig. 2. γ_{11} has been changed from the value of 1.1, Fig. 2a where the solutions exhibit limit cycle, also presented at [10], to the value 1.089, Fig. 2b where yielding damped oscillations converging to the nontrivial steady states $\overline{u}_i$ is presented. The values for the other parameter are: $\gamma_{12} = g_{cb} = -0.5$, $\gamma_{22} = g_{bb} = 0$, $\gamma_{21} = g_{bc} = 1$ are dimensionless parameters and $\alpha_1 = \alpha_c = 3$, $\alpha_2 = \alpha_b = 4$, $\beta_1 = \beta_c = 0.2$, $\beta_2 = \beta_b = 0.02$ have dimension day^{-1}. In [10] it has been claimed that this parameter is the primary factor in the regulation of bone remodeling dynamics. The following parameter‘s range values are used in this research: $\gamma_{11} = g_{cc} \in (0; 1.2)$, $\gamma_{12} = g_{cb} \in (-1; 0)$, $\gamma_{22} = g_{bb} \in (0; 1)$, $\gamma_{21} = g_{bc} \in (0; 1)$, $\alpha_1 = \alpha_c \in (0; 5)$, $\alpha_2 = \alpha_b \in (0; 5)$, $\beta_1 = \beta_c \in (0.1; 0.5)$, $\beta_2 = \beta_b \in (0.01; 1)$. Initial conditions are constantly [OC(0);OB(0)]=(11, 231) in all simulations. However, having in mind the form of function ψ it is obvious that changing, for instance value of parameter $\beta_2 = \beta_b = 0.028$ the response of the system dynamic no longer depend on the small changes of the value of $\gamma_{11} = g_{cc}$ with the same sensitivity, Fig. 2e, d. This is what underlines the importance of such a multi-parametric analysis presented herein. As an overall outcome, we can conclude that the right relation between parameters, not only value range of one primary parameter, is valuable to be analyzed. This on the other side confirmed that continuous therapies that targeted only one parameter most likely would not be appropriate for a long-term period treatment.

Similar parameter analysis has been done in [11] for a more advanced model that incorporates osteocyte signalling and two different stages of osteoblast maturity, which gives five dependent equations in system (1) with at least eighteen different parameters that describe autocrine and paracrine signaling more realistically. Authors have analyzed the steady state of the bone mass and presented it as a function of two parameters in the 3D diagram in parameter space which is a very useful way of presentation but it is constrained with representation of dependence on two parameters at most. With the multi-parametric analysis these findings are confirmed, but also the influence of the other parameters is possible to be revealed.

It is possible to perform the changes of any parameter and even of all of them simultaneously with this procedure in a very functional way. By performing similar simulations with all involved parameters it is straightforward to decide which of the parameter‘s relation is the most influential and most responsible for changes of model dynamics, even if the number of parameters or equations in the model is enhanced. Obtained conclusions and discussions for parameter values and ranges are applicable for the justification of effectiveness of mathematical models and their compliance with in vivo experiments of bone cells. Nevertheless, one should bear in mind that the presented range of parameters for this chapter are chosen only intuitively and further readouts from in vitro experiments could be extremely important in order to further validate the model. Mention should also be made of the fact that known histopathological samples suggest limited number of cells per one cycle of remodeling what puts constraint to the maximal number of OC and OB present in the simulations, e.g., $OC_{max} < 20$ and $80 < OB_{max} < 120$ per one cycle of bone remodeling. Parameters α_i and β_i, are responsible for bounding the number of cells as they are cell’s growth and death rate coefficients. If the value of parameter $\alpha_2 = 1$, then it is obvious from the Fig. 2c that number of OB is bonded between 200 and 500 cells.

This allows probability that the transition number of cells, in the beginning and end of cycle, become small than 10 cells in which case predator–prey system yields to population extinction after only one cycle. Reasonably, our current research is the stochastic analogue of the system of equations where we run 1000 simulations to explore these effects and perform cross-correlation simulation of parameters to find out appropriate intervals of expected value of parameters. Both the stochastic and deterministic trajectories can be found. Any individual stochastic trajectory does not match the deterministic solution. However, the average of these 1000 stochastic trajectories does correspond extremely well with the first peak of the deterministic solution. These results will be published in the future and are not presented herein as they are far beyond the scope of this chapter that has the aim to present the practical and applicable side of the multi-parametric simultaneous analysis. However, comparative analysis of different mathematical approaches shows that biological real event can be interpreted with a useful model depending on the different real constraints from the biological experiment. If the simulated population are large enough then the continuum deterministic approximation is appropriate, while if the population number tends to fall below 10 cells then the stochastic description is more apt.

4 Conclusions

Different types of external signals were observed in [6] and the nonlinear stability analysis [8] was used for obtaining parameter's ranges for steady state solutions, the authors have proved that the modified model yields a positive non-oscillatory solution. This behavior of the solution is consistent with the bone remodeling cycle [5] returning to a quiescent state after 3 or 5 months. Additional quantitative and qualitative analysis of model sensibility to parameter's changes have been presented with this chapter results. Based on a simultaneous multi-parametric analysis it is possible to provide a parametric range where model exhibits periodic solutions and to analyze their effects on model dynamics. The usage of Mathematica program provides different visualization and animation technique of bone cell population model dynamics. As being nonlinear with power law terms γ_{ij} representing autocrine and paracrine signaling pathways in bone cell communication process system of ODEs (1) is highly sensitive on small changes of parameters (e.g., changes in system dynamics for only one parameter shifting is presented on Fig. 2). The models explored in numerous numerical (in silico) experiments also provide the more realistic approaches to understanding the development of interventions for patients with bone trauma and diseases. The good example is in [9] where this model is adjusted and enhanced to description and explanation of bone myeloma diseases.

Acknowledgements These results are part of research carried within project MMoBEER (Nov. 2017-Nov. 2019) supported by the European Union H2020 MGA MSCA-IF-2016 under grant agreement No. 752793.

References

1. Hedrih (Stevanović), K.R., Ivanović-Šašić, A., Simonović, J., Kolar-Anić, L.J., Čupić, Ž.: Oscillators: phenomenological mappings and analogies—first part: mathematical analogy and chains. Sci. Tech. Rev. **65**(3), 27–38 (2015). http://www.vti.mod.gov.rs/ntp/eindex.htm
2. Hedrih (Stevanović), K.R., Simonović, J.: Structural analogies on systems of deformable bodies coupled with non-linear layers. Int. J. Nonlinear Mech. **73**, 18–24 (2015). https://doi.org/10.1016/j.ijnonlinmec.2014.11.004

3. Kovacic, I., Brennan, M.J., Lineton, B.: Asymmetric Duffing oscillator: hysteretic behaviour and jump phenomena. In: 12th Conference on Nonlinear Vibrations, Dynamics, and Multibody Systems, pp. 1–5. Virginia Polytechnic Institute and State University, Blacksburg (2008)
4. Noori, H.R.: Hysteresis phenomena in biology. Springer, Berlin (2014). ISBN 978-3-642-38217-8
5. Pivonka, P., Komarova, S.: Mathematical modeling in bone biology: from intracellular signaling to tissue mechanics. Bone. **47**, 181–189 (2010)
6. Chen-Charpentier, B.M., Diakite, I.: A mathematical model of bone remodeling with delays. J. Comput. Appl. Math. **291**, 76–84 (2016)
7. Hollnagel, E.: FRAM—The Functional Resonance Analysis Method. Ashgate, Farnham (2012)
8. Jerez, S., Chen, B.: Stability analysis of a Komarova type model for the interactions of osteoblast and osteoclast cells during bone remodeling. Math. Biosci. **264**, 29–37 (2015)
9. Ayati, B.P., Edwards, C.M., Webb, G.F., Wikswo, J.P.: A mathematical model of bone remodeling dynamics for normal bone cell populations and myeloma bone disease. Biol. Direct. **5**(1), 4–17 (2010)
10. Komarova, S.V., Smith, R.J., Dixon, S.J., Sims, S.M., Wahl, L.M.: Mathematical model predicts a critical role for osteoclast autocrine regulation in the control of bone remodeling. Bone. **33**(2), 206–215 (2003)
11. Graham, J.M., Ayati, B.P., Holstein, S.A., Martin, J.A.: The role of osteocytes in targeted bone remodeling: a mathematical model. PLoS One. **8**(5), 10–14 (2013). https://doi.org/10.1371/journal.pone.0063884

Nonlinear Dynamics of RRc Lyrae Stars

Sandip V. George, Ranjeev Misra, and G. Ambika

Abstract In this work we consider the nonlinear dynamics of RRc Lyrae variable stars, using intensity data from the Kepler space telescope. RRc Lyrae variables are shown to exhibit rich nonlinear dynamical behavior, including strange nonchaotic and quasiperiodic behaviors. We examine the intensity variations (or light curves) of four confirmed RRc Lyrae stars (KIC 4064484, 5520878, 8832417, and 9453114) in the Kepler field of view. Earlier work has classified two of the four stars as quasiperiodic with noise contamination and two as exhibiting strange nonchaotic dynamics George et al. (Nonlinear Dyn 89(1):465, 2017). We explore the recurrence network properties and bicoherence of these stars. We find the groups of stars that show the same dynamical behavior also show similar nonlinear properties. We also do a comparison of the metallicities and pulsation periods of these stars to their nonlinear properties. Our results support the existence of two distinct subcategories of RRc Lyrae stars and indicate a link between the nonlinear and astrophysical properties of RRc Lyrae variable stars.

Keywords RR Lyrae stars · Chaos · Recurrence plots · Recurrence networks · Bicoherence

S. V. George (✉)
Indian Institute of Science Education and Research, Pune, India

University Medical Center Groningen, Groningen, the Netherlands
e-mail: sandip.varkey@students.iiserpune.ac.in

R. Misra
Inter-University Centre for Astronomy and Astrophysics, Pune, India
e-mail: rmisra@iucaa.in

G. Ambika
Indian Institute of Science Education and Research, Tirupati, Tirupati, India
e-mail: g.ambika@iisertirupati.ac.in

W. Lacarbonara et al. (eds.), *New Trends in Nonlinear Dynamics*,
https://doi.org/10.1007/978-3-030-34724-6_25

1 Introduction

Pulsating variable stars are stars whose light variation is primarily due to oscillations in the star. A prominent class of pulsating variables is the RR Lyrae class. These are evolved He burning stars located in the boundary between the instability strip and horizontal branch, on the H-R diagram. The pulsations are thought to be due to the κ mechanism operating in the partial ionization zone. RR Lyrae stars are further classified based on the mode in which they pulsate in. RRab Lyrae stars pulsate in the radial fundamental mode, whereas the RRc Lyrae stars oscillate in the first overtone and RRd variables pulsate simultaneously in the radial fundamental and first overtone modes [1]. RRab stars also have slightly longer periods than RRc Lyrae variables, with the former having pulsation periods ranging between 0.3 and 1 days, and the latter having periods between 0.2 and 0.5 days. Many RRc Lyrae variables are shown to have a secondary pulsation mode that is about 0.6 times the fundamental [2]. The closeness of this ratio to the golden ratio has invoked considerable interest in these stars. RRc Lyrae variable stars are also the first astrophysical objects where strange nonchaotic dynamics were detected [3].

We calculate the nonlinear quantifiers of the time series of intensity variations (light curves) of the stars that were measured using the Kepler space telescope [4]. Nonlinear time series analysis techniques are useful in uncovering the underlying dynamics of the system from which the time series is derived. Knowing the underlying dynamics can help in modeling and classification of astrophysical systems. A very popular nonlinear time series measure is the bicoherence function, which is a higher order spectrum. It examines the extent of quadratic coupling between the different frequencies present in the time series. Unlike the phase blind power spectrum, the ability to retain phase information makes the bicoherence an attractive tool to analyze nonlinear time series [5]. This has led to the widespread use of the bicoherence function in a large number of fields [6, 7]. The use of the bicoherence to analyze dynamical states arising in nonlinear dynamical systems has been demonstrated previously [8, 9]. In this work, we use two measures, namely the mean bicoherence over the plane and the fraction of frequency pairs having significant bicoherence, to quantify the bicoherence plane.

A recurrence plot is a method that quantifies the underlying patterns in a phase space trajectory. The recurrence plot represents the extent to which phase space vectors are adjacent to each other. A network constructed using the recurrence plot as an adjacency matrix is called an ϵ recurrence network. The properties of these networks, constructed from time series, are used to determine the underlying dynamical properties of the time series themselves. Both recurrence plots and networks have been put to considerable use to analyze datasets from a large number of fields [10, 11].

One of the most popular methods used to identify strange nonchaotic behavior in real systems is through the scaling behavior of peaks in the strobed power spectrum. Recent work has, however, shown that it is difficult to distinguish between noise contaminated quasiperiodicity and strange nonchaotic behavior using this approach.

A bicoherence based filter was shown to be effective in distinguishing between the two states. It has been shown that the four RRc Lyrae variable stars in the Kepler field of view may be classified into two distinct groups showing noise contaminated quasiperiodic and strange nonchaotic behaviors [12]. We analyze the recurrence network and bicoherence plane measures of the two groups. Finally, we look for common astrophysical characteristics of the two RRc Lyrae groups. The effective temperature, period, and metallicities seem to be distinctly different for the two sets of stars.

2 Bicoherence Analysis

The bicoherence is a higher order spectrum which checks the extend of quadratic coupling between every pair of frequencies present in the measured light curve. The light curve is initially segmented into k evenly sampled segments of 1024 points each, the global mean is then reduced from the time series. The bicoherence is defined as

$$b(f_1, f_2) = \frac{\left|\sum_{i=1}^{k} A_i(f_1)A_i(f_2)A_i^*(f_1 + f_2)\right|}{\sum_{i=1}^{k} \left|A_i(f_1)A_i(f_2)A_i^*(f_1 + f_2)\right|}, \tag{1}$$

where $A(f)$ is the Fourier transform of the signal at frequency, f, and $A^*(f)$ is its complex conjugate [5]. The full bicoherence planes, corresponding to the four stars considered, are shown in Fig. 1.

The bicoherence plane is quantified using the mean bicoherence of the plane and the fraction of significant pairs in the plane. The former is calculated by averaging over the bicoherence in the entire plane and the latter is found by counting the relative number of pairs with bicoherence greater than the 99% significance level. For k-segments, the 99% significance level is given by $\sqrt{\frac{9.2}{2k}}$ [9]. These quantifiers are listed in Table 1. We see that the characteristics of the bicoherence plane is distinctly different for the two categories identified using the spectral scaling with bicoherence based filtering technique developed in [12].

3 Recurrence Analysis

In this section, we construct ϵ recurrence networks from the measured light curves of the RRc Lyrae variable stars. We start by reconstructing the phase space using the method of delay embedding. To eliminate differences that arise from the amplitude distributions of the different light curves, we first take the uniform deviate of the light curves. A delay time (τ) equal to the time at which the auto-correlation falls to $\frac{1}{e}$ is chosen. If $I_u(t)$ is the light curve after taking uniform deviates, a vector in an M-dimensional reconstructed phase space is given as

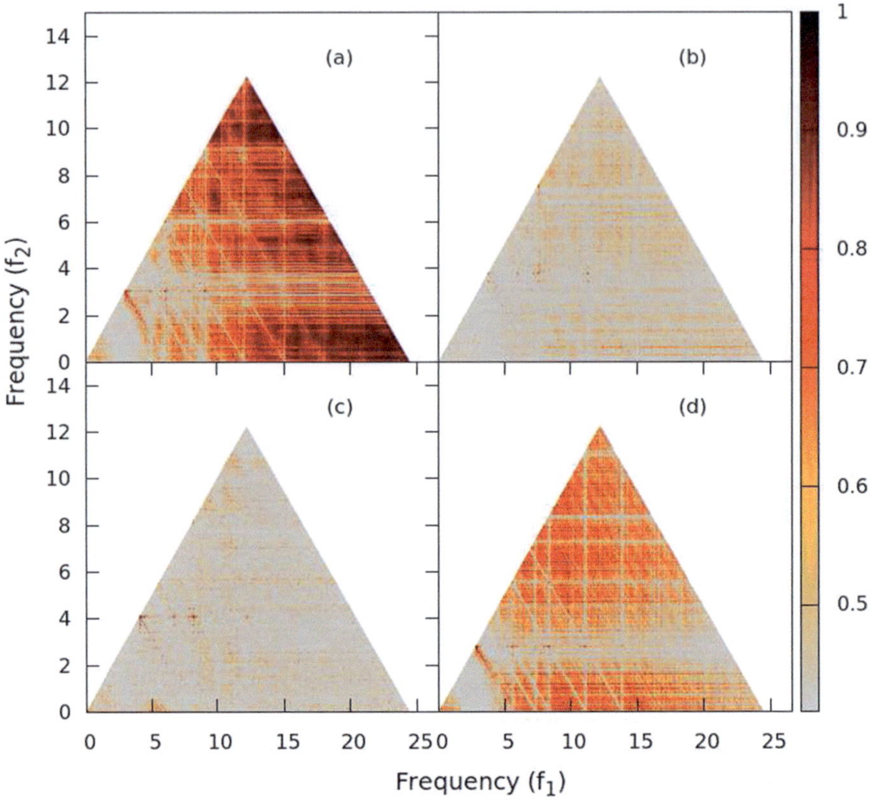

Fig. 1 Full bicoherence plots corresponding to the four RRc Lyrae stars (**a**) KIC 4064484, (**b**) KIC 5520878, (**c**) KIC 8832417, and (**d**) KIC 9453114. The planes for the strange nonchaotic stars shown in (**a**) and (**d**) seem to be distinctly different from the quasiperiodic stars in (**b**) and (**c**)

$$\mathbf{v}_i = [I_u(t_i), I_u(t_i + \tau), \ldots, I_u(t_i + M\tau)].$$

This embedding dimension M is chosen as the maximum of the nearest integer greater than the correlation dimension (D_2). In our case the dimension is chosen as 4, which is larger than the value for D_2 for all four stars. The delay time (τ), saturated correlation dimension (D_2), and deviation of data from surrogates [13] (square difference) are shown in Table 2. The square difference is calculated as follows:

$$sq.diff. = \frac{1}{M_{\max} - 1} \sum_{M=2}^{M_{\max}} D_2(M) - < D_2^{\text{surr}}(M) >^2 . \tag{2}$$

Here, M is the embedding dimension, $M_{\max}$ is the maximum embedding dimension till which the saturation of the correlation dimension was checked, $D_2(M)$ is the

Table 1 Number of segments (k), mean bicoherence (B_{avg}), and fraction of significant number of bicoherence pairs (SPF) for the RRc Lyrae stars considered

Kepler ID	k	B_{avg}	SPF
4064484	22	0.64	0.77
5520878	27	0.40	0.31
8832417	27	0.33	0.24
9453114	27	0.55	0.70

The two subcategories have distinctly different values for B_{avg} and SPF. The mean for all the cases is calculated with 22 segments only, the minimum number of segments present across the four datasets

Table 2 Table showing the delay time (τ), saturated correlation dimension (D_2), and mean square difference ($sq.diff.$) for the four RRc Lyrae stars

KID	τ (days)	D_2	$sq.diff.$
4064484	0.08	3.05	0.43
5520878	0.06	3.58	0.25
8832417	0.06	2.87	0.52
9453114	0.08	2.89	0.54

All the stars show D_2 less than 4, which is the embedding dimension used for constructing the recurrence plots and networks

correlation dimension at embedding dimension M for the data, and $D_2^{surr}(M)$ is the correlation dimension at embedding dimension M for the surrogates. The delay time (τ), saturated correlation dimension (D_2), and mean square difference ($sq.diff.$) for the four RRc Lyrae stars are listed in Table 2.

Every pair of vectors in this reconstructed 4 dimensional phase space is a point in the recurrence plot. If the vectors are within ϵ distance of each other, the recurrence point is 1, otherwise it is 0. In our case ϵ is chosen to be 0.14 following the criteria described in [14] for embedding dimension, $M = 4$. This is chosen to be large enough such that above 95% of the nodes in the network form part of its largest component, and small enough such that the network does not become over connected. The recurrence plots are shown in Fig. 2. Two important characterizers of the recurrence plot are the laminarity (LAM) and the determinism(DET). The former is a measure of the vertical structures in the recurrence plot, whereas the latter measures diagonal lines. The vertical lines are indicative of the tangential motion in phase space, while the diagonal lines are associated with greater correlations in the process [10]. These characteristics are listed in Table 3. We see distinctly different laminarities and laminarity to determinism ratios for the two groups of RRc Lyrae stars identified in [12].

One can use the recurrence plot as an adjacency matrix to construct a network (the diagonal elements are set to zero to avoid self-loops). Such a network is called a recurrence network [11]. We look at three main properties of the recurrence network, namely the average degree (d_{avg}), the characteristic path length (CPL), and the average clustering coefficient (CC_{avg}). The average degree is defined at the number

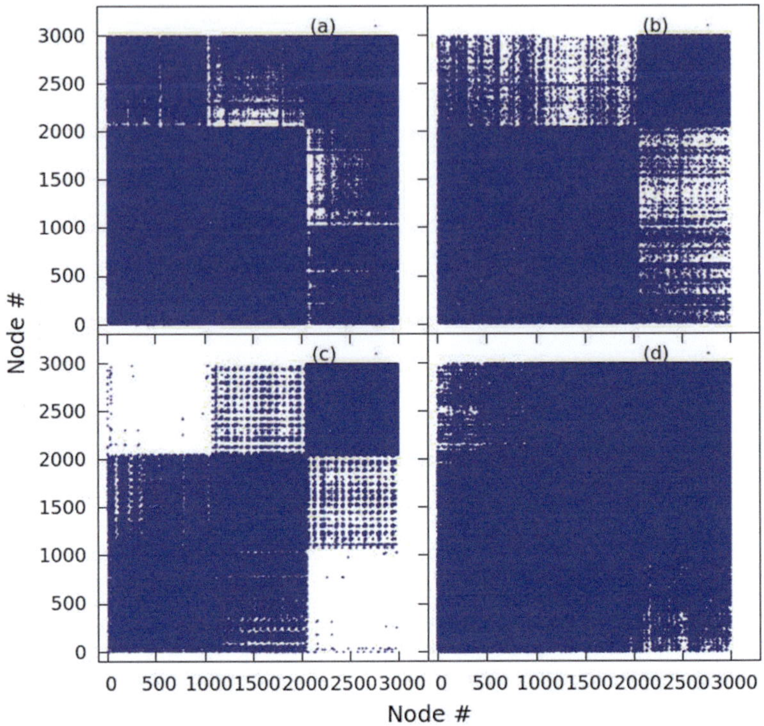

Fig. 2 Recurrence plots corresponding to the four RRc Lyrae stars (**a**) KIC 4064484, (**b**) KIC 5520878, (**c**) KIC 8832417, and (**d**) KIC 9453114. While difficult to distinguish visually, the plots are quantified using recurrence quantification analysis (RQA) in Table 3

of neighbors a node has on an average. The characteristic path length (CPL) is calculated to be the average of shortest paths between every possible node pair.

$$CPL = \frac{1}{N \cdot (N-1)} \cdot \sum_{i \neq j} d(v_i, v_j). \tag{3}$$

Here v_i and v_j are nodes in the network and $d(v_i, v_j)$ is the shortest path between them. The average clustering coefficient is the average of the local clustering coefficients of individual nodes. Let N be the set of all nodes and E be the set of all links. The local clustering for a node i is then given by

$$C_i = \frac{2|\{e_{jk} : v_j, v_k \in N_i, e_{jk} \in E\}|}{(k_i)(k_i - 1)}. \tag{4}$$

Here k_i is the degree of the node v_i and $e_{i,j}$ is the link between two nodes v_i and v_j. This local clustering is a measure of how close a node is to being a complete graph [15]. The values for the four light curves for these properties are listed in Table 3.

Table 3 Table showing the recurrence plot and recurrence network quantifiers for the four RRc Lyrae stars

KID	RR	DET	LAM	$\frac{LAM}{DET}$	d_{avg}	CPL	CC_{avg}
4064484	0.02	0.71	0.07	0.1	92.6	10.22	0.73
5520878	0.02	0.61	0.02	0.04	69.8	15.0	0.75
8832417	0.01	0.60	0.02	0.04	60.5	15.77	0.75
9453114	0.02	0.61	0.08	0.13	104	7.95	0.80

Recurrence rate (*RR*), Determinism (*DET*), Laminarity (*LAM*), Laminarity determinism ratio $(\frac{LAM}{DET})$, Average degree (d_{avg}), Characteristic path length (CPL), and average clustering (CC_{avg}) for the RRc Lyrae stars considered. We see that the two subcategories of the RRc Lyrae stars can be identified using LAM, $\frac{LAM}{DET}$, d_{avg}, and CPL

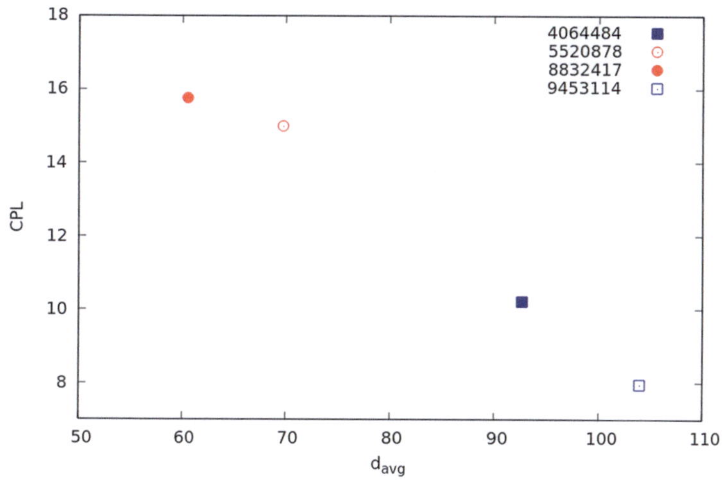

Fig. 3 CPL-d_{avg} plot for the four RRc Lyrae stars considered. We see that the noisy quasiperiodic stars and strange nonchaotic stars fall into distinct regions of the CPL-d_{avg} plane

The strange nonchaotic stars, KIC 4064484 and KIC 9453114, show higher average degree and lower path length while the noisy quasiperiodic stars, KIC 5520878 and KIC 8832417, are characterized by lower average degree and higher path length (see Fig. 3).

4 Results and Summary

We conduct a detailed nonlinear time series analysis of the measured light curves of the four RRc Lyrae stars in the Kepler field of view, using the bicoherence, recurrence plots, and recurrence networks. Earlier work had suggested that RRc Lyrae stars may be classified into two different categories, depending on the underlying dynamical state [12]. This work probes to see whether other nonlinear

Table 4 Astrophysical and nonlinear parameters that show distinct difference for the two categories of RRc Lyrae stars

Kepler ID	P_0	T_{eff}	$[Fe/H]$	B_{avg}	SPF	LAM	$\frac{LAM}{DET}$	d_{avg}	CPL
4064484	**0.34**	**6500**	**−1.58**	**0.64**	**0.77**	**0.07**	**0.01**	**92.6**	**10.22**
5520878	0.27	7250	−0.18	0.40	0.31	0.02	0.04	69.8	15.0
8832417	0.25	7000	−0.27	0.33	0.24	0.02	0.04	60.5	15.77
9453114	**0.37**	**6500**	**−2.13**	**0.55**	**0.70**	**0.08**	**0.13**	**104**	**7.95**

The strange nonchaotic stars are shown in boldface

time series measures also show differences between the two categories of RRc Lyrae stars. The mean of the bicoherence across the plane (B_{avg}) and the significant pair fraction (SPF) both are high for the strange nonchaotic stars, while they are low for the noise contaminated quasiperiodic stars. The recurrence quantification analysis (RQA) using the recurrence plots shows that the laminarity (LAM) and laminarity-determinism ratio $\left(\frac{LAM}{DET}\right)$ are distinctly different for the two groups. Finally, for the corresponding recurrence networks we find that strange nonchaotic stars show higher average degree and lower path length while the noisy quasiperiodic stars show lower average degree and higher path length.

The two categories identified through nonlinear time series analysis correspond to similar properties in the pulsation period, effective temperature, and metallicity [2]. The quasiperiodic stars appear to be more metal rich than the strange nonchaotic stars. Moreover, the former have lower pulsation periods and higher effective temperature than the latter. A comprehensive table of nonlinear and astrophysical properties that differ for the four stars considered is shown in Table 4. These results seem to conclusively establish the existence of two separate subcategories of RRc Lyrae variable stars that differ in nonlinear and astrophysical properties, and provide a number of quantifiers that can be used to identify these subcategories.

5 Conclusions

Using the bicoherence and the properties of the recurrence plots and networks constructed from the measured light curves, we can clearly distinguish between the different categories of RRc Lyrae stars identified previously in [12]. KIC 4064484 and KIC 9453114 were identified as being strange nonchaotic, whereas KIC 5520878 and KIC 8832417 were shown to be exhibiting noise contaminated quasiperiodicity. This paper goes beyond the previous work by explicitly identifying differences in the nonlinear properties and the astrophysical properties of these two classes. The differences that were shown in the nonlinear properties of the RRc Lyrae stars cement the idea that there are indeed two subgroups with distinct physical and dynamical behaviors. Further, this paper also demonstrates that these differences can be established using relatively fewer datapoints using recurrence based analysis. The estimates of the bicoherence were made using datasets that contained between 20,000 and 30,000 datapoints, whereas all the calculations for recurrence plots and networks were done with just 3000 points. This order of magnitude improvement suggests that the two subcategories of RRc Lyrae stars can be distinguished using much smaller datasets. Finally this paper explicitly shows that the astrophysical properties of RRc Lyrae stars in different dynamical states are different. While this may not be entirely surprising, this link between astrophysics and nonlinear dynamics has been largely under-utilized. As newer space missions like the Transiting Exoplanet Survey Satellite (TESS) sends richer data, nonlinear time series analysis must be exploited to explore the links between dynamical behavior and astrophysical properties in greater depth.

Acknowledgements SVG acknowledges with gratitude financial support from the Council of Scientific & Industrial Research(CSIR) and The Infosys Foundation for presentation of this work at NODYCON 2019.

References

1. Moskalik, P., Smolec, R., Kolenberg, K., Molnár, L., Kurtz, D.W., Szabó, R., Benko, J.M., Nemec, J.M., Chadid, M., Guggenberger, E., Ngeow, C.C., Jeon, Y.B., Kopacki, G., Kanbur, S.M.: Mon. Not. R. Astron. Soc. **447**(3), 2348 (2015)
2. Nemec, J.M., Cohen, J.G., Ripepi, V., Derekas, A., Moskalik, P., Sesar, B., Chadid, M., Bruntt, H.: Astrophys. J. **773**(2), 181 (2013)
3. Lindner, J.F., Kohar, V., Kia, B., Hippke, M., Learned, J.G., Ditto, W.L.: Phys. Rev. Lett. **114**(5), 054101 (2015)
4. Borucki, W., Koch, D., Basri, G., Batalha, N., Brown, T., Caldwell, D., Christensen-Dalsgaard, J., Cochran, W., Dunham, E., Gautier, T.N., et al., Proc. Int. Astron. Union **3**(S249), 17 (2007)
5. Totsky, A., Lukin, V, Zelensky, A., Astola, J., Egiazarian, K., Khlopov, G., Morozov, V.Y., Kurbatoc, I., Molchanov, P., Roenko, A., et al.: Tampereen teknillinen yliopisto. TICSP series (2008)
6. Gajraj, R., Doi, M., Mantzaridis, H., Kenny, G.: Br. J. Anaesth. **80**(1), 46 (1998)
7. Fergusson, J., Shellard, E.: Phys. Rev. D **80**(4), 043510 (2009)
8. Pezeshki, C., Elgar, S., Krishna, R.: J. Sound Vib. **137**(3), 357 (1990)
9. Chandran, V., Elgar, S., Pezeshki, C.: Int. J. Bifurcation Chaos **3**(03), 551 (1993)
10. Marwan, N., Romano, M.C., Thiel, M., Kurths, J.: Phys. Rep. **438**(5–6), 237 (2007)
11. Donner, R.V., Zou, Y., Donges, J.F., Marwan, N., Kurths, J.: New J. Phys. **12**(3), 033025 (2010)
12. George, S.V., Ambika, G., Misra, R.: Nonlinear Dyn. **89**(1), 465 (2017)
13. Hegger, R., Kantz, H., Schreiber, T.: Chaos: Interdiscip. J. Nonlinear Sci. **9**(2), 413 (1999)
14. Jacob, R., Harikrishnan, K., Misra, R., Ambika, G.: Phys. Rev. E **93**(1), 012202 (2016)
15. Newman, M.: Networks. Oxford University Press, Oxford (2018)

Part IV
Chaos in Electronic Systems

Multijump Resonance with Chua's Circuit

Maide Bucolo, Arturo Buscarino, Carlo Famoso, Luigi Fortuna, and Mattia Frasca

Abstract In this paper the existence of multijump resonance is described for the driven Chua circuit. Multijump resonance is a phenomenon observable in driven nonlinear systems which leads to a multi-valued frequency response, showing several hysteresis windows with respect to the increase/decrease of the driving frequency. The phenomenon is here investigated in order to characterize its onset as a function of system parameter values. Finally, a physical implementation of the driven Chua circuit, in which jump and multijump resonance occur, is presented and discussed allowing to verify the robustness of such nonlinear phenomenon.

Keywords Jump resonance · Chua's circuit · Nonlinear dynamics

1 Introduction

The phenomenon of jump resonance is an appealing topic in the field of nonlinear oscillators. It identifies a hysteretic behavior which occurs in the frequency response of forced nonlinear systems leading to sudden jumps in the amplitude of the output signal [1]. The jump resonance has been widely investigated in mechanical systems [2] and in control applications. In particular, the phenomenon is classically considered as negative since its occurrence is detrimental for the performance of the control actions. Therefore techniques to mitigate jump resonance have been introduced. From a different perspective, looking at jump resonance from a system

---

M. Bucolo · C. Famoso
Dipartimento di Ingegneria Elettrica Elettronica e Informatica, University of Catania, Catania, Italy

A. Buscarino (✉) · L. Fortuna · M. Frasca
Dipartimento di Ingegneria Elettrica Elettronica e Informatica, University of Catania, Catania, Italy

IASI, Consiglio Nazionale delle Ricerche (CNR), Roma, Italy
e-mail: arturo.buscarino@unict.it

W. Lacarbonara et al. (eds.), *New Trends in Nonlinear Dynamics*,
https://doi.org/10.1007/978-3-030-34724-6_26

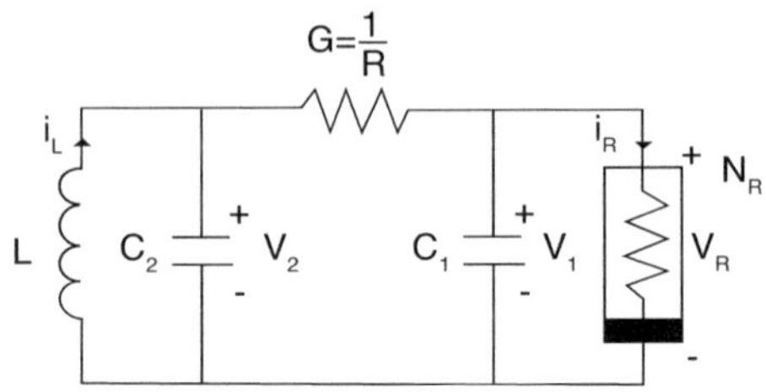

Fig. 1 The Chua circuit

theory point, it may be represented as a "memory" effect, due to presence of the hysteresis. Under this perspective it is necessary a strategy to intentionally design electronic circuits with jump resonance [3]. Furthermore, with the term "multijump" resonance, we indicate a frequency response where more frequency hysteresis windows are present, even in different ranges or within the same range leading to multiple jump paths [4]. In this paper we show that the Chua's circuit allows the realization of electronic circuits with this peculiarity.

2 The Chua Circuit

The Chua's circuit [5] is represented in Fig. 1. Let x indicate the voltage v_1 across C_1, y the voltage v_2 across C_2, z the current i_L in the inductor, and $\psi(x)$ the Chua diode nonlinear characteristic. The classical normalized equations of the Chua circuit are

$$\begin{aligned} \dot{x} &= \alpha\,(y - \psi(x)), \\ \dot{y} &= x - y + z, \\ \dot{z} &= -\beta y, \end{aligned} \tag{1}$$

where α and β are system parameters, and $\psi(x)$ may assume either a piece-wise linear or continuous form. We assume a forcing signal $r(t) = A \sin \omega t$ acting on the first equation of the system [6].

3 Describing Function Approach to Determine the Frequency Response

The system in Eq. (1) can be rewritten according to the closed-loop scheme shown in Fig. 2, better known as Lur'e form. In this representation, $F(s)$ is the transfer function of the linear part of the Chua's circuit, which can be calculated as:

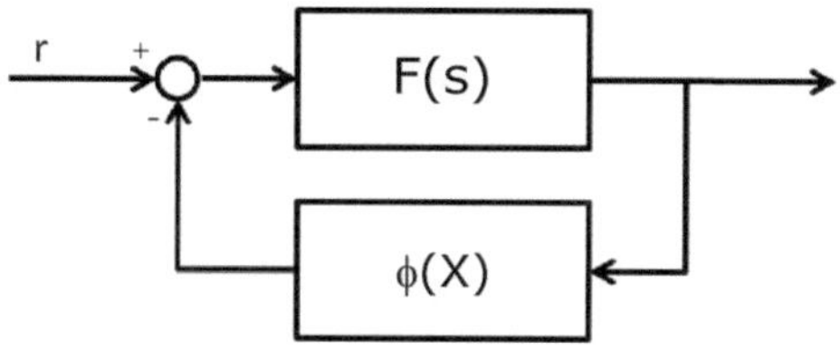

Fig. 2 Lur'e representation of the Chua circuit dynamics

$$F(s) = \frac{\alpha\left(s^2 + s + \beta\right)}{s^3 + (1+\alpha)\,s^2 + \beta s + \alpha\beta} \tag{2}$$

and $\phi(X)$ represents an approximation of the nonlinearity $\psi(x)$ in the sinusoidal domain, i.e., the so-called describing function [1]. The describing function approach assumes that each dynamical variable of the system can be approximated as a finite sum of sinusoidal terms. Under this assumption, and the further condition of the presence of filtering effect in the loop, the nonlinear elements of the system are then represented by blocks that are functions of the input amplitude X and, in case, its frequency ω, as their action on each sinusoidal term were linear. The describing function approach, thus, introduces a quasi-linear approximation of the system under the assumption of the periodicity of the involved signals. Assuming $\psi(x) = ax^5 + bx^3$, being an odd symmetric polynomial, we can use the following describing function:

$$\phi(X) = AX^4 + BX^2, \tag{3}$$

where $A = \frac{5}{8}a$ and $B = \frac{3}{4}b$ [7].

4 Jump Resonance in Chua's Circuit

Taking into consideration the scheme of Fig. 2, the input signal $r(t) = \bar{r}\sin(\omega t)$ is a solution for closed-loop system of Fig. 2 if it satisfies:

$$\left[\phi(X) + F^{-1}(j\omega)\right] X = \bar{r}e^{-j\psi}. \tag{4}$$

Assuming $\psi = 0$, and defining R and I as the real and imaginary part of $F^{-1}(j\omega)$, calculating the modulus we have

$$[X\phi(X) + RX]^2 + I^2 = r^2. \tag{5}$$

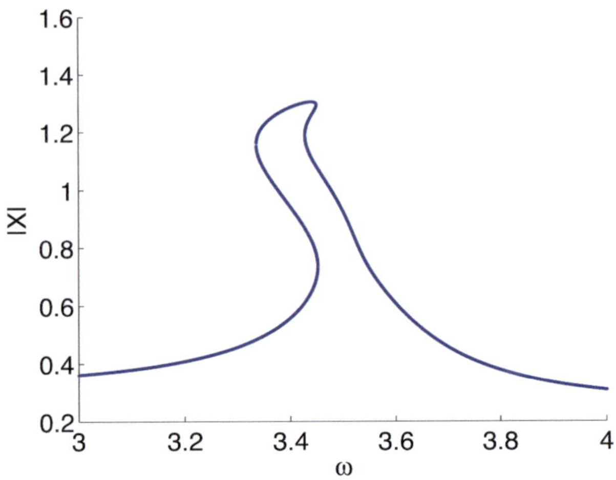

Fig. 3 Multijump resonance in Chua's circuit: frequency response for $\alpha = 6.881$, $\beta = 14$, $A = 0.1$, $B = -0.378$

A polynomial of tenth order with only even coefficients in X is thus obtained. A jump resonance window is obtained searching for values of α, β, A, B for which the polynomial in Eq. (5) assumes five positive real solution for a given ω. As an example, setting $\alpha = 6.881$, $\beta = 14$, $A = 0.1$, $B = -0.378$ the frequency response shown in Fig. 3 is obtained assuring us that in the range of $3.3 < \omega < 3.45$ a multijump resonance window is achieved.

In order to establish the region of $\alpha - \beta$ parameter space in which jump or multijump resonance occurs, we performed extensive numerical simulations, investigating the behavior of the circuit varying the parameters α and β. In Fig. 4 we report the number of positive real solutions retrieved for Eq. (5) with respect to different values of α and β. The area of the parameter space in which three solutions are retrieved, depicted in green, corresponds to the region of single-jump resonance, i.e., a single hysteresis window can be observed and two jump occurs in the frequency response. The red dark area in Fig. 4 indicates the occurrence of multijump resonance, i.e., more nested hysteresis windows within the same range of frequency. It appears clearly that the jump resonance can be retrieved in a wide area of the parameter space.

In the example reported in Fig. 3, the multijump range is narrow. Therefore, the width of the frequency range in which multiple solutions can be retrieved is a further numerical indication of the robustness of the phenomenon. In order to investigate such aspect, we define the parameters $\Delta\omega_3$ and $\Delta\omega_5$ as the width of the range corresponding to three and five positive real solutions, respectively. The two parameters are reported in the maps shown in Figs. 5 and 6, color-coded according to the reported color bar.

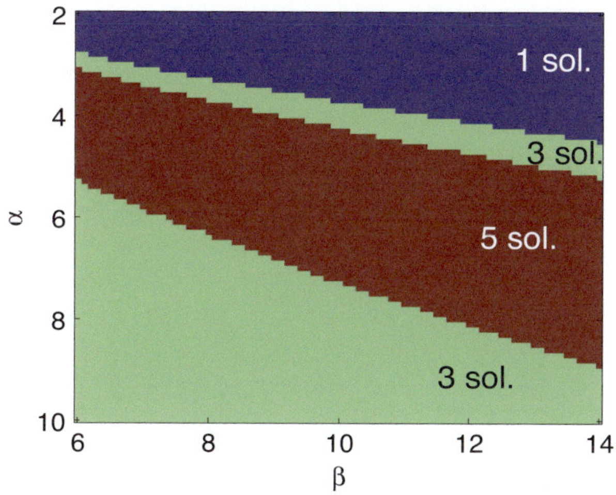

Fig. 4 Multijump resonance in Chua's circuit: number of multiple solutions for Eq. (5) in the parameter space $\alpha - \beta$. Other parameters as: $A = 0.1$, $B = -0.378$

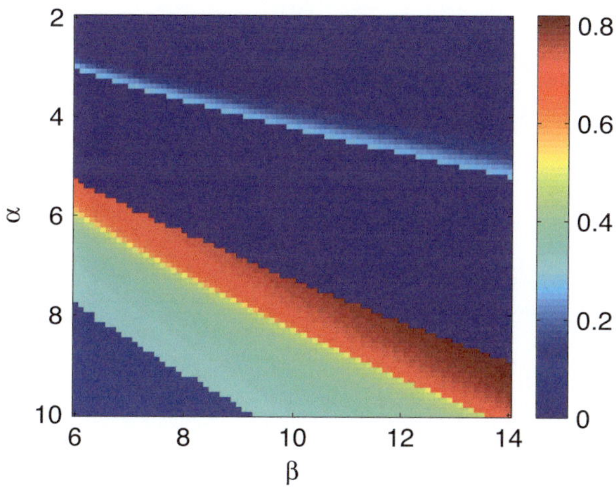

Fig. 5 Width of the jump resonance (three solutions) frequency range in the parameter space $\alpha - \beta$. Other parameters as: $A = 0.1$, $B = -0.378$

The trend of hysteresis width are also nontrivial with respect to α and β. To account for this, we report the trend of $\Delta\omega_3$ and $\Delta\omega_5$ in two cross sections of Fig. 5 and 6. In Fig. 7a, the dependence on α, fixing $\beta = 10$, is reported, while in Fig. 7b, the dependence on β, when $\alpha = 6$, is shown. In both cases, we can observe non-monotone behavior of $\Delta\omega$.

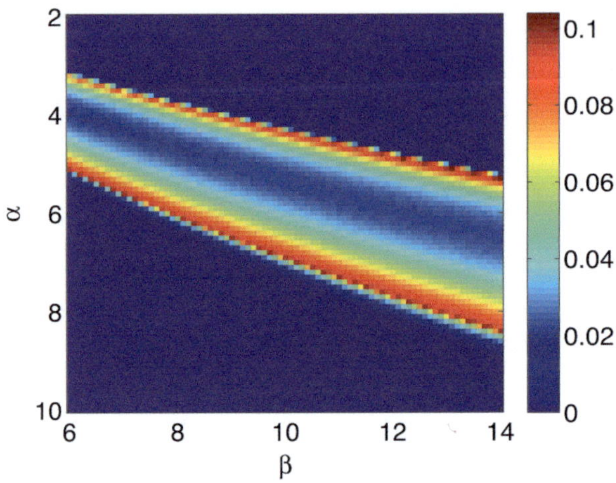

Fig. 6 Width of the multijump resonance (five solutions) frequency range in the parameter space $\alpha - \beta$. Other parameters as: $A = 0.1$, $B = -0.378$

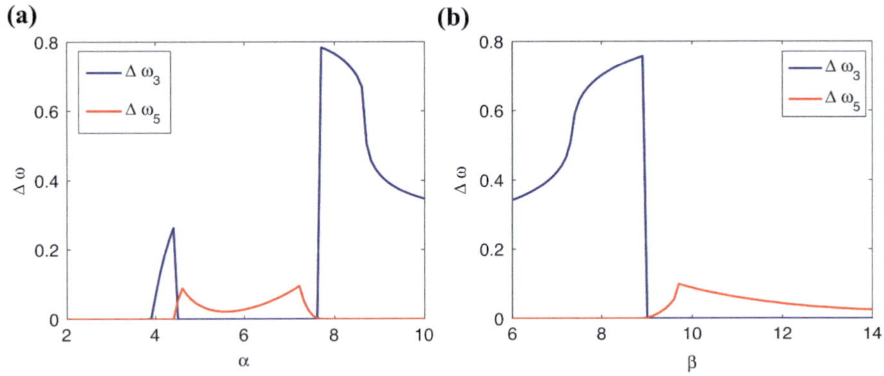

Fig. 7 (**a**) Width of the jump resonance (three solutions) frequency range in the parameter space $\alpha - \beta$; (**b**) Width of the multijump resonance (five solutions) frequency range in the parameter space $\alpha - \beta$. Other parameters as: $A = 0.1$, $B = -0.378$

5 Experimental Analysis

The investigation described in the previous section allows to assess that jump and multijump resonance can be retrieved in the Chua circuit in a wide region of the parameter space. This robustness to parametric mismatched encouraged us to retrieve jump and multijump resonance in the physical implementation of Chua's circuit.

Two main points have to be faced when dealing with the experimental observation of nonlinear resonance phenomena. First of all, the frequency range in which jump resonance is retrieved is scaled according to the temporal rescaling introduced by integrators implemented to realize the dynamical equations of the circuit. Therefore, choosing the time-constant of integrators, frequency range is suitably rescaled. Moreover, the analysis carried out is based on the harmonic balance method which approximates the solution of the closed-loop system neglecting higher frequency. Although the linear part of the Chua circuit has a low-pass effect, the approximation may fail for some values of the amplitude and the frequency of the driving signal. Despite this, jump and multijump resonance can be still observed in the physical implementation according to the prediction based on the harmonic balance.

We adopted the State Controlled Cellular Nonlinear Network (SC-CNN) implementation of the Chua circuit [5], so that temporal rescaling can be obtained fixing the capacitors value. In Fig. 8, the schematic of the realized system is reported, component values can be found in the figure caption. The two potentiometers allow to change the parameter α and β values. As concerns the nonlinearity $\psi(x)$, it has been implemented by using three analog multipliers, as reported in the schematic of Fig. 9.

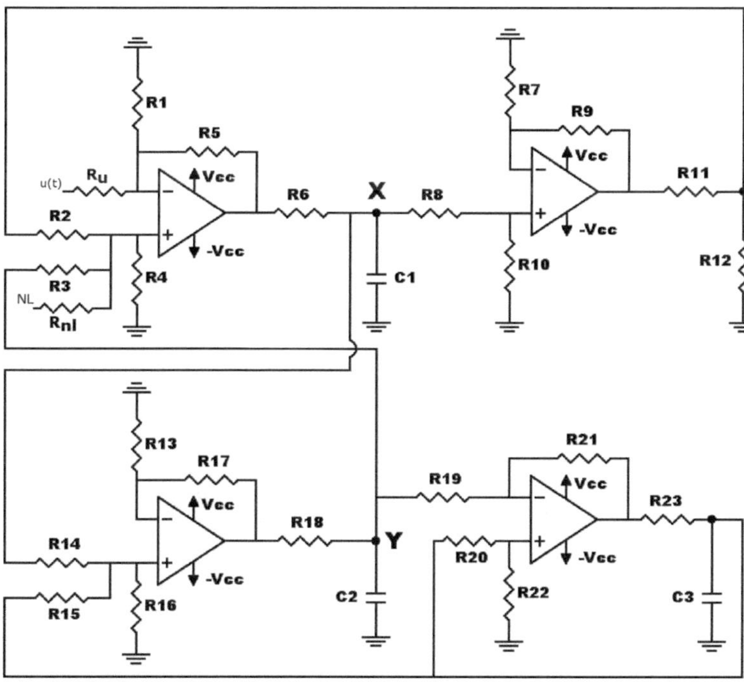

Fig. 8 SC-CNN based circuit implementation of the Chua circuit. Components: $R_1 = 4\,\text{k}\Omega$, $R_2 = 13.3\,\text{k}\Omega$, $R_3 = 5.6\,\text{k}\Omega$, $R_4 = 20\,\text{k}\Omega$, $R_5 = 20\,\text{k}\Omega$, $R_nl = 20\,\text{k}\Omega$, $R_u = 20\,\text{k}\Omega$, $R_6 = 380\,\Omega$ (potentiometer), $R_7 = 112\,\text{k}\Omega$, $R_8 = 112\,\text{k}\Omega$, $R_9 = 1\,\text{M}\Omega$, $R_{10} = 1\,\text{M}\Omega$, $R_{11} = 8.2\,\text{k}\Omega$, $R_{12} = 1\,\text{k}\Omega$, $R_{13} = 51.1\,\text{k}\Omega$, $R_{14} = 100\,\text{k}\Omega$, $R_{15} = 100\,\text{k}\Omega$, $R_{16} = 100\,\text{k}\Omega$, $R_{17} = 100\,\text{k}\Omega$, $R_{18} = 1\,\text{k}\Omega$, $R_{19} = 8.2\,\text{k}\Omega$, $R_{20} = 100\,\text{k}\Omega$, $R_{21} = 100\,\text{k}\Omega$, $R_{22} = 7.8\,\text{k}\Omega$, $R_{23} = 1\,\text{k}\Omega$, $C_1 = C_2 = C_3 = 100\,\text{nF}$, $V_{cc} = 9\,\text{V}$

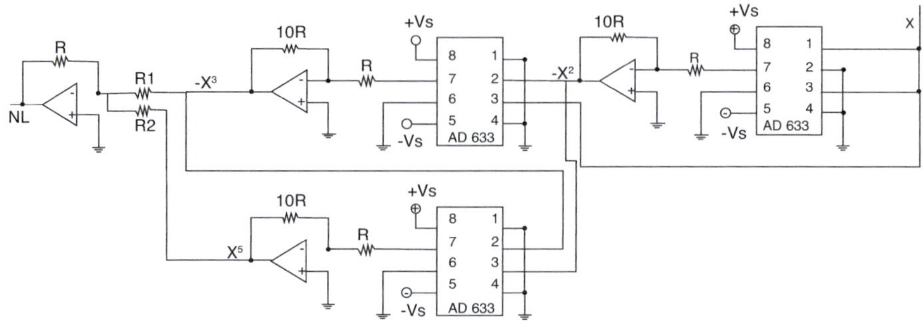

Fig. 9 Implementation of the nonlinearity $\psi(x)$. Components: $R = 10\,\text{k}\Omega$, $R_1 = 20\,\text{k}\Omega$, $R_2 = 62.5\,\text{k}\Omega$, AD633 analog multipliers, $V_{cc} = 9\,\text{V}$

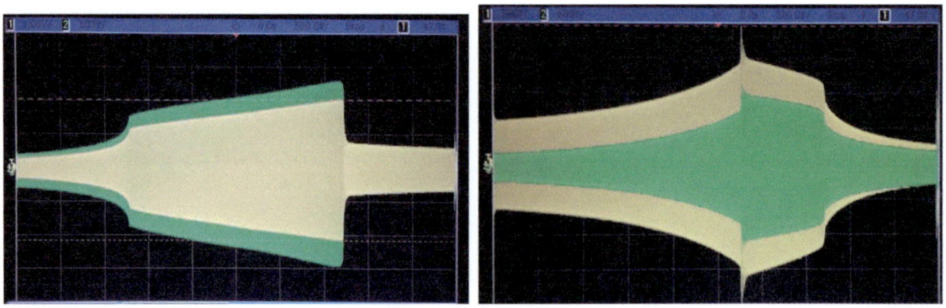

(a) (b)

Fig. 10 Experimental observation of multijump resonance in the driven Chua circuit: state variables x and y when driving the system with a sinusoidal signal $u(t) = R\sin(2\pi f(t)t)$ with $R = 200\,\text{mV}$ and frequency f sweeping upwards (**a**), or downwards (**b**), between 700 Hz and 1.3 kHz in a time window of $T = 5\,\text{s}$. Oscilloscope scales: (**a**) 500 ms/div (X-axis), 500 mV/div (Y-axis), (**a**) 500 ms/div (X-axis), 500 mV/div (Y-axis, x variable), 1 V/div (Y-axis, y variable)

The circuit is now driven by a sinusoidal signal $u(t) = R\sin(2\pi f(t)t)$ whose amplitude $R = 200\,\text{mV}$ is kept constant, while its frequency $f(t)$ varies in time performing a sweep upwards and downwards between 700 Hz and 1.3 kHz. The state variables x and y are shown in Fig. 10a, b, reporting the response when the frequency sweeps upwards and downwards, respectively. It is possible to retrieve that the amplitude of the output signal undergoes sudden jumps, either up or down, indicating the existence of jump resonance. Moreover, the presence of multiple jumps in Fig. 10b indicates the existence of a multijump resonance.

In order to further assess the validity of the experimental results, we report here the numerical simulation of the Chua circuit as in Fig. 2 when driven by a sinusoidal input signal $u(t) = R\sin(2\pi f(t)t)$ with $R = 0.2$ and $f(t)$ varies in time performing a sweep upwards and downwards. The output of the system, namely state variable x of the Chua circuit, is reported in Fig. 11. The evidence of multijump resonance behavior are clearly confirmed, moreover the experimental realization is able to reproduce the peculiar features of the driven Chua circuit.

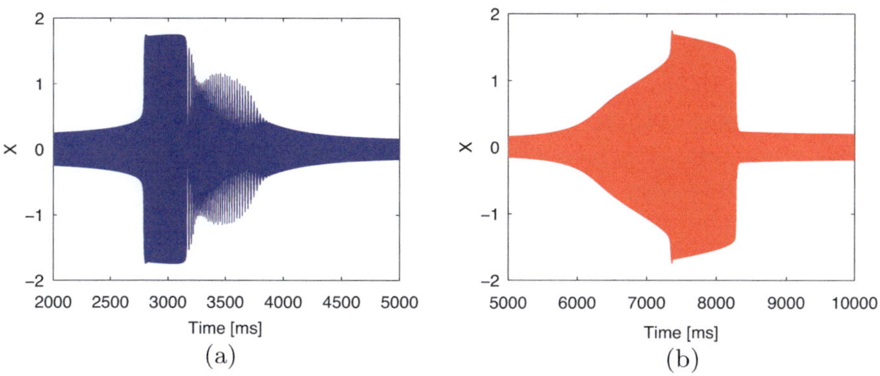

Fig. 11 Numerical simulation of the driven Chua circuit with $\alpha = 6.881$, $\beta = 14$, $A = 0.1$, $B = -0.378$: state variable x when driving the system with a sinusoidal signal $u(t) = R\sin(2\pi f(t)t)$ with $R = 0.2$ and frequency f sweeping upwards (**a**), or downwards (**b**), between 0.3 and 1 Hz in a time window of $T = 10$ s

6 Conclusions

In this paper we presented the first evidence of multijump resonance in the driven Chua circuit. Furthermore, we discuss the robustness of the observed phenomenon proposing a circuit implementation able to show a multijump resonance behavior.

A complete characterization of the occurrence of the phenomenon allows to determine the existence of family of curves with the same peculiarity as that shown in Fig. 3 in a wide range on the parameter space. The existence of jump and multijump resonance in the Chua's circuit make this simple device suitable for the implementation of analog multi-state memory. In fact, providing to the circuit input signals with specific sweeping frequencies, it is possible to drive the system towards different states. This novel scenario makes further impressive the capabilities of the Chua circuit.

References

1. Atherton, D.P.: Nonlinear Control Engineering. Van Nostrand Reinhold, Wokingham (1982)
2. Cook, P.A.: Nonlinear Dynamical Systems. Prentice Hall, UK (1994)
3. Buscarino, A., Caponetto, R., Famoso, C., Fortuna, L.: Jump resonance in fractional order circuits. Int. J. Bifurcation Chaos **28**(1), 1850016 (2018)
4. Buscarino, A., Famoso, C., Fortuna, L., Frasca, M.: Multi-jump resonance systems. Int. J. Control 1–11 (2018)
5. Fortuna, L., Frasca, M., Xibilia, M.G.: Chua's Circuit Implementations: Yesterday, Today and Tomorrow. World Scientific, Singapore (2009)
6. Buscarino, A., Fortuna, L., Frasca, M.: Jump resonance in driven Chua's circuit. Int. J. Bifurcation Chaos **19**(8), 2557–2561 (2009)
7. Vander Velde, W.E.: Multiple-Input Describing Functions and Nonlinear System Design. McGraw-Hill, New York (1968)

Experimental Observation of Robust Chaos in a 3D Electronic Circuit

Soumyajit Seth

Abstract Robust chaos occurring in piecewise smooth dynamical systems is very important in practical applications. It is defined by the absence of periodic windows and coexisting attractors in some neighborhood of the parameter space. In earlier works, the conditions of stability and region of parameter space of robust chaos were reported in the context of piecewise linear 1D and 2D maps, and have been demonstrated in 1D and 2D switching circuits. In this brief, the first experimental observation of this phenomenon in a 3D electronic switching system is reported and the region of its parameter space is determined experimentally.

Keywords Border collision bifurcation · Piecewise smooth dynamical systems · Robust chaos · Electronic switching systems

1 Introduction

Chaos has many practical applications nowadays in different physical and engineering systems [1]. Therefore, it is very important to use a system which generates chaos. But due to the existence of coexisting periodic attractors, the system toggles between the attractors [2], and due to the periodic windows, any slight fluctuation of the parameter values may take the system out of chaos [3]. So, a robust chaos generator system is very much required.

A chaotic attractor is *robust* if, for its parameter values, there exists a neighborhood in the parameter space with no periodic attractor as well as coexisting attractors and the chaotic attractor is unique in that neighborhood [4]. Robust chaos generally occurs in piecewise smooth dynamical systems [5] where the so-called border collision bifurcation occurs.

S. Seth (✉)
Indian Institute of Science Education and Research Kolkata, Mohanpur, India
e-mail: ss14rs057@iiserkol.ac.in

W. Lacarbonara et al. (eds.), *New Trends in Nonlinear Dynamics*,
https://doi.org/10.1007/978-3-030-34724-6_27

The possibilities of occurrence of this phenomenon have been explored theoretically in case of 1D and 2D piecewise linear maps and have been shown experimentally in case of 1D and 2D switching electronic circuits. Gardini et al. [6] have shown the conditions of stability and region of parameter space of this phenomenon theoretically in a 1D piecewise smooth map. Mandal et al. [7] have shown the occurrence of robust chaos in a 1D switching circuit. Banerjee et al. [4] have obtained the conditions of stability and the region of occurrences of robust chaos in a 2D piecewise linear map and have demonstrated it experimentally in boost and buck converters [8]. One needs to check whether robust chaos can occur in higher dimensional systems. In this brief, it is shown that with the proper arrangements of the circuit parameter values, this phenomenon does occur in a 3D switching system.

2 System Description

Figure 1a is the electronic circuit which is used to check the occurrence of the robust chaos in a 3D electronic system. Here, the current i, voltages V_1 and V_2 across the capacitors C_1 and C_2, respectively, are the three state variables. The clock pulse (CLK) with small ON period is applied at the RESET input of the S–R latch. When the system starts, the current i is very low. So, the voltage drop across R_L is $V_{R_L} = iR_L < V_{ref}$. This makes the comparator output zero and at this time, the ON period of clock pulse just arrives, the Q output of the S–R latch is zero and $\bar{Q}$ is high. This makes the analog switches S_1, S_3 close and S_2 opens. At this time, as shown in Fig. 1b, the current in the L increases and energy is stored in it. When the current i reaches $i_{ref} = \frac{V_{ref}}{R_L}$, the comparator output changes the state; therefore, the outputs of the latch reverse, and the switches S_1, S_3 become OFF and S_2 becomes ON (in Fig. 1c). So, the stored energy in the inductor drops and also current decreases unless the next clock pulse arrives. After that, the current in the inductor rises again and the system repeats its operation.

Here, V_{in} and R_2 are used as bifurcation parameters in order to observe robust chaos in this system.

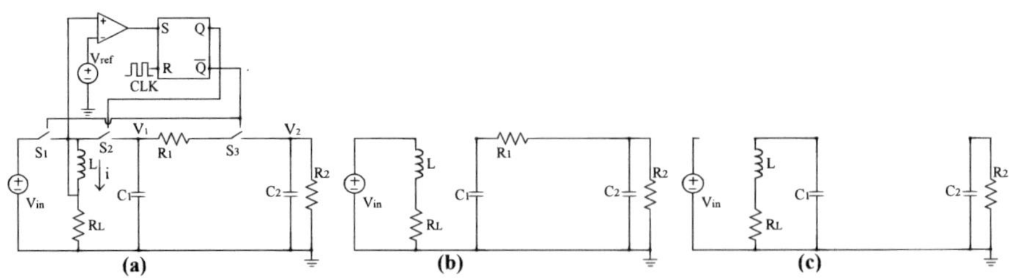

Fig. 1 (**a**) The switching circuit under consideration. (**b**) The subsystem for $iR_L < V_{ref}$, (**c**) The subsystem for $iR_L = V_{ref}$

3 Mathematical Model

Assuming all the components to be ideal, the system can be described by ODEs, given by

$$\left.\begin{aligned}\frac{di}{dt} &= \frac{1}{L}(V_{\text{in}} - iR_{\text{L}})\\ \frac{dV_1}{dt} &= \frac{V_2 - V_1}{C_1 R_1}\\ \frac{dV_2}{dt} &= \frac{V_1}{C_2 R_1} - \frac{V_2}{C_2}\left(\frac{1}{R_1} + \frac{1}{R_2}\right)\end{aligned}\right\} \quad \text{if } i < i_{\text{ref}} \tag{1}$$

$$\left.\begin{aligned}\frac{di}{dt} &= \frac{1}{L}(V_1 - iR_{\text{L}})\\ \frac{dV_1}{dt} &= -\frac{i}{C1}\\ \frac{dV_2}{dt} &= -\frac{V_2}{C_2 R_2}\end{aligned}\right\} \quad \text{otherwise} \tag{2}$$

Note that, in this system, when the switches S_1 and S_3 are ON and S_2 is OFF (Fig.1b), the system obeys a set of differential equations which is denoted as Eq. (1) and for the other condition, the system obeys Eq. (2).

4 Simulation Results

The bifurcation diagrams for the different parameter values of the above system (Fig. 1a) can be observed in "switch-on sampling" where the peaks of the output waveform of one state variable are identified and depending upon the peaks, the periodicity of the system can be determined.

4.1 V_{in} as Bifurcation Parameter

The bifurcation diagram of the circuit (Fig. 1a) with input voltage as parameter is shown in Fig. 2a. Input voltage is varied from 7.0 V to 20.0 V with a step of 0.04 V with other parameters fixed (as shown in Fig. 2b). The circuit shows bifurcation from period 1 to chaos as the input voltage is decreased from 20.0 V. The first bifurcation takes place around 19.0 V, where period 1 bifurcates to period 2 through normal bifurcation. Period 2 orbit goes to period 2 through a border collision bifurcation around 16.8 V. Then period 2 orbit again bifurcates to 4-piece chaotic orbit around 16.4 V and the system gradually goes to chaos. The dense set of points in the bifurcation diagram signifies chaos in the system.

Time series waveform for the chaotic attractor at $V_{\text{in}} = 8.0$ V is shown in Fig. 2b. The frequency spectrum at input voltage 8.0 V clearly shows that the orbit is chaotic. From the bifurcation diagram it can be said that chaos is robust because the chaotic attractor will exist up to 7.0 V without any co-existing attractor (as from Fig. 2a).

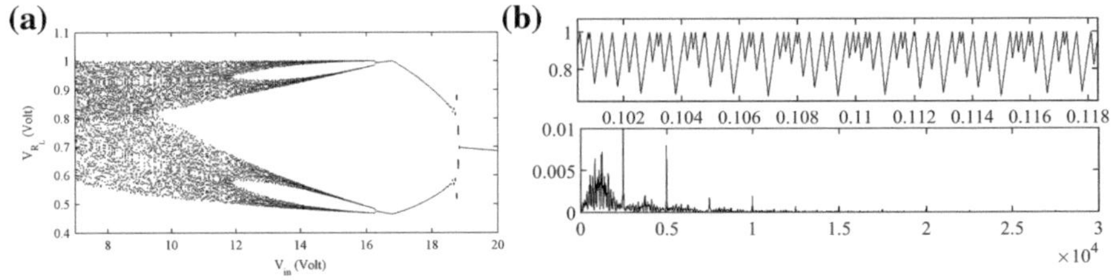

Fig. 2 (**a**) Bifurcation diagram for the system with input voltage as parameter, and (**b**) Time series waveform of switch-on sampling from simulation: the parameters are $V_{\text{ref}} = 1$ V, $L = 100$ mH, $C_1 = 100\,\mu$F, $C_2 = 220\,\mu$F, $R_1 = R_L = 50\,\Omega$, $R_2 = 125\,\Omega$, and clock frequency $f = 5000$ Hz. Chaotic time series occurred at $V_{\text{in}} = 8.0$ V. The upper trace is the voltage across R_L in volt, where x-axis is the time in second and y-axis is the voltage in volt, and the lower trace is the frequency spectrum where x-axis is the frequency in Hz and y-axis is the no of counts, i.e., distribution of counts of the frequencies and plot them in frequency axes

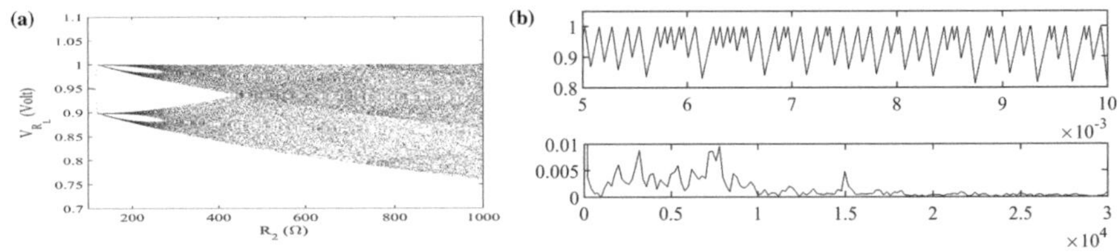

Fig. 3 (**a**) Bifurcation diagram for the system with resistance R_2 as parameter, and (**b**). Time series waveform from simulation; x-axis is the time in second and y-axis is the voltage in volt. The parameters are $V_{\text{ref}} = 1$ V, $V_{\text{in}} = 4.0$ V, $L = 100$ mH, $C_1 = 100\,\mu$F, $C_2 = 220\,\mu$F, $R_1 = R_L = 50\,\Omega$, and clock frequency $f = 5000$ Hz, $R_2 = 1000\,\Omega$. The upper trace is the voltage across R_L in volt, and the lower trace is the frequency spectrum where x-axis is the frequency in Hz and y-axis is the no of counts

4.2 R_2 as Bifurcation Parameter

The bifurcation diagram of the above system (Fig. 1a) with load resistance R_2 as parameter are shown in the Fig. 3a. The load resistance is to be varied from 100 Ω to 1000 Ω with a step of 3 Ω with other parameters fixed which is shown in Fig. 3b. The dense set of points in the bifurcation diagram confirms that only the chaotic orbit exists in the whole parameter range. In that sense, we can say that chaotic orbit is robust after R_2 is 145 Ω. Figure 3b shows the time series as well as the frequency spectra of the chaotic attractor at $R_2 = 1000\,\Omega$.

4.3 Conditions of Occurrence of Robust Chaos

The time period of the clock pulse needs to be very less than the characteristic times of the two capacitors C_1 and C_2 in order to achieve the robust chaos phenomenon of the inductor current for a particular parameter range. Furthermore, the characteristic time of C_1 should be less than the characteristic time of C_2.

5 Experimental Results

In order to validate the aforementioned results, an experimental setup was fabricated with approximately the same parameters (this cannot be done exactly to specification). Figure 4 gives the circuit implementation. Here, we have used a 555-timer as an "astable multivibrator" to generate a square wave pulse. R_A, R_B, and C have been chosen with proper values (as mentioned in the caption of Fig. 4) to generate a frequency 5000 Hz. The output of the timer is passed through a comparator comparing with a DC signal V_{ref1}. The output of the comparator is the clock pulse of small ON period (10% of the total waveform). There is also another comparator which is connected with the inductor which compares voltage (V_{R_L}) across R_L and V_{ref}. When $V_{ref} = V_{R_L}$, the comparator goes into the ON state. HEF4013BP and HEF4016BP are used as S–R latch and an analog switch, respectively. The voltage across the R_L is taken as V_{R_L}.

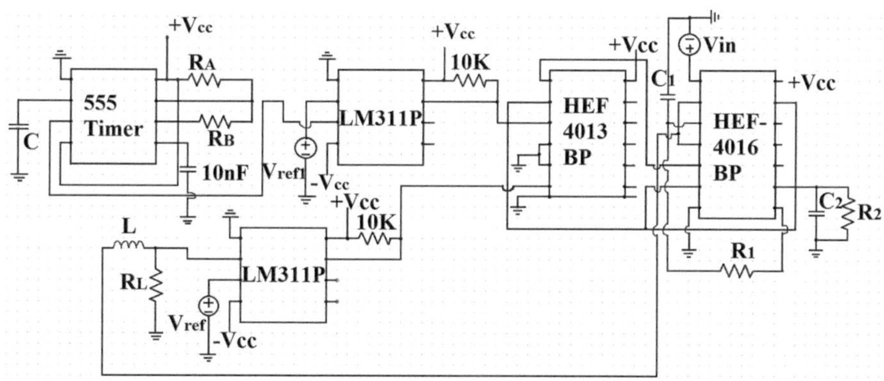

Fig. 4 Circuit implementation of the experimental system. LM311P is comparator and HEF4013BP is S–R Flip Flop, HEF4016BP is an analog switch. The parameters are: $R_A = 23\,580\,\Omega$, $R_B = 2620\,\Omega$, $C = 9.75\,\text{nF}$, $C_1 = 97.8\,\mu\text{F}$, $C_2 = 218.6\,\mu\text{F}$, $R_1 = R_L = 50\,\Omega$, $V_{ref} = 1.0\,\text{V}$, $V_{ref1} = 5.0\,\text{V}$. (1) For the first bifurcation, $R_2 = 125\,\Omega$, V_{in} is to be varied from 6.0 V to 20.0 V, (2) For the second bifurcation, $V_{in} = 4.0\,\text{V}$ and R_2 is to be varied from $100\,\Omega$ to $1000\,\Omega$

5.1 V_{in} as Bifurcation Parameter

The experimentally obtained bifurcation diagram is shown in the Fig. 5a when the input voltage is used as the bifurcation parameter. One can notice that there is only higher density of dots corresponding to a chaotic orbit. The experimentally obtained waveform and its corresponding FFT presented in Fig. 5b show that the chaotic orbit exists at $V_{in} = 8.0$ V.

5.2 R_2 as Bifurcation Parameter

As the bifurcation parameter here is the load resistance R_2, we cannot draw the bifurcation diagram directly in experiment. But we can verify the bifurcation diagram which has been obtained numerically as shown in Fig. 3 by analyzing the time series in the parameter ranges.

The experimentally obtained time series waveforms and their corresponding FFT presented in Fig. 6 show that for $R_2 = 100\,\Omega$, there is period 1 orbit (Fig. 6a). As the load resistance R_2 is increased further, the border collision bifurcation occurs and a four-piece chaotic orbit (Fig. 6b) emerges at R_2 is $200\,\Omega$. At $R_2 = 1000\,\Omega$, the chaotic orbit is generated. The FFT confirms that the orbit is chaotic.

(a) **(b)**

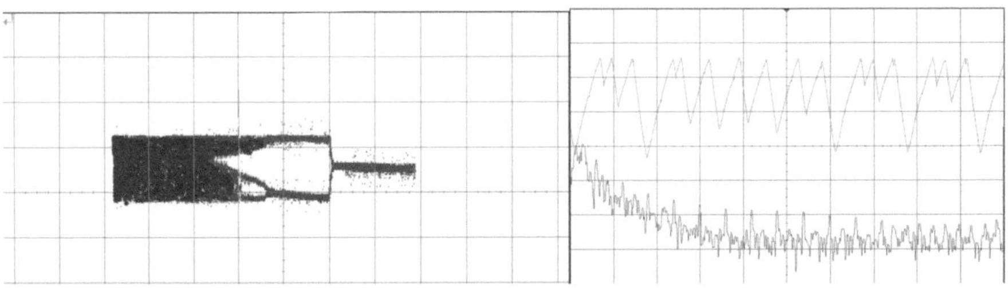

Fig. 5 (**a**) Experimentally obtained bifurcation diagram of the system with input voltage V_{in} as parameter: x-axis is the input voltage in volt and y-axis is the voltage across R_L in volt. V_{in} is varied from 7 V to 20 V. Along the x-coordinate, the grid division is 500 mV and along the y-coordinate it is 200 mV, and (**b**) Upper trace is the time series waveforms: x-axis is time in second and y-axis is the voltage in volt and lower trace is the Fast Fourier Transform. Chaotic orbit at $V_{in} = 8.0$ V. Green (grid division 200 mV)—voltage across R_L in volt and purple—fast Fourier transform obtained from digital oscilloscope. Time division 200 μs

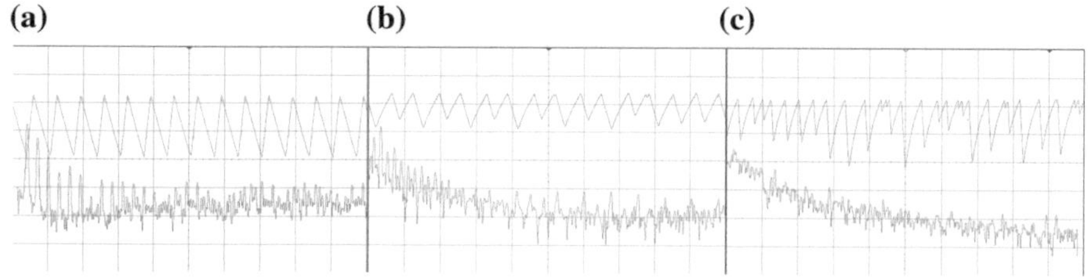

Fig. 6 Upper trace is the time series waveforms: x-axis is time in second and y-axis is the voltage in volt and lower trace is the Fast Fourier Transform. (**a**). Period 1 orbit at $R_2 = 100\,\Omega$, (**b**) Four piece chaotic orbit at $R_2 = 200\,\Omega$, and (**c**) Chaotic orbit at $R_2 = 1000\,\Omega$. Green (grid division 100 mV)—voltage across R_L in volt and purple—fast Fourier transform obtained from digital oscilloscope. Time division for (**a**) and (**b**) are 100 μs, and for (**c**), it is 500 μs.

6 Conclusions

The occurrence of robust chaos was so far predicted theoretically and observed experimentally in case of 1D and 2D piecewise smooth systems. It was predicted using piecewise linear maps and the conditions of stability and region of parameter space have been demonstrated in case of the 1D and 2D electronic switching circuits. The condition of occurrences and the region of parameter spaces of this phenomenon have not yet been investigated in case of a 3D system. In this paper, an electronic switching system has been constructed whose discrete time gives rise to a 3D piecewise smooth map which can be approximated as a 3D piecewise linear map. This system exhibits “robust” chaos for a particular parameter setting after bifurcation. This paper offers the first experimental observation of this phenomenon in case of a 3D system.

In order to understand this phenomenon in a 3D piecewise linear map, it is important to know how it occurs in a physical system. The system described in this paper will be helpful in understanding the occurrence of the robust chaos from a theoretical point of view. Also, the effect of smoothness of robust chaos for this system can be shown later.

References

1. Hayes, S., Grebogi, C., Ott, E.: Communicating with chaos. Phys. Rev. Lett. **70**(20), 3031 (1993)
2. Seth, S., Banerjee, S.: Experimental observation of multiple attractor bifurcation in an electronic circuit. IEEE Trans. Circuits Syst. Express Briefs **65**(9), 1254–1258 (2018)
3. Graczyk, J., Swiatek, G.: Generic hyperbolicity in the logistic family. Ann. Math. **146**(1), 1–52 (1997)
4. Banerjee, S., Yorke, J.A., Grebogi, C.: Robust chaos. Phys. Rev. Lett. **80**(14), 3049 (1998)
5. Banerjee, S., Grebogi, C.: Border collision bifurcations in two-dimensional piecewise smooth maps. Phys. Rev. E **59**(4), 4052 (1999)

6. Gardini, L., Tramontana, F., Banerjee, S.: Bifurcation analysis of an inductorless chaos generator using 1D piecewise smooth map. Math. Comput. Simul. **95**, 137–145 (2014)
7. Mandal, S., Banerjee, S.: Analysis and CMOS implementation of a chaos-based communication system. IEEE Trans. Circuits Syst. Regul. Pap. **51**(9), 1708–1722 (2004)
8. Banerjee, S., Kastha, D., Das, S., Vivek, G., Grebogi, C.: Robust chaos-the theoretical formulation and experimental evidence. In: Proceedings of the 1999 IEEE International Symposium on Circuits and Systems, ISCAS'99, vol. 5, pp. 293–296. IEEE, Piscataway (1999)

Homogenous Multistability in Memristive System

Chunbiao Li

Abstract Many dynamical systems may have special parameters to control the amplitude sometimes concomitantly with frequency or offset. In this chapter, we select a chaotic system with amplitude-frequency parameter, which controls the scale and speed of the oscillation without changing its basic property of chaos. By introducing a memristor into the feedback for amplitude/frequency control, a special regime of homogenous multistability shows up in memristive system where the initial condition of the internal variable only determines the amplitude (here combined with frequency) of the variables without changing the essential oscillation of chaos. This phenomenon can exist in other systems with scale parameter. Unlike other regular multistability, the coexisting attractors share the same shape of phase trajectory except with different scales. Following this routine, a new pattern of homogenous multistability is also demonstrated where the stable oscillation stands in phase space with different offset in the dimension of x without changing the fundamental oscillation. To our best knowledge, it is reported firstly on homogenous multistability in memristive system systematically.

Keywords Amplitude control · Homogenous multistability · Memristive system · Offset boosting · Regular multistability

1 Introduction

Multistability has received great attention in nonlinear science since it may pose great threat or merit in engineering applications. Several studies on multistability show such systems with different attractors [1–6]. However, in a memristive system

---

C. Li
Jiangsu Collaborative Innovation Center of Atmospheric Environment and Equipment Technology (CICAEET), Nanjing University of Information Science and Technology, Nanjing, China

School of Electronic and Information Engineering, Nanjing University of Information Science and Technology, Nanjing, China
e-mail: chunbiaolee@nuist.edu.cn

W. Lacarbonara et al. (eds.), *New Trends in Nonlinear Dynamics*,
https://doi.org/10.1007/978-3-030-34724-6_28

different initial conditions may lead to different attractors of the same shape (sometimes with different amplitude and/or frequency or offset). This special regime of multistability is defined as homogenous multistability [7]. In this chapter, we discuss this newly-found regime of homogenous multistability, which appears in memristive systems and is relatively rare compared with other regular multistability.

In this work, a chaotic system is applied for hosting memristor(s) for hatching coexisting attractors with growing amplitude/frequency or growing offset. In Sect. 2, a new memristive model is proposed for demonstrating homogenous multistability. In Sect. 3, regular multistability is discussed for comparison. Conclusions are given in the last section.

2 Special Regime of Multistability

2.1 *Homogenous Multistability*

As suggested in [7], here we select a system with single parameter for amplitude and frequency control for hosting memristor [8],

$$\begin{aligned} \dot{x} &= y - xz - yz, \\ \dot{y} &= axz, \\ \dot{z} &= y^2 - bz^2, \end{aligned} \tag{1}$$

When $a = 4$, $b = 0.3$, system (1) has a chaotic solution. System (1) has three equilibrium points, two of which are $\left(0, \pm\sqrt{b}, 1\right)$, and another is a line equilibria $(x, 0, 0)$. When a memristor is introduced in the coefficient for amplitude-frequency control, the corresponding system can exhibit chaotic signal with increasing frequency and amplitude according to the time [7]. A memristor is introduced into the coefficient for amplitude control as,

$$\begin{aligned} \dot{x} &= W(u)y - xz - yz, \\ \dot{y} &= axz, \\ \dot{z} &= y^2 - bz^2, \\ \dot{u} &= y, \end{aligned} \tag{2}$$

As predicted, when $a = 4$, $b = 0.3$, $W(u) = 0.1\,|u| - 0.1$, the above memristive system has coexisting attractors of the same type since the memristor brings the information of the variable y. And therefore, the initial condition can be used to control the amplitude and frequency of the chaotic signals.

As shown in Fig. 1, when the initial condition varies, most of the coexisting attractors are of the same type but with different frequency and amplitude. System (2) has a plane of equilibria now $(x, 0, 0, u)$, whose eigenvalue is $(0, 0, 0, 0)$ showing the equilibria is of critical stability and further study shows that the system

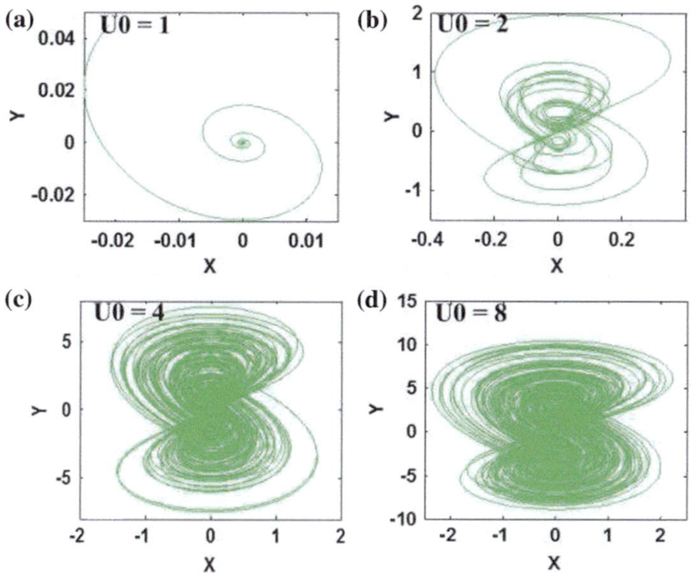

Fig. 1 Phase trajectories with different initial condition $[1, 0, -1, u_0]$ under the same time duration $(T = 500)$

is nonlinearly unstable. Figure 1 also shows that system (2) also has coexisting point attractor. System (2) is of inversion symmetry, which can be proved by $x \rightarrow -x$, $y \rightarrow -y$, $z \rightarrow z$, $u \rightarrow -u$. System (2) retains the rotational symmetry with a rate of hypervolume contraction given by the Lie derivative,

$$\nabla V = \frac{\partial \dot{x}}{\partial x} + \frac{\partial \dot{y}}{\partial y} + \frac{\partial \dot{z}}{\partial z} + \frac{\partial \dot{u}}{\partial u} = -(2b + 1)z \tag{3}$$

To further investigate how the initial condition of u_0 changes the amplitude and frequency of the chaotic signal, we obtain the average value of signals and Lyapunov exponents when u_0 varies in $[-8, 8]$, which is similar to what was shown in [7]. The average of system variables grows with the absolute value of u_0. There is a wide window of non-chaos where the initial condition draws the system to a stable point which is a state of suspended animation. The initial condition of other variables can also change the amplitude and frequency of the signals since all system variables are connected as a whole by the governing equation [7]. However, as compared, when random disturbance is applied in the flux-controlled or charge-controlled memristor in other cases, transition in formation of attractors and mode transition from chaotic to periodical oscillation occurs since in those cases memristor changes the value of bifurcation parameter [9].

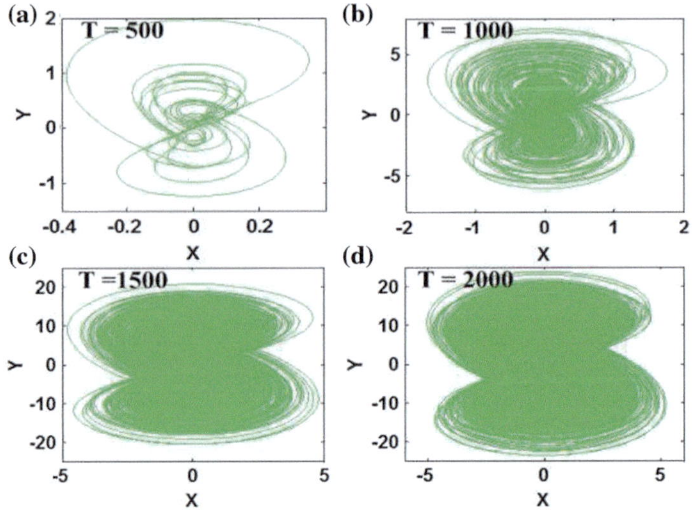

Fig. 2 Phase trajectory evolvement with various durations of time under the same initial condition $[1, 0, -1, 2]$

2.2 *Growing Scale of Attractor*

More distinctively the amplitude and frequency of the variables are also modified according to the time duration, as shown in Fig. 2. The mechanism is from the fact that longer time duration may lead to larger value of $W(u)$. Here the increase of variable u leads to an increase of y. Larger y gives larger signal of u and m in turn, yields larger variables with higher frequency. It is proven that there is an identical trend where the time duration and the initial condition are two effective ways to influence the integral value.

The increasing speed of the amplitude and frequency over the time duration is not constant. It is due to the fact that the hysteresis property of the memristor is associated with the frequency of the input signal. Here the flux-controlled memristor is defined as

$$\begin{aligned} i &= W(u)y, \\ W(u) &= 0.1 \mid u \mid -0.1, \\ \dot{u} &= y, \end{aligned} \tag{4}$$

The flux-controlled memductance is a function of internal state of u, which is associated with the voltage y,

$$W(u) = 0.1\,|u| - 0.1 = 0.1\left|\int_{-\infty}^{t} y\right| - 0.1 = W_0 + 0.1\left|\int_{0}^{t} y ds\right| - 0.1 \tag{5}$$

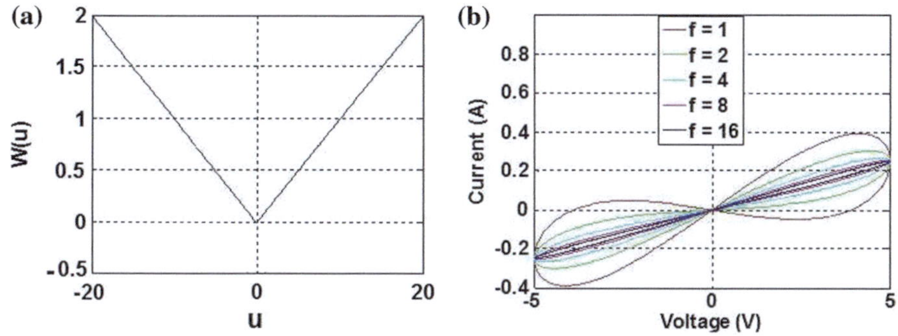

Fig. 3 The memductance and pinched hysteresis loop

where,

$$W_0 = 0.1\left(\left|\int_{-\infty}^{t} yds\right| - \left|\int_{0}^{t} yds\right|\right) \tag{6}$$

The memductance and the theoretical pinched hysteresis loop are shown in Fig. 3. Here the input sinusoidal function reads

$$v(t) = 5\sin(2\pi f t) \tag{7}$$

We can see the mechanism of homogenous multistability clearly by revising the memristor to be a simple function as the following:

$$\begin{aligned} \dot{x} &= 0.1yu - xz - yz, \\ \dot{y} &= axz, \\ \dot{z} &= y^2 - bz^2, \\ \dot{u} &= 1, \end{aligned} \tag{8}$$

In Eq. (8), the variable u is linearly increasing with a constant slope. The initial condition of the variable u changes the oscillation starting point, and therefore produce the same type of attractors with different scale of amplitude and frequency, as shown in Fig. 4. Note that the variable $u(t)$ here is different from the signal in system (2), and so results in a continuous increase of amplitude and frequency in the signals x, y, z.

Offset boosting is another helpful control for chaotic signal in electronic engineering [10–13]. Following the routine above mentioned, we predict that homogenous multistability may also be induced by offset boosting in memristive system. Suppose there are two memristors being introduced into the above system (1) as,

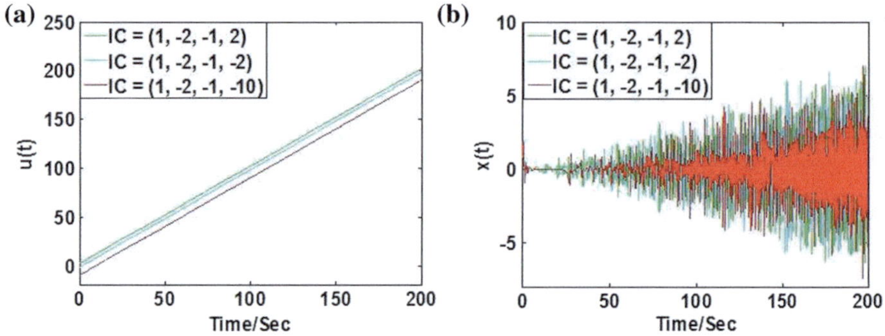

Fig. 4 Signals of system (8) with different initial condition over a time duration of 200

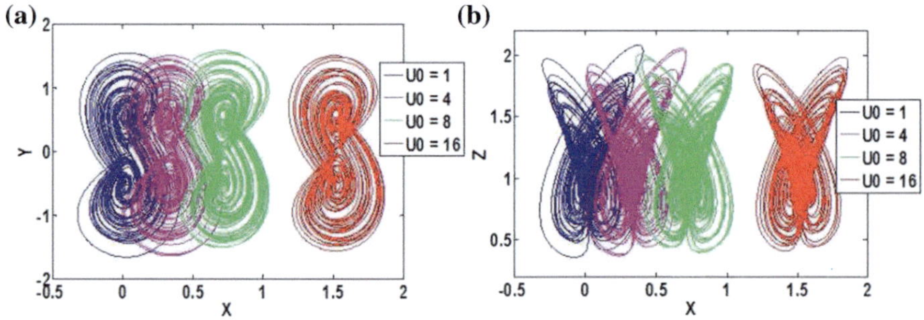

Fig. 5 Memristor-resulted offset boosting from different initial condition $[1, 0, -1, u_0]$ over a time duration of 500

$$\begin{aligned} \dot{x} &= y - (x - W(u))\, z - yz, \\ \dot{y} &= a\,(x - W(u))\, z, \\ \dot{z} &= y^2 - bz^2, \\ \dot{u} &= 0.001z, \end{aligned} \tag{9}$$

Here the flux-controlled memductance is also a function of internal state of u, which is defined as

$$\begin{aligned} i &= W(u)z, \\ W(u) &= 0.1 \mid u \mid -0.1, \\ \dot{u} &= 0.001z, \end{aligned} \tag{10}$$

Now the chaotic oscillation in system (9) shows different offset in the dimension of x even under the same parameters, which is another pattern of homogenous multistability. As shown in Fig. 5, different initial conditions of the variable u lead the oscillation towards different location space in the dimension of x.

3 Regular Multistability

3.1 Different Types of Coexisting Attractors

To understand the homogenous multistability in memristive system further, here we try to introduce the memristor into a different position of feedback. Suppose a memristor is introduced in the position of bifurcation parameter in the following,

$$\begin{aligned} \dot{x} &= y - xz - yz, \\ \dot{y} &= axz + W(u)x, \\ \dot{z} &= y^2 - bz^2, \\ \dot{u} &= x, \end{aligned} \tag{11}$$

When $a = 4$, $b = 0.3$, $W(u) = 4u^2 + 1$, system (11) has coexisting limit cycles and strange attractors with different Lyapunov exponents, as shown in Fig. 6. Note that compared with the above system all these coexisting attractors are no longer changing with the time which shows that it belongs to a class of regular multistability. When the initial condition of u_0 varies in $[-8, 8]$, corresponding Lyapunov exponents and bifurcation diagram are shown in Fig. 7.

System (11) is of symmetric structure, which can be proved by $x \to -x$, $y \to -y$, $z \to z$, $u \to -u$. Unlike other memristive system, now system (11) has three lines of equilibria $(0, 0, 0, u)$ and $\left(0, \pm\sqrt{b}, 1, u\right)$. Here coexisting strange attractors are common but are still rare in other symmetric systems [14–16].

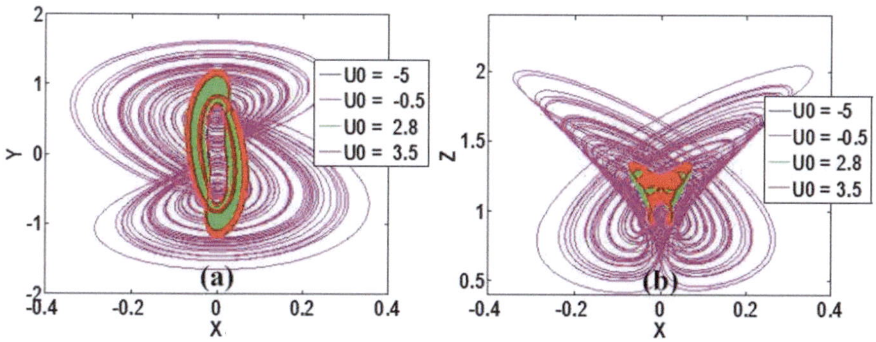

Fig. 6 Coexisting attractors with different initial condition $[1, 0, -1, u_0]$

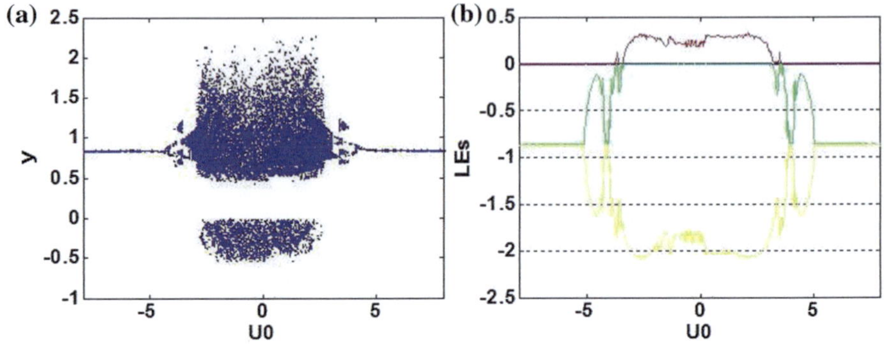

Fig. 7 Bifurcation diagram and the corresponding Lyapunov exponents of system (11) with the initial condition of $[1, 0, -1, u_0]$, u_0 varies in $[-8, 8]$

3.2 *Multistability Comparison*

Coexisting attractors may have different properties, some of which share the same type or shape but with different amplitude (sometimes with frequency) or offset, while some of which have totally different shape of phase trajectories. These two types of multistability come from different mechanisms. Memristor can be applied as a special feedback, and its existence can definitely define the dynamics in system. The integration of the internal variable will change the bifurcation parameter or non-bifurcation parameter, and therefore breeding different types of multistabilities. Further exploration shows that sometimes two types of coexisting attractors may exist in a normal system, but it is difficult to divide them clearly and find which the dominant factor is. When the system is constructed in circuit [17–22], the increasing process of the chaotic signal may be hidden for the saturation of the IC chip.

4 Conclusions

A memristor may exhibit different multistability depending on the system parameters. Two positions are selected to host the memristor for producing different coexisting attractors with a similar shape of phase trajectories. Homogenous multistability is a new regime of multistability, where coexisting attractors stay at different stages of development. It is clear that the occurrence of multistability in the proposed memristive system can account for multiple modes in neural activities. Furthermore, to a certain extent, a memristor-involved neuron model may explain the effect of electromagnetic induction and even radiation which deserves further research.

Acknowledgments The author sincerely thank the anonymous reviewers for their valuable comments that led to the present improved version of the original manuscript. This work was supported financially by the National Nature Science Foundation of China (Grant No.: 61871230), the Natural Science Foundation of Jiangsu Province (Grant No.: BK20181410), and a Project Funded by the Priority Academic Program Development of Jiangsu Higher Education Institutions.

References

1. Bao, B., Xu, Q., Bao, H.: Extreme multistability in a memristive circuit. Electron. Lett. **52**(12), 1008–1010 (2016)
2. Yuan, F., Wang, G., Wang, X.: Extreme multistability in a memristor-based multi-scroll hyper-chaotic system. Chaos. **26**(7), 073107 (2016)
3. Wang, G., Shi, C., Wang, X., Yuan, F.: Coexisting oscillation and extreme multistability for a memcapacitor-based circuit. Math. Probl. Eng. **2017**(5), 6504969 (2017)
4. Robinson, A., Calov, R., Ganopolski, A.: Multistability and critical thresholds of the Greenland ice sheet. Nat. Clim. **2**, 429–432 (2012)
5. Lai, Q., Chen, S.: Generating multiple chaotic attractors from Sprott B system. Int. J. Bifurcat. Chaos. **26**, 1650177 (2016)
6. Zhou, L., Wang, C., Zhou, L.: Generating four-wing hyperchaotic attractor and two-wing, three-wing, and four-wing chaotic attractors in 4D memristive system. Int. J. Bifurcat. Chaos. **27**(2), 1750027 (2017)
7. Li, C., Thio, W., Iu, H.H.C., Lu, T.: A memristive chaotic oscillator with increasing amplitude and frequency. IEEE Access. **6**, 12945–12950 (2018)
8. Li, C., Sprott, J.C.: Chaotic flows with a single nonquadratic term. Phys. Lett. A. **378**(3), 178–183 (2014)
9. Ma, J., Wu, F., Ren, G., Tang, J.: A class of initials-dependent dynamical systems. Appl. Math. Comput. **298**, 65–76 (2017)
10. Li, C., Sprott, J.C., Xing, H.: Constructing chaotic systems with conditional symmetry. Nonlinear Dyn. **87**, 1351–1358 (2017)
11. Li, C., Sprott, J.C.: Variable-boostable chaotic flows. Optik. **127**, 10389–10398 (2016)
12. Li, C., Xu, Y., Chen, G., Liu, Y., Zheng, J.: Conditional symmetry: bond for attractor growing. Nonlinear Dyn. **95**, 1245–1256 (2019)
13. Li, C., Sprott, J.C., Kapitaniak, T., Lu, T.: Infinite lattice of hyperchaotic strange attractors. Chaos Soliton. Fract. **109**, 76–82 (2018)
14. Li, C., Sprott, J.C., Yuan, Z., Li, H.: Constructing chaotic systems with total amplitude control. Int J Bifurcation Chaos. **25**, 1530025 (2015)
15. Jafari, S., Sprott, J.C.: Simple chaotic flows with a line equilibrium. Chaos Soliton. Fract. **57**, 79–84 (2013)
16. Bao, H., Wang, N., Bao, B., Chen, M., Jin, P., Wang, G.: Initial condition-dependent dynamics and transient period in memristor-based hypogenetic jerk system with four line equilibria. Commun. Nonlinear Sci. Numer. Simulat. **57**, 264–275 (2018)
17. Zhou, L., Wang, C., Zhou, L.: Generating hyperchaotic multi-wing attractor in a 4D memristive circuit. Nonlinear Dyn. **85**(4), 2653–2663 (2016)
18. Zhou, L., Wang, C., Zhou, L.: A novel no-equilibrium hyperchaotic multi-wing system via introducing memristor. Int. J. Circ. Theor. Appl. **46**(1), 84–98 (2018)
19. Wang, C., Liu, X., Xia, H.: Multi-piecewise quadratic nonlinearity memristor and its 2N-scroll and 2N+1-scroll chaotic attractors system. Chaos. **27**(3), 033114 (2017)

20. Li, C., Thio, W., Sprott, J.C., Iu, H.H.C., Xu, Y.: Constructing infinitely many attractors in a programmable chaotic circuit. IEEE Access. **6**, 29003–29012 (2018)
21. Li, C., Thio, W., Sprott, J.C., Zhang, R., Lu, T.: Linear synchronization and circuit implementation of chaotic system with complete amplitude control. Chin. Phys. B. **26**(12), 120501 (2017)
22. Li, C., Akgul, A., Sprott, J.C., Iu, H.H.C., Thio, W.: A symmetric pair of hyperchaotic attractors. Int. J. Circ. Theor. Appl. **1-10**, 2018 (2018)

Part V
Fractional-Order Systems

Numerical Study of Nonlinear Vibrations of Fractionally Damped Cylindrical Shells Under the Additive Combinational Internal Resonance

Marina V. Shitikova and Basem Ajarmah

Abstract In the present chapter, nonlinear damped vibrations of a cylindrical shell embedded into a fractional derivative medium are investigated for the case of the combinational internal resonance, resulting in modal interaction, using two different numerical methods with further comparison of the results obtained. The damping properties of the surrounding medium are described by the fractional derivative Kelvin–Voigt model utilizing the Riemann–Liouville fractional derivatives. Within the first method, the generalized displacements of a coupled set of nonlinear ordinary differential equations of the second order are estimated using numerical solution of nonlinear multi-term fractional differential equations. Adopting the second procedure based on the generalized method of multiple time scales, the governing nonlinear differential equations describing amplitude-and-phase modulations for the case of the combinational internal resonance have been derived, and the envelopes of amplitudes have been constructed.

Keywords Cylindrical shell · Fractional derivative · Combinational internal resonance · Nonlinear damped vibrations · Fractional differential equations

M. V. Shitikova (✉)
Research Center on Dynamics of Solids and Structures, Voronezh State Technical University, Voronezh, Russia
e-mail: mvs@vgasu.vrn.ru

B. Ajarmah
Research Center on Dynamics of Solids and Structures, Voronezh State Technical University, Voronezh, Russia

Al-istiqlal University, Jericho, Palestine
e-mail: basem@pass.ps

W. Lacarbonara et al. (eds.), *New Trends in Nonlinear Dynamics*,
https://doi.org/10.1007/978-3-030-34724-6_29

1 Introduction

Cylindrical shells are widely used as structural elements in pipes and ducts, bodies of cars, space shuttles, aircraft fuselages, ship hulls, submarines, and building structures. Recent investigations on dynamic response of cylindrical shells are based on nonlinear theories, and a comprehensive review can be found in [1–3].

In recent years, much attention is given to the damping features of mechanical systems subjected to the conditions of different internal resonances. Damping properties of nonlinear systems are described mainly by the first-order time-derivative of a generalized displacement [3, 4]. However, the analysis of free damped vibrations of a suspension combined system under the conditions of the one-to-one internal resonance has revealed that for good fit of the theoretical investigations with the experimental results it is better to describe the damping features of nonlinear mechanical systems in terms of fractional time-derivatives of the generalized displacements [5].

Nowadays, fractional calculus models are widely spread in mechanics of solids and structures [5, 6] for modeling viscoelastic properties of such structures as beams, pipes, plates, and bridges. However, there are a few papers dealing with fractionally damped shells. The first attempt was made in 2001 to study the linear dynamic response of an elastic circular cylindrical shell in a fractional derivative medium [7]. Recently the fractional derivative standard linear solid model has been utilized to model the viscoelastic features of a core in a cylindrical [8] and conical [9] shells.

As for the nonlinear response of fractionally damped cylindrical shells, then a new approach has been suggested recently [10] for the case when the dynamic response of the shell is described by a set of three coupled nonlinear differential equations of the Donnell–Mushtari–Vlasov type [11]. The procedure resulting in decoupling linear parts of equations is proposed with the further use of the generalized method of multiple scales for solving nonlinear governing equations of motion involving fractional derivative terms.

This procedure has been utilized for the analytical analysis of free vibrations of cylindrical shells subjected to the conditions of the different internal resonances, resulting in the interaction of two (in the case of 2:1, 1: or 3:1 internal resonance [10]) or three modes corresponding to the mutually orthogonal displacements (in the case of combinational internal resonances of additive or difference type [12, 13]).

In this chapter, we first analyze parameter values resulting in nonlinear vibrations of a fractionally damped cylindrical shell subjected to the combinational additive internal resonance, when a certain natural frequency of vibration is equal to the sum of two other natural frequencies. Then the generalized displacements are found via the numerical solution of nonlinear multi-term fractional differential equations by the procedure based on the reducing of the problem to a system of fractional differential equations [14, 15]. Utilizing the generalized method of multiple timescales [5], the governing nonlinear differential equations defining the amplitude-and-phase modulations for the case of the combinational internal resonance are solved by the fourth order Runge–Kutta algorithm, and the envelopes of amplitudes are constructed.

2 Problem Formulation

The dynamic response of a free supported nonlinear elastic circular cylindrical shell of radius R and length l vibrating in a fractionally damped surrounding medium are described by the Donnell–Mushtari–Vlasov equations in terms of three mutually orthogonal displacements [11, 12]:

$$\begin{aligned} &u_{xx} + \tfrac{1-\nu}{2}\ \beta_1^2 u_{\varphi\varphi} + \tfrac{1+\nu}{2}\ \beta_1 v_{x\varphi} - \nu\beta_1 w_x + w_x\left(w_{xx} + \tfrac{1-\nu}{2}\ \beta_1^2 w_{\varphi\varphi}\right) \\ &\quad + \tfrac{1+\nu}{2}\ \beta_1^2 w_\varphi w_{x\varphi} = \ddot{u} + æ_1\left(\tfrac{d}{dt}\right)^\gamma u, \end{aligned} \tag{1}$$

$$\begin{aligned} &\beta_1^2 v_{\varphi\varphi} + \tfrac{1-\nu}{2}\ v_{xx} + \tfrac{1+\nu}{2}\ \beta_1 u_{x\varphi} - \beta_1^2 w_\varphi + \beta_1 w_\varphi\left(\beta_1^2 w_{\varphi\varphi} + \tfrac{1-\nu}{2}\ w_{xx}\right) \\ &\quad + \tfrac{1+\nu}{2}\ \beta_1 w_x w_{x\varphi} = \ddot{v} + æ_2\left(\tfrac{d}{dt}\right)^\gamma v, \end{aligned} \tag{2}$$

$$\begin{aligned} &\tfrac{\beta_2^2}{12}\left(w_{xxxx} + 2\ \beta_1^2 w_{xx\varphi\varphi} + \beta_1^2 w_{\varphi\varphi\varphi\varphi}\right) - \nu\ \beta_1\ u_x - \beta_1^2\ v_\varphi + \beta_1^2 w \\ &\quad + \tfrac{1}{2}\ \nu\beta_1 (w_x)^2 + \tfrac{1}{2}\ \beta_1^3 (w_\varphi)^2 - w_x\left(u_{xx} + \tfrac{1-\nu}{2}\beta_1^2 u_{\varphi\varphi} + \tfrac{1+\nu}{2}\beta_1 v_{x\varphi}\right) \\ &\quad - \beta_1\ w_\varphi\left(\beta_1^2 v_{\varphi\varphi} + \tfrac{1-\nu}{2} v_{xx} + \tfrac{1+\nu}{2}\beta_1 u_{x\varphi}\right) - w_{xx}\left(u_x + \nu\beta_1 v_\varphi - \nu\beta_1 w\right) \\ &\quad - \beta_1^2 w_{\varphi\varphi}\left(\nu u_x + \beta_1 v_\varphi - \beta_1 w\right) - (1-\nu)\,\beta_1 w_{x\varphi}\left(\beta_1 u_\varphi + v_x\right) \\ &= -\ddot{w} - æ_3\left(\tfrac{d}{dt}\right)^\gamma w, \end{aligned} \tag{3}$$

subjected to the initial

$$u|_{t=0} = v|_{t=0} = w|_{t=0} = 0, \quad \dot{u}|_{t=0} = \dot{v}|_{t=0} = \dot{w}|_{t=0} = 0, \tag{4}$$

and the boundary conditions

$$\begin{aligned} &w|_{x=0} = w|_{x=l} = 0, \quad v|_{x=0} = v|_{x=l} = 0, \quad u_x|_{x=0} = u_x|_{x=l} = 0, \\ &\qquad w_{xx}|_{x=0} = w_{xx}|_{x=l} = 0, \end{aligned} \tag{5}$$

where the x-axis is directed along the axis of the cylinder, r and φ are the polar radius and angle, respectively, $u = u(x, \varphi, t)$, $v = v(x, \varphi, t)$, and $w = w(x, \varphi, t)$ are the displacements of points in the shell's middle surface in three mutually orthogonal directions x, φ, r, $\beta_1 = l/R$ and $\beta_2 = h/l$ are parameters defining the dimensions of the shell, h is its thickness, t is time, ν is the Poisson's ratio, $æ_i = \varepsilon\mu_i\tau_i^\gamma$ ($i = 1,2,3$) are coefficients of small viscosity, $0 < \varepsilon < 1$, μ_i are modal damping coefficients, τ_i are modal retardation times, over dot denotes the time-derivative, lower indices label the derivatives with respect to the corresponding coordinates, and γ is the fractional order of the operator of differentiation $\left(\frac{d}{dt}\right)^\gamma$ which is equivalent to the Riemann–Liouville fractional derivative of the γ-order [5]

$$D^{\gamma} F = \frac{d}{dt} \int_0^t \frac{F\left(t - t'\right)}{\Gamma\left(1 - \gamma\right)} dt'. \tag{6}$$

For solving the partial differential equations (1)–(3) subjected to the initial (4) and boundary (5) conditions, we will adopt the procedure [10] for the decoupling linear parts of Eqs. (1)–(3) with further utilization of the generalized method of multiple timescales [12], utilizing the solution of the Navier type

$$u\left(x, \varphi, t\right) = \sum_{m=1}^{\infty} \sum_{n=1}^{\infty} x_{1mn}(t) \eta_{1mn}\left(x, \varphi\right), \tag{7}$$

$$v\left(x, \varphi, t\right) = \sum_{m=1}^{\infty} \sum_{n=1}^{\infty} x_{2mn}(t) \eta_{2mn}\left(x, \varphi\right), \tag{8}$$

$$w\left(x, \varphi, t\right) = \sum_{m=1}^{\infty} \sum_{n=1}^{\infty} x_{3mn}(t) \eta_{3mn}\left(x, \varphi\right), \tag{9}$$

where $x_{imn}(t)$, $\eta_{imn}(x, \varphi)$ $(i = 1, 2, 3)$ are, respectively, the generalized time-dependent displacements and eigen functions corresponding to the free supported shell.

Following the idea [10] to model nonlinear damped vibrations of shells in a viscoelastic medium by representing viscous resistance forces via fractional order time derivatives, as distinct to the traditional way [1–3] when damping forces are assumed to be proportional to first order time-derivatives of displacements, i.e., to utilize the fractional derivative Kelvin–Voigt model and the procedure [10] for decoupling the linear parts of fractional-order differential equations of motion of the cylindrical shell, the set of equations of three predominating modes of vibrations, which could be coupled by some conditions of internal resonance, have been derived in [12] and written in the following dimensionless form:

$$\ddot{X}_{1mn} + æ_1 D^{\gamma} X_{1mn} + \Omega_{1mn}^2 X_{1mn} = -\sum_{i=1}^{3} F_{imn} L_{imn}^{I}, \tag{10}$$

$$\ddot{X}_{2mn} + æ_2 D^{\gamma} X_{2mn} + \Omega_{2mn}^2 X_{2mn} = -\sum_{i=1}^{3} F_{imn} L_{imn}^{II}, \tag{11}$$

$$\ddot{X}_{3mn} + æ_3 D^{\gamma} X_{3mn} + \Omega_{3mn}^2 X_{3mn} = -\sum_{i=1}^{3} F_{imn} L_{imn}^{III}, \tag{12}$$

where m and n are integers, F_{imn} are nonlinear terms involving cubic nonlinearities [12], Ω_{1mn}, Ω_{2mn} and Ω_{3mn} are the square roots of eigenvalues of the matrix S_{ij}^{mn} and L_{imn}^{I}, L_{imn}^{II} and L_{imn}^{III} are eigenvectors of the same matrix with the elements

$$S_{ij}^{mn} = \begin{bmatrix} \left(\pi^2 m^2 + \frac{1-v}{2}\beta_1^2 n^2\right) & \frac{1+v}{2}\beta_1 \pi mn & v\beta_1 \pi m \\ \frac{1+v}{2}\beta_1 \pi mn & \left(\frac{1-v}{2}\pi^2 m^2 + \beta_1^2 n^2\right) & \beta_1^2 n \\ v\beta_1 \pi m & \beta_1^2 n & \frac{\beta_2^2}{12}\left(\pi^2 m^2 + \beta_1^2 n^2\right)^2 + \beta_1 \end{bmatrix} \tag{13}$$

3 Methods of Solution

Equations (10)-(12) for the case of the combinational internal resonance $\Omega_3 = \Omega_1 + \Omega_2$, when the frequency of flexural vibration is approximately equal to the sum of frequencies of two other orthogonal modes, could be solved numerically by two different methods.

The first method is based on the Diethelm's discretizing procedure [14] of all derivatives involved in (10)–(12). Introducing the following notation

$$\begin{gathered} Y_1 = X_1, \quad Y_2 = D^\gamma X_1 = D^\gamma Y_1, \quad Y_3 = DX_1 = DY_1, \\ \ddot{X}_1 = DDX_1 = DY_3 = -\sum_{i=1}^{3} F_{1\,mn} L_{i\,mn}^{I} - æ_1 Y_2 - \Omega_1^2 Y_1, \end{gathered} \tag{14}$$

the first time-derivative is discretized using the trapezoidal rule

$$DY_1 = Y_3 \rightarrow Y_{1\,i} = Y_{1\,i-1} + \frac{1}{2}h\left(Y_{3\,i} + Y_{3\,i-1}\right), \tag{15}$$

or

$$Y_{1\,i} - \frac{1}{2}hY_{3\,i} = Y_{1\,i-1} + \frac{1}{2}h\left(Y_{3\,i-1}\right) = s_{2i-1}. \tag{16}$$

As for the fractional derivative, then it could be represented as [14, 15]

$$D^\gamma Y_1 = \frac{1}{{}^{\alpha}\chi_i}\left(\sum_{k=0}^{i} {}^{\gamma}\omega_{k,i}\, Y_{1\,i-k} + \frac{Y_{1\,0}}{\gamma}\right) = Y_2, \tag{17}$$

therefore

$${}^{\gamma}\chi_i Y_2 - {}^{\gamma}\omega_{0,i}\, Y_{1\,i} = \left(\sum_{k=1}^{i} {}^{\gamma}\omega_{k,i}\, Y_{1\,i-k} + \frac{Y_{1\,0}}{\gamma}\right). \tag{18}$$

Letting $s_{1i-1} = \sum_{k=1}^{i} {}^{\gamma}\omega_{k,i} Y_{1i-k} + \frac{Y_{10}}{\gamma}$, and adopting the above procedure to other two equations, Eqs (10)–(12) could be written in the following matrix form:

$$\begin{bmatrix} -{}^{\gamma}\omega_{0,i} & {}^{\gamma}\chi_i & 0 & 0 & 0 & 0 & 0 & 0 & 0 \\ 1 & 0 & -\frac{h}{2} & 0 & 0 & 0 & 0 & 0 & 0 \\ \frac{h}{2}\Omega_1^2 & \frac{h}{2}æ_1 & 1 & 0 & 0 & 0 & 0 & 0 & 0 \\ 0 & 0 & 0 & -{}^{\gamma}\omega_{0,i} & {}^{\gamma}\chi_i & 0 & 0 & 0 & 0 \\ 0 & 0 & 0 & 1 & 0 & -\frac{h}{2} & 0 & 0 & 0 \\ 0 & 0 & 0 & \frac{h}{2}\Omega_2^2 & \frac{h}{2}æ_2 & 1 & 0 & 0 & 0 \\ 0 & 0 & 0 & 0 & 0 & 0 & -{}^{\gamma}\omega_{0,i} & {}^{\gamma}\chi_i & 0 \\ 0 & 0 & 0 & 0 & 0 & 0 & 1 & 0 & -\frac{h}{2} \\ 0 & 0 & 0 & 0 & 0 & 0 & \frac{h}{2}\Omega_3^2 & \frac{h}{2}æ_3 & 1 \end{bmatrix} * \begin{bmatrix} Y_{1\,i} \\ Y_{2\,i} \\ Y_{3\,i} \\ Y_{4\,i} \\ Y_{5\,i} \\ Y_{6\,i} \\ Y_{7\,i} \\ Y_{8\,i} \\ Y_{9\,i} \end{bmatrix} = \begin{bmatrix} s_{1\,i-1} \\ s_{2\,i-1} \\ s_{3\,i-1} \\ s_{4\,i-1} \\ s_{5\,i-1} \\ s_{6\,i-1} \\ s_{7\,i-1} \\ s_{8\,i-1} \\ s_{9\,i-1} \end{bmatrix} \tag{19}$$

To find the **Y** column, it is a need to inverse the left matrix and multiply it by **s** (the right column which represents the already known values or initial conditions from the previous iteration step).

The second method of solution is based on the fractional derivative expansion procedure [5], resulting in a set of six nonlinear differential equations in terms of amplitudes a_i and phases φ_i for the case $\Omega_3 = \Omega_1 + \Omega_2$:

$$\dot{\left(a_1^2\right)} + s_1 a_1^2 = -\Omega_1^{-1} a_{23}^{I} a_1 a_2 a_3 \sin\delta, \quad \dot{\varphi}_1 - \frac{1}{2}\sigma_1 - \frac{1}{2}\frac{a_{23}^{I}}{\Omega_1}\frac{a_2 a_3}{a_1}\cos\delta = 0, \tag{20}$$

$$\dot{\left(a_2^2\right)} + s_2 a_2^2 = -\Omega_2^{-1} a_{13}^{II} a_1 a_2 a_3 \sin\delta, \quad \dot{\varphi}_2 - \frac{1}{2}\sigma_2 - \frac{1}{2}\frac{a_{13}^{II}}{\Omega_2}\frac{a_1 a_3}{a_2}\cos\delta = 0, \tag{21}$$

$$\dot{\left(a_3^2\right)} + s_3 a_3^2 = \Omega_3^{-1} a_1 a_2 a_3 \sin\delta, \quad \dot{\varphi}_3 - \frac{1}{2}\sigma_3 - \frac{1}{2}\frac{a_{12}^{III}}{\Omega_3}\frac{a_1 a_2}{a_3}\cos\delta = 0, \tag{22}$$

where $\delta = \varphi_3 - (\varphi_1 + \varphi_2)$ is the phase difference, $s_i = \mu_i \tau_i^{\gamma} \Omega_i^{\gamma-1} \sin\psi$, $\psi = \pi\gamma/2$, and $\sigma_i = \mu_i \tau_i^{\gamma} \Omega_i^{\gamma-1} \cos\psi$.

Reference to Eqs. (20)–(22) shows that damping coefficients of nonlinear vibrations s_i depend on natural frequencies Ω_i and retardation times τ_i, and only at $\gamma = 1$, i.e., in the case of the conventional viscosity, they are frequency-independent.

4 Numerical Results

Since our aim is to study the additive combinational internal resonance $\Omega_3 = \Omega_1+\Omega_2$, then first it is necessary to reveal the geometrical parameters of the shell that could lead to such a resonance. In other words, it is a need to find parameters β_2 and β_1 which could initiate the condition of the additive combinational resonance $\Omega_3 = \Omega_1+\Omega_2$, and the results are shown in Fig. 1. Figure 2 represents the interdependency between the shell's parameters and the natural frequencies satisfying the condition $\Omega_3 = \Omega_1+\Omega_2$ at fixed magnitudes of interacting modes numbers: $m_1 = 1$, $n_1 = 1$, $m_2 = 1$, $n_2 = 4$, $n_2 = 1$, $m_3 = 4$, $n_3 = 1$, i.e., for one curve from Fig. 1. From Fig. 2a it is evident that the shell parameters β_2 and β_1 are exponentially interrelated, while from Fig. 2b, c it is seen that magnitudes of β_2 vary linearly with every frequency of combinational internal resonance in the case of $\Omega_3 = \Omega_1+\Omega_2$, and β_1 are exponentially related to them.

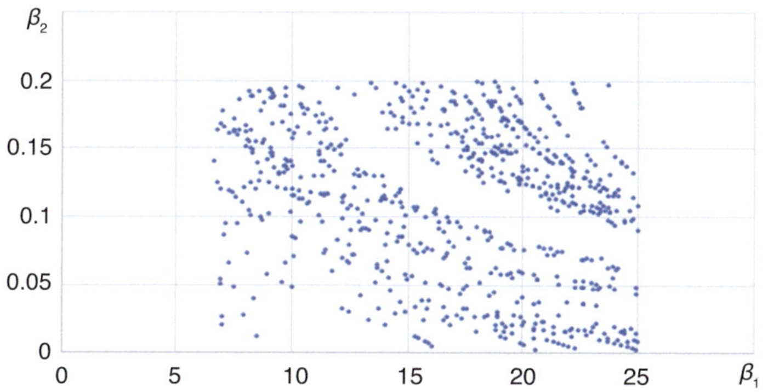

Fig. 1 Shell's parameters satisfying the condition $\Omega_3 = \Omega_1+\Omega_2$

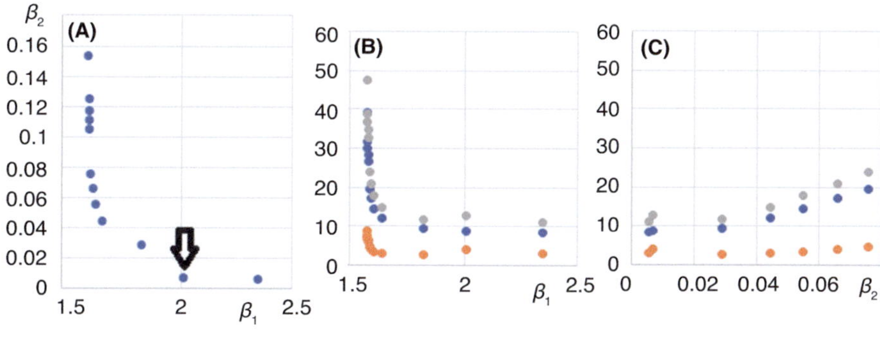

Fig. 2 Correlation between parameters β_1, β_2 and frequencies Ω_1, Ω_2, Ω_3: (**a**) $\beta_1 - \beta_2$ dependence, (**b**) β_2-dependence of frequencies, (**c**) β_1-dependence of frequencies, where orange filled circle—Ω_1, blue filled circle—Ω_2, black filled circle—Ω_3

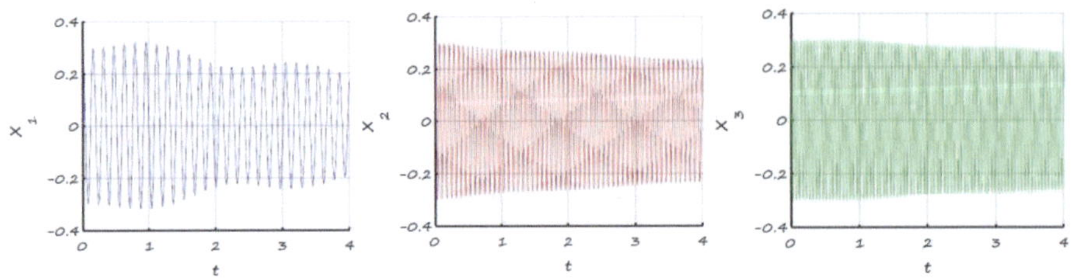

Fig. 3 Time-history of the generalized displacements X_i ($i = 1, 2, 3$) calculated via the Diethelm's multi-term method according to Eq. (19)

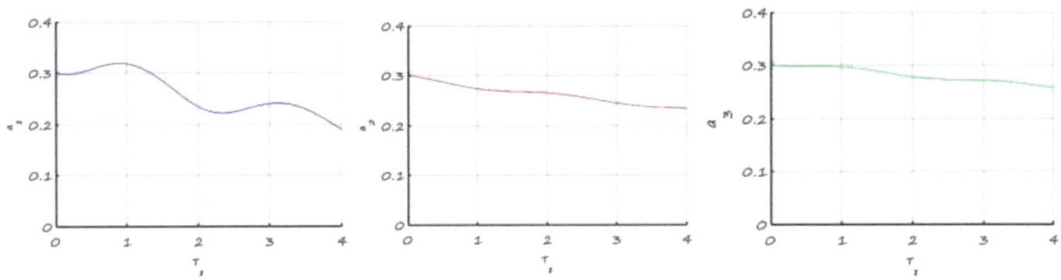

Fig. 4 T_1-dependence of amplitudes' envelopes via Eqs. (20)–(22)

The numerical solution of (19) via the multi-step method has been done for the following dimensionless parameters: $\beta_1 = 2.0095$, $\beta_2 = 0.00672$, $\Omega_1 = 3.9151$, $m_1 = n_1 = 1$, $\Omega_2 = 8.8242$, $m_2 = 1$, $n_2 = 4$, $\Omega_3 = 12.7393$, m_3=4, $n_3 = 1$, æ $= 0.25$, and $\nu = 0.33$, which correspond to the point indicated by an arrow in Fig. 2a. The time-history curves for the generalized displacements X_1, X_2, and X_3 are constructed at the fractional parameter $\gamma = 0.25$ in Fig. 3.

The solution of (20)–(22) has been carried out numerically using the Runge–Kutta fourth-order algorithm at $\tau = 1$, $\mu = 2.5$, $\gamma = 0.25$, and the envelopes of amplitudes corresponding to the same case of the combinational resonance are shown in Fig. 4.

From Figs. 3 and 4, it is seen that energy transfer takes place between the three coupled modes of vibration, and the amplitude of flexural vibrations attenuates more rapidly as compared with other two modes. The comparison of the results obtained by two methods shows their good correlation.

5 Conclusions

In the present chapter, the nonlinear dynamic response of a fractionally damped cylindrical shell was studied, when its motion is governed by a set of three coupled nonlinear fractional differential equations subjected to the occurrence of the additive combinational internal resonance, resulting in the interaction of three coupled orthogonal modes. The nonlinear set of resolving fractional-order differential equations was obtained in terms of the generalized displacements and in terms of the amplitudes and phases. The two systems were solved numerically by two different methods. A good agreement between the results obtained by the two methods was found.

The numerical procedures presented could be used for the analysis of all types of internal resonances which may occur during nonlinear vibrations of rods, beams, plates, shells, bridges, and other structures.

Acknowledgements This research was partially supported by Grant No. 9.5138.2017/8.9 as a Government task from the Ministry of Education and Science of the Russian Federation.

References

1. Kubenko, V.D., Koval'chuk, P.S.: Int. Appl. Mech. **34**, 703–728 (1998)
2. Amabili, M., Paidoussis, M.P.: Appl. Mech. Rev. **56**, 349–381 (2003)
3. Breslavsky, I.D., Amabili, M.: Nonlinear Dyn. **93**, 53–62 (2018)
4. Nayfeh, A.: Nonlinear Interaction: Analytical, Computational, and Experimental Methods. Wiley, New York (2000)
5. Rossikhin, Y.A., Shitikova, M.V.: Appl. Mech. Rev. **63**, Paper ID 010801 (2010)
6. Rossikhin, Y.A., Shitikova, M.V.: Fractional calculus in structural mechanics. In: Baleanu, D., Lopes, A.M. (eds.) Handbook of Fractional Calculus with Applications. Applications in Engineering, Life and Social Sciences, Part A, vol. 7, pp. 159–192. De Gruyter, Berlin (2019)
7. Rossikhin, Y.A., Shitikova, M.V., Ovsjannikova, E.I.: Proceedings of the 5th International Conference on Vibration Problems, Moscow, Russia, October 8-10, 2001, pp. 378–381 (2001)
8. Mokhtari, M., Permoon, M.R., Haddadpour, H.: Compos. Struct. **186**, 165–174 (2018)
9. Permoon, M.R., Shakouri, M., Haddadpour, H.: Compos. Struct. **241**, 62–72 (2019)
10. Rossikhin, Y.A., Shitikova, M.V.: Appl. Math Comput. **257**, 498–525 (2015)
11. Volmir, A.: Nonlinear Dynamics of Plates and Shells (in Russian). Nauka, Moscow (1972)
12. Rossikhin, Y.A., Shitikova, M.V.: Chapter 3 in Computational problems in science and engineering. In: Mastorakis, N., et al. (eds.) Lecture Notes in Electrical Engineering, vol. 343, pp. 59–107. Springer, Berlin (2015)
13. Ajarmah, B., Shitikova, M.V.: J. Phys. Conf. Ser. **1203**, 012008 (2019)
14. Diethelm, K.: Elect. Trans. Num. Anal. **5**, 1–6 (1997)
15. Baleanu, D., Diethelm, K., Scalas, E., Trujillo, J.J.: Fractional Calculus: Models and Numerical Methods, pp. 88–114. World Scientific, Singapore (2012)

Stability of Caputo-Type Fractional Variable-Order Biquadratic Difference Equations

Oana Brandibur, Eva Kaslik, Dorota Mozyrska, and Małgorzata Wyrwas

Abstract The problem of stability of the Caputo-type linear fractional variable-order discrete-time equations known as biquadratic equations is discussed. We study linear equations with constant coefficients and variable-order differences defined by functions with values from the interval $(0, 2]$. For the stability analysis of the considered equations, the $\mathcal{Z}$-transform is used. Sufficient conditions for the asymptotic stability are presented.

Keywords Fractional order difference equations · Variable-order · Stability · Biquadratic equations

1 Introduction

In this paper we discuss the stability property of linear fractional variable-order discrete-time equations with the Caputo-type operator. A thorough understanding of the stability of dynamic systems is essential, because it provides important relationships among process dynamics, controller tuning, and achievable performance. These relationships are used in a variety of ways, such as selecting controller modes, tuning controllers, and designing processes that are easier to control.

The following new aspects are discussed: fractional differences, discrete-time operators. About fractional derivatives or differences we can think that they are operators introducing into the system/equation part of the memory effects. However, the paper is about stability issue, and it is not based on data and does not fit models to data. The reader can treat the proposed model as one of the possible

---

O. Brandibur · E. Kaslik
Department of Mathematics and Computer Science, West University of Timisoara, Timisoara, Romania
e-mail: oana.brandibur@e-uvt.ro; eva.kaslik@e-uvt.ro

D. Mozyrska (✉) · M. Wyrwas
Faculty of Computer Science, Bialystok University of Technology, Bialystok, Poland
e-mail: d.mozyrska@pb.edu.pl; m.wyrwas@pb.edu.pl

W. Lacarbonara et al. (eds.), *New Trends in Nonlinear Dynamics*,
https://doi.org/10.1007/978-3-030-34724-6_30

to be used in the modelling. The pioneer paper introducing fractional calculus in the area of viscoelastically damped structures is [1]. During the last decades fractional calculus has been noticed as a useful tool in modelling various processes and in hereditary physics. There are plenty of examples of deep analysis presented for damping behaviour of viscoelastic media, see, for example, [4–6, 12]. Mainly, there are used continuous-time models with fractional derivative changing the first order derivatives, for example, as the following equation: $\left(D^2 + aD^\nu + b\right)x(t) = f(t)$, where ν is fractional order from interval $(0, 1)$ or $(0, 2)$. And where D^ν introduces memory into the system, changing the operator D^1 in the ordinary differential equations conventional biquadratic equations. In [4] the authors claim that some viscoelastic materials, such as silicone gel and rubber, show a particular weak frequency dependence and they do not obey classical viscoelastic models with integer-order derivative operators. For that they proposed to use fractional differential operators.

In this paper, we show in the sequel the descriptions of the linear Caputo-type difference fractional variable-order equations and we analyse the stability of the considered equations. We formulate and prove stability conditions for generalised form of equations known in the literature as biquadratic systems/equations. Stability conditions for difference systems of Caputo-type or Grünwald–Letnikov-type with positive constant orders were previously analysed in paper [7, 8] and for variable orders with values from $(0, 1]$ in [9, 11, 14, 15]. Very interesting attempts have also been made regarding stability concepts in two-dimensional systems involving Caputo-type differences for incommensurate fractional-order, see [2]. We study linear equations with constant coefficients and fractional variable-order differences with variable-order defined by a function with values $\nu(k) \in (0, 1]$, while the higher order is $1 + \nu(k) \in (1, 2]$. We consider discrete-time equations with the Caputo-type fractional variable-order difference operator of convolution type. In the discrete-time case, the Grünwald–Letnikov fractional operator is usually used, with variable-order backward difference. Here, we propose the convolution operator of the Caputo-type, defined originally in [11]. However, if we consider zero initial condition, both definitions (Grünwald–Letnikov-type and Caputo-type) are then equivalent. The Caputo-type operator allows nonzero stationary points to equations, while equations or systems with the Grünwald–Letnikov operator can have only zero as the stationary point. As a sequence of coefficients as the oblivion function defined for variable-order we use coefficients defined, for example, in [10, 13]. Moreover, considering the convolution type operators it is possible to solve initial value problems for linear equations using the $\mathcal{Z}$-transform.

The paper is organized as follows: Preliminary information about the fractional variable-order operators is given in Sect. 2. The main results of the paper are presented in Sect. 3, where the solutions of the considered equations are provided. Moreover, in Sect. 3 the conditions that guarantee the stability of the considered equation are presented. Finally, Sect. 4 provides brief conclusions.

2 Preliminaries

Firstly, let us introduce definitions of oblivion function with variable-order and the Caputo-type difference operator.

Definition 1 Let $k, l \in \mathbb{Z}$ and $\nu(\cdot)$ be a given order function. Then, we define the oblivion function by the following:

$$a^{[\nu(l)]}(k) = \begin{cases} 0 & \text{for } k < 0 \\ 1 & \text{for } k = 0 \\ (-1)^k \frac{\nu(l)[\nu(l)-1]\cdots[\nu(l)-k+1]}{k!} & \text{for } k > 0 \end{cases} . \tag{1}$$

In this paper we use order functions $\nu(\cdot)$ with values in $(0, 1]$.

Let $y : \mathbb{Z} \to \mathbb{R}$, then the difference operator on $y(\cdot)$ we understand as ([3])

$$(\Delta y)(k) = y(k+1) - y(k) .$$

Let $q \in \mathbb{N}_0 := \{0, 1, 2, \ldots\}$ and $\Delta^q := \Delta \circ \cdots \circ \Delta$ is q-folded operator Δ. Then,

$$(\Delta^q y)(k) = \sum_{i=0}^{q} (-1)^{q-i} \binom{q}{i} y(k+i) .$$

Definition 2 Let $\nu : \mathbb{Z} \to \mathbb{R}_+ \cup \{0\}$. For a function $y : \mathbb{Z} \to \mathbb{R}$ the *fractional variable-order sum of convolution type* is defined by

$$\left(\Delta^{-\nu(\cdot)} y\right)(k) := \left(a^{[-\nu(\cdot)]} * y\right)(k) = \sum_{i=0}^{k} a^{[-\nu(i)]}(i) y(k-i) ,$$

where $k \in \mathbb{N}_0$ and "$*$" means the convolution operator.

The definition, that uses a convolution operator, it is possible to use $\mathcal{Z}$-method. Hence one gets

$$\mathcal{Z}\left[\Delta^{-\nu(\cdot)} y\right](z) = Y(z) \mathcal{Z}\left[a^{[-\nu(\cdot)]}\right](z) , \tag{2}$$

where $Y(z) := \mathcal{Z}[y](z)$ and $\mathcal{Z}\left[a^{[-\nu(\cdot)]}\right](z) = \sum_{i=0}^{\infty} (-1)^i \binom{-\nu(i)}{i} z^{-i}$ are the $\mathcal{Z}$-transforms of the sequences y and $a^{[-\nu(\cdot)]}$, respectively. Then, one can show that the following properties hold

(a) For $\nu(k) \equiv \alpha$ Eq. (2) takes form

$$\mathcal{Z}\left[\Delta^{-\nu(\cdot)} y\right](z) = \left(\frac{z}{z-1}\right)^{\alpha} Y(z).$$

(b) Let $\nu_1, \nu_2 : \mathbb{Z} \to \mathbb{R}_+ \cup \{0\}$. Then,

$$\left(\Delta^{-\nu_1(\cdot)}\left(\Delta^{-\nu_2(\cdot)}y\right)\right)(k) = \left(\Delta^{-\nu_2(\cdot)}\left(\Delta^{-\nu_2(\cdot)}y\right)\right)(k)\,.$$

In what follows, we define the Caputo fractional variable-order difference operator of convolution type.

Definition 3 Let $\nu : \mathbb{Z} \to (q-1, q], q \in \mathbb{N}_1$. Then, *the Caputo fractional variable-order difference operator of convolution type* $\Delta^{\nu(\cdot)}$ with order function $\nu(\cdot)$, for a function $y : \mathbb{Z} \to \mathbb{R}$ is defined by

$$\left(\Delta^{\nu(\cdot)}y\right)(k) = \left(\Delta^{-(q-\nu(\cdot))}\left(\Delta^{q}y\right)\right)(k). \tag{3}$$

It is easy to see that

(a) If $q = 1$, then $\left(\Delta^{\nu(\cdot)}y\right)(k) = \left(\Delta^{-(1-\nu(\cdot))}\left(\Delta y\right)\right)(k)$;
(b) If $\nu(k) \equiv q \in \mathbb{N}_1$, then we have $\left(\Delta^{\nu(\cdot)}y\right)(k) = \left(\Delta^{q}y\right)(k)$.
(c) For $q = 1$: $\mathcal{Z}\left[\Delta^{\nu(\cdot)}y\right](z) = ((z-1)Y(z) - zy(0))\,\mathcal{Z}\left[a^{[\nu(\cdot)-1]}\right](z)$, where $Y(z) = \mathcal{Z}[y](z)$ and $\mathcal{Z}\left[a^{[\nu(\cdot)-1]}\right](z) = \sum_{i=0}^{\infty}(-1)^i\binom{\nu(i)-1}{i}z^{-i}$.

3 Biquadratic Equations and Their Solutions

Let $\nu : \mathbb{Z} \to (0, 1]$. Then, $(1+\nu) : \mathbb{Z} \to (1, 2]$. In the considered equations we use two different operators of higher orders: $\Delta^{1+\nu(\cdot)}$ and $\Delta^{\nu(\cdot)}$. Then, $\left(\Delta^{1+\nu(\cdot)}y\right)(k) = \left(\Delta^{\nu(\cdot)-1}\left(\Delta^2 y\right)\right)(k)$.

Let us consider equations with a variable-order in the form similar to the classical biquadratic equations mentioned in the Introduction:

$$\left(\Delta^{1+\nu(\cdot)}y\right)(k) + 2\xi\omega\left(\Delta^{\nu(\cdot)}y\right)(k) + \omega^2 y(k) = \omega^2 u(k)\,,\ k \geq 0, \tag{4}$$

with initial values $y(0), y(1) \in \mathbb{R}$, where $\nu : \mathbb{Z} \to (0, 1]$ is a variable-order function, function $u : \mathbb{N}_0 \to \mathbb{R}$ is an input function, $y : \mathbb{N}_0 \to \mathbb{R}$ is a response and $\xi > 0, \omega > 0$ are given real constants. The following proposition shows that Eq. (4) has the unique solution for the given initial values $y(0)$ and $y(1)$:

Proposition 1 *The recursive solution to Eq. (4) is as follows:*

$$\begin{aligned}
y(2) &= \omega^2 u(0) + (2\xi\omega - \omega^2 - 1)y(0) + 2(1-\xi\omega)y(1)\\
y(k) &= \omega^2 u(k-2) - 2\xi\omega\left(\Delta^{\nu(\cdot)}y\right)(k-2) - (\omega^2+1)y(k-2) + 2y(k-1)\\
&\quad - \sum_{i=0}^{k-3} a^{[\nu(k-2-i)-1]}(k-2-i)\left(\Delta^2 y\right)(i)\,,\ k \geq 3,
\end{aligned} \tag{5}$$

where values $y(0), y(1) \in \mathbb{R}$ *are given.*

Proof Let the values of order function $\nu(\cdot)$ be from (0, 1]. As we have that

$$\left(\Delta^{1+\nu(\cdot)}y\right)(k) = \left(\Delta^{\nu(\cdot)-1}\left(\Delta^2 y\right)\right)(k) = \sum_{i=0}^{k} a^{[\nu(k-i)-1]}(k-i)\left(\Delta^2 y\right)(i)$$

and

$$\left(\Delta^{\nu(\cdot)}y\right)(k) = \left(\Delta^{\nu(\cdot)-1}\left(\Delta y\right)\right)(k) = \sum_{i=0}^{k} a^{[\nu(k-i)-1]}(k-i)\left(\Delta y\right)(i),$$

then to prove the recurrence formula we need to rewrite both operators in Eq. (4) and then solve the equation according to $y(k+2)$ and recalculate the formula for $y(k)$.

Dealing with linear difference equations, one of the best tools is the $\mathcal{Z}$-transform method. It is also possible to use it in the field of fractional variable-order operators. As the definitions of summations or differences are based on the convolution of the oblivion function, let us state the following notation:

$$\mathcal{A}(z) := \mathcal{Z}\left[a^{[\nu(\cdot)-1]}\right](z) = \sum_{i=0}^{\infty}(-1)^i \binom{\nu(i)-1}{i} z^{-i}. \tag{6}$$

Then since $\nu(k) \in (0, 1]$ for $k \in \mathbb{Z}$, we have that

$$\mathcal{Z}\left[\Delta^{\nu(\cdot)}y\right](z) = ((z-1)Y(z) - zy(0))\,\mathcal{A}(z). \tag{7}$$

Moreover,

$$\mathcal{Z}\left[\Delta^2 y\right](z) = (z-1)^2 Y(z) - zy(1) - z(z-2)y(0) \tag{8}$$

and then

$$\begin{aligned}\mathcal{Z}\left[\Delta^{1+\nu(\cdot)}y\right](z) &= \mathcal{Z}\left[\Delta^2 y\right]\mathcal{A}(z) = \\ &= (z-1)^2 Y(z)\mathcal{A}(z) - (zy(1) + z(z-2)y(0))\,\mathcal{A}(z).\end{aligned} \tag{9}$$

From Eqs. (7)–(9) we see that generally

$$\left(\Delta\left(\Delta^{\nu(\cdot)}y\right)\right)(k) \neq \left(\Delta^{1+\nu(\cdot)}y\right)(k). \tag{10}$$

The equality holds only for zero initial conditions. Hence, taking into account only such situations, we can obtain the same conditions of stability for two cases of higher order operators.

Now, taking $y(0) = y(1) = 0$ and applying $\mathcal{Z}$-transform to Eq. (4) we get the following image:

$$Y(z) = \omega^2 W^{-1}(z) U(z), \tag{11}$$

where $W(z) = (z-1)^2 \mathcal{A}(z) + 2\xi\omega(z-1)\mathcal{A}(z) + \omega^2$. Observe that $W(1) = \omega^2$, $W(-1) = 4(1-\xi\omega)\sum_{i=0}^{\infty}\binom{\nu(i)-1}{i} + \omega^2$, $W(0) = (\omega-\xi)^2 + 1 - \xi^2$.

Proposition 2 *Solutions to (4) are*

(a) *unstable if* $\{z \in \mathbb{C} : W(z) = 0, |z| > 1\} \neq \emptyset$;
(b) *asymptotically stable if* $\{z \in \mathbb{C} : W(z) = 0, |z| > 1\} = \emptyset$.

Proof Using the same arguments as in [8] one can show that because $\{z \in \mathbb{C} : W(z) = 0\}$ is the set of all poles of $(z-1)Y(z)$, so one gets

(a) If there is an element from $\{z \in \mathbb{C} : W(z) = 0\}$ that $|z| > 1$, then $\limsup_{k\to\infty} y(k) = \infty$ and system (4) is unstable.
(b) If all elements from $\{z \in \mathbb{C} : W(z) = 0\}$ are strictly inside the unit circle, then $\lim\limits_{k\to\infty} y(k) = \lim\limits_{z\to 1}(z-1)Y(z) = 0$.

Proposition 3 *If in Eq. (4) parameters* $\xi > 0, \omega > 0$ *satisfy conditions*

$$\xi > \frac{1}{\sqrt{\mathcal{A}(-1)}}, \quad 0 < \omega < 2\left(\xi\mathcal{A}(-1) - \sqrt{\xi^2\mathcal{A}^2(-1) - \mathcal{A}(-1)}\right), \tag{12}$$

then the solutions are asymptotically stable.

Proof Let us consider zeros of $W(z)$, then we can treat equation $W(z) = 0$, i.e.

$$\omega^2 + 2\xi(z-1)\mathcal{A}(z)\omega + (z-1)^2\mathcal{A}(z) = 0 \tag{13}$$

as quadratic equation with respect to ω. Taking into account that $\omega \in \mathbb{R}_+$, we need to take the assumption for ξ, that $\xi \geq 1$ as $\delta_\omega = 4(z-1)^2(\xi^2\mathcal{A}^2 - \mathcal{A})$, and the roots of quadratic equation are $\omega_{1,2} = (z-1)\left(\xi\mathcal{A}(z) \pm \sqrt{\xi^2\mathcal{A}^2(z) - \mathcal{A}(z)}\right)$.

Then, for $z = -1$ we obtain the maximum for ω, for which solutions are stable in the form $\omega_{\max} = 2\left(\xi\mathcal{A}(-1) - \sqrt{\xi^2\mathcal{A}^2(-1) - \mathcal{A}(-1)}\right)$.

Example 1 Let us consider a response of unit signal from Eq. (4) with parameters for stable responses $\xi = \frac{1}{\sqrt{A}} + 1$, $\omega = 2\left(\xi A - \sqrt{A^2\xi^2 - A}\right) * 0.9$ and for the following different order functions:

(1) $\nu_1(k) = 0.1$, see Fig. 1, $\xi = 2.130367$, $\omega = 0.457303$.
(2) $\nu_2(k) = 1 - \frac{1}{k+1}$, see Fig. 2, $\xi = 2.242281$, $\omega = 0.4380649$.
(3) $\nu_3(k) = 0.5 + 0.5\exp(-0.01k)$, see Fig. 3, $\xi = 1.997183$, $\omega = 0.482884$.
(4) $\nu_4(k) = 0.7 + 0.3\cos^2(k/6)$, see Fig. 4, $\xi = 2.000846$, $\omega = 0.482136$.

Calculations of $\mathcal{A}(-1)$ and parameters ξ, ω are made for $T = 1200$ steps.

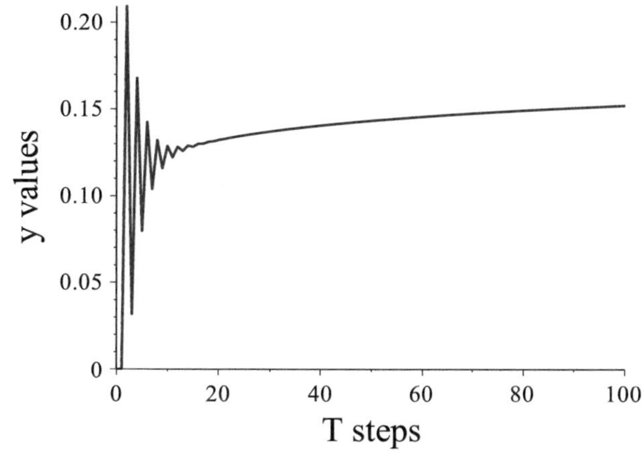

Fig. 1 Graphs of unit response of Eq. (4) for variable-order function $\nu_1(k) = 0.1$, $T = 100$ steps

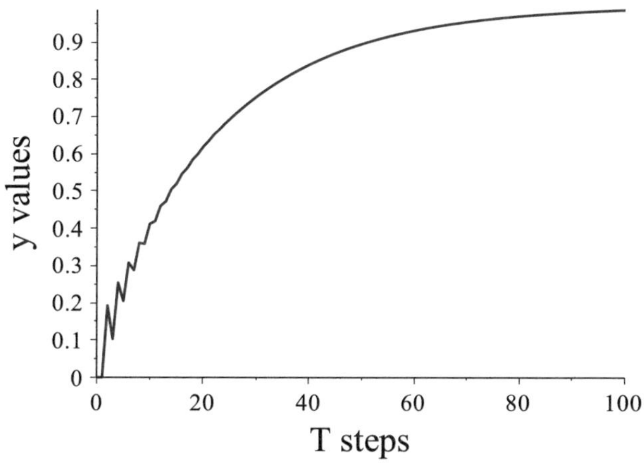

Fig. 2 Graphs of unit response of Eq. (4) for variable-order function $\nu_2(k) = 1 - \frac{1}{k+1}$, $T = 100$ steps

The example illustrates the behaviour of the solutions for different order functions. We can observe that behaviours of four considered order functions are slightly different, as well as values of the border of proposed parameters, as this depends on monotonicity of order functions. It is not possible to compare to another methods, as our definitions are new and it is not connected with direct discretization of factional derivative of continuous-time. We use directly discrete-time version of operators. In Figs. 1, 2, 3 and 4, the unit step for t is equal to 1, as we work only with differences.

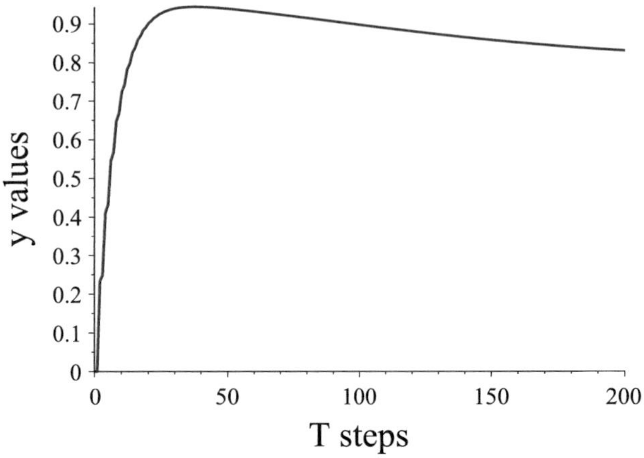

Fig. 3 Graphs of unit response of Eq. (4) for variable-order function $\nu_3(k) = 0.5 + 0.5 \exp(-0.01k)$, $T = 200$ steps

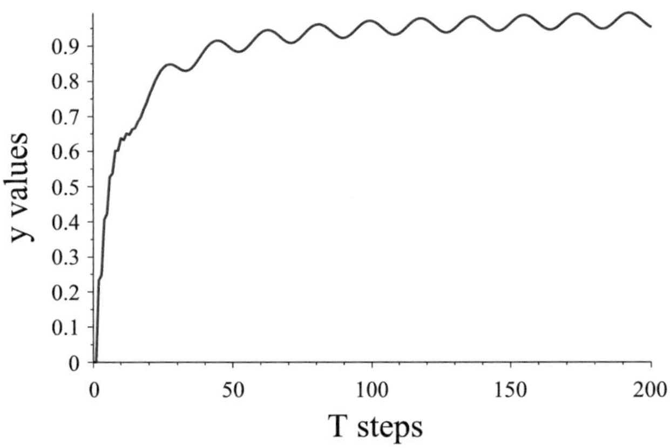

Fig. 4 Graphs of unit response of Eq. (4) for variable-order function $\nu_4(k) = 0.7 + 0.3\cos^2(k/6)$, $T = 200$ steps

4 Conclusions

The goal of this chapter was to study the behavior of the Caputo-type linear fractional variable-order biquadratic difference equations. We showed the existence of the solutions of the considered equations and subseqeuntly, based on the $\mathcal{Z}$-transform, we presented the conditions that guarantee asymptotic stability.

Acknowledgements This article is based upon work from COST Action CA15225, a network supported by COST (European Cooperation in Science and Technology). The work was supported by Polish funds of National Science Center, granted on the basis of decision DEC-2016/23/B/ST7/03686.

References

1. Bagley, R.L., Torvik, J.: Fractional calculus—a different approach to the analysis of viscoelastically damped structure. AIAA J. **21**(5), 741–748 (1983)
2. Brandibur, O., Kaslik, E.: Stability of two-components incommensurate fractional-order systems and applications to the investigations of a Fitzhugh–Nagumo neuronal model. Math. Methods Appl. Sci. **41**(17), 7182–7194 (2018). https://doi.org/10.1002/mma.4768
3. Ferreira, R.A.C., Torres, D.F.M.: Fractional h-difference equations arising from the calculus of variations. Appl. Anal. Discrete Math. **5**(1), 110–121 (2011)
4. Fukunaga, M., Shimizu, N.: Role of prehistories in the initial value problems of fractional viscoelastic equations. Nonlinear Dyn. **38**, 207–220 (2004)
5. Kempfle, S., Schäfer, I., Beyer, H.: Fractional calculus via functional calculus: theory and applications. Nonlinear Dyn. **29**, 99–127 (2002)
6. Kempfle, S., Schäfer, I., Beyer, H.: Impulse responses of fractional damped systems. Nonlinear Dyn. **38**, 61–68 (2004)
7. Mozyrska, D., Wyrwas, M.: Advances in the thoery and applications of non-integer order systems. In: Latawiec, K.J., Łukaniszyn, M., Stanisławski, R. (eds.) Lecture Notes in Electrical Engineering, vol. 320, chap. Fractional Linear Equations with Discrete Operators of Positive Order, pp. 47–58. Springer, Berlin (2014)
8. Mozyrska, D., Wyrwas, M.: The Z-transform method and delta type fractional difference operators. Discrete Dyn. Nat. Soc. **2015**, 852734 (2015). https://doi.org/10.1155/2015/852734
9. Mozyrska, D., Ostalczyk, P.: Variable-fractional-order Grünwald-Letnikov backward difference selected properties. In: Proceedings of the 39th International Conference on Telecommunications and Signal Processing (2016)
10. Mozyrska, D., Ostalczyk, P.: Generalized fractional–order discrete–time integrator. Complexity **2017**, 3452409 (2017)
11. Mozyrska, D., Wyrwas, M.: Stability of linear systems with Caputo fractional-, variable-order difference operator of convolution type. In: 2018 41st International Conference on Telecommunications and Signal Processing (TSP). IEEE, Athens (2018). https://doi.org/10.1109/TSP.2018.8441360
12. Najafi, H.S., Toroqi, E.A., Divishali, A.J.: A new fractional model of single degree of freedom system, by using generalized differential transform method. Rom. J. Math. Comput. Sci. **6**(1), 93–105 (2016)
13. Ostalczyk, P.: Discrete fractional calculus. Applications in control and image processing. Series in Computer Vision, vol. 4. World Scientific, Singapore (2016)
14. Sierociuk, D., Malesza, W., Macias, M.: On a new definition of fractional variable-order derivative. In: Proceedings of the 2013 14th International Carpathian Control Conference (ICCC), pp. 340–345 (2013)
15. Valério, D., Sáda Costa, J.: Variable-order fractional derivatives and their numerical approximations. Signal Process. **91**(3), 470–483 (2011). https://doi.org/10.1016/j.sigpro.2010.04.006

Stability of Systems of Fractional-Order Difference Equations and Applications to a Rulkov-Type Neuronal Model

Oana Brandibur, Eva Kaslik, Dorota Mozyrska, and Małgorzata Wyrwas

Abstract Necessary and sufficient conditions for the asymptotic stability and instability of two-dimensional linear autonomous incommensurate systems of fractional-order Caputo difference equations are presented. Moreover, the occurrence of discrete Flip and Hopf bifurcations is also discussed, choosing the fractional orders as bifurcation parameters. The theoretical results are then applied to the investigation of the stability and instability properties of a fractional-order version of the Rulkov neuronal model. Numerical simulations are further presented to illustrate the theoretical findings, revealing complex bursting behavior in the fractional-order Rulkov model.

Keywords Fractional-order difference equation · Caputo fractional difference · Fractional-order Rulkov model · Neuronal model · Incommensurate fractional-order system

1 Introduction

As fractional-order derivatives are known to reflect both memory and hereditary properties of different processes, numerous recent results have proven that fractional-order systems provide more realistic results in practical applications. Stability properties and linearization theorems for both continuous-time fractional systems [2, 3] and discrete-time fractional systems [6, 7] were recently obtained. Regarding the stability analysis of linear discrete-time fractional-order systems, the most used and effective tool is the $\mathcal{Z}$-transform. Due to the fact that fractional

O. Brandibur (✉) · E. Kaslik
Department of Mathematics and Computer Science, West University of Timisoara, Timisoara, Romania
e-mail: oana.brandibur@e-uvt.ro

D. Mozyrska · M. Wyrwas
Faculty of Computer Science, Bialystok University of Technology, Bialystok, Poland
e-mail: d.mozyrska@pb.edu.pl; m.wyrwas@pb.edu.pl

W. Lacarbonara et al. (eds.), *New Trends in Nonlinear Dynamics*,
https://doi.org/10.1007/978-3-030-34724-6_31

derivatives can be approximated by fractional h-differences of corresponding types, we can establish the relationship between the stability of fractional-order differential systems and the stability of their discrete-time counterparts.

In this work, we present general stability and instability results for two-dimensional systems of fractional-order difference equations. As an application to the theoretical findings, we investigate the stability of a fractional-order version of the Rulkov neuronal model [8, 9], providing extensive numerical simulations, where discrete Hopf bifurcations that occur can also be observed, in conjunction with complex and chaotic bursting behavior.

Experimental neuronal research [1, 5] has recently suggested that the mathematical modeling of neuronal dynamics should incorporate fractional-order derivatives or fractional-order differences. Fractional-order membrane potential dynamics have been shown to introduce capacitive memory effects [10], proving their utility in reproducing the electrical activity of neurons. Furthermore, [4] suggests that the index of memory could be a possible physical interpretation of the fractional order, which is in accordance with the use of fractional-order derivatives and differences in mathematical models arising from neuroscience.

2 Preliminaries on Fractional-Order Difference Operators

Let $h > 0$, $(h\mathbb{N})_0 = \{0, h, 2h, \ldots\}$ and a function $x : (h\mathbb{N})_0 \to \mathbb{R}$. The operator

$$(\Delta_h x)(t) := \frac{x(t+h) - x(t)}{h}, \quad t \in (h\mathbb{N})_0$$

is called the forward h-difference operator.

Let us introduce the family of binomial functions on $\mathbb{Z}$ parameterized by $q > 0$ and given by the values

$$\tilde{\varphi}_q(j) = \binom{j+q-1}{j} = (-1)^j \binom{-q}{j}, \text{ for } j \in \mathbb{N}_0, \quad \tilde{\varphi}_q(j) = 0, \text{ if } j < 0.$$

Definition 1 For a function $x : (h\mathbb{N})_0 \to \mathbb{R}$, the fractional h-sum of order $q > 0$ is given by $(\Delta_h^{-q} x)(t) := h^q(\tilde{\varphi}_q * \bar{x})(j), t = jh, \bar{x}(s) := x(sh), s \in \mathbb{N}$, where "$*$" denotes the convolution operator.

Definition 2 Let $q \in (0, 1]$. The Caputo-type h-difference operator ${}^c\Delta^q$ of order q for a function $x : (h\mathbb{N})_0 \to \mathbb{R}$ is defined by

$$({}^c\Delta^q x)(t) = (\Delta_h^{-(1-q)}(\Delta_h x))(t), \quad t \in (h\mathbb{N})_0.$$

Proposition 1 *For $q \in (0, 1]$, let us define $y(j) = ({}^c\Delta^q x)(t)$, where $t \in (h\mathbb{N})_0$ and $t = jh$. Then the $\mathcal{Z}$-transform of $y(j)$ is*

$$\mathcal{Z}[y](z) = h^{-q}\left(\frac{z}{z-1}\right)^{1-q}\Big((z-1)X(z) - zx(0)\Big),$$

where $X(z) = \mathcal{Z}[\bar{x}](z)$ *and* $\bar{x}(j) = x(jh)$.

3 General Stability Results for Two-dimensional Systems of Fractional-Order Difference Equations

Let us consider the n-dimensional fractional-order system

$$^{c}\Delta^{\mathbf{q}}\mathbf{x}(nh) = f(nh, \mathbf{x}(nh)), \tag{1}$$

where $\mathbf{q} = (q_1, q_2, \ldots, q_n) \in (0, 1)^n$ and $f : (h\mathbb{N})_0 \times \mathbb{R}^n \to \mathbb{R}^n$ is continuous on the whole domain of definition and Lipschitz-continuous with respect to the second variable, such that $f(nh, 0) = 0$ for any $nh \in (h\mathbb{N})_0$. Let $\varphi(nh, x_0)$ denote the unique solution of (1) which satisfies the initial condition $x(0) = x_0$.

Definition 3

(a) The trivial solution of (1) is called *stable* if for any $\varepsilon > 0$ there exists $\delta = \delta(\varepsilon) > 0$ such that for every $x_0 \in \mathbb{R}^n$ satisfying $\|x_0\| < \delta$ we have $\|\varphi(n, x_0)\| \leq \varepsilon$ for any $n \geq 0$.
(b) The trivial solution of (1) is called *asymptotically stable* if it is stable and there exists $\rho > 0$ such that $\lim_{n\to\infty} \varphi(n, x_0) = 0$ whenever $\|x_0\| < \rho$.
(c) The trivial solution of (1) is called $\mathcal{O}(n^{-q})$*-asymptotically stable* if it is stable and there exists $\rho > 0$ such that for any $\|x_0\| < \rho$ one has $\|\varphi(n, x_0)\| = \mathcal{O}(n^{-q})$ as $n \to \infty$.

Consider the following two-dimensional linear autonomous incommensurate fractional-order system:

$$\begin{cases} {}^{c}\Delta^{q_1} x(nh) = a_{11}x(nh) + a_{12}y(nh) \\ {}^{c}\Delta^{q_2} y(nh) = a_{21}x(nh) + a_{22}y(nh) \end{cases}, \tag{2}$$

where $A = (a_{ij})$ is a real two-dimensional matrix, $q_1, q_2 \in (0, 1)$ are the fractional orders of Caputo forward difference operators, and h is the discretization step.

Applying the $\mathcal{Z}$-transform to system (2) and denoting $\mathcal{Z}[x] = X$ and $\mathcal{Z}[y] = Y$ the $\mathcal{Z}$-transforms of x and y, respectively, we deduce

$$\left(\begin{bmatrix} h^{-q_1}z(1-z^{-1})^{q_1} & 0 \\ 0 & h^{-q_2}z(1-z^{-1})^{q_2} \end{bmatrix} - A\right)\begin{bmatrix} X(z) \\ Y(z) \end{bmatrix} = \begin{bmatrix} F_1(z) \\ F_2(z) \end{bmatrix},$$

where

$$F_1(z) = x(0)h^{-q_1}z(1-z^{-1})^{q_1-1} \quad \text{and} \quad F_2(z) = y(0)h^{-q_2}z(1-z^{-1})^{q_2-1}.$$

The following characteristic equation is obtained:

$$\det\left(\mathrm{diag}\left(h^{-q_1}z(1-z^{-1})^{q_1}, h^{-q_2}z(1-z^{-1})^{q_2}\right) - A\right) = 0.$$

Hence, the characteristic function of the system (2) is

$$\Delta_A(z) = z^2(1-z^{-1})^{q_1+q_2} - a_{11}h^{q_1}z(1-z^{-1})^{q_2} - a_{22}h^{q_2}z(1-z^{-1})^{q_1} + h^{q_1+q_2}\det(A).$$

As in the continuous-time case [2, 3], the following basic result is obtained for the characterization of the stability and instability properties of system (2), in terms of the distribution of the roots of the characteristic function.

Theorem 1

1. *Denoting* $q = \min\{q_1, q_2\}$, *system (2) is* $\mathcal{O}(n^{-q})$*-globally asymptotically stable if and only if all the roots of* $\Delta_A(z)$ *are inside the unit circle* ($|z| < 1$).
2. *If* $\det(A) \neq 0$ *and* $\Delta_A(z)$ *has a root outside the closed unit circle, system (2) is unstable.*

The following sufficient conditions for the instability of system (2), which do not depend on the fractional orders q_1 and q_2 can be obtained using basic mathematical tools:

Theorem 2 (Fractional-Order Independent Instability Results) *System (2) is unstable regardless of the fractional orders* q_1 *and* q_2 *if one of the following conditions hold*

1. $\det(A) < 0$;
2. $\det(A) > 0$ *and* $(1-h)(a_{11}+a_{22}) \geq (1-h)^2\det(A) + 1$;
3. $a_{11} > 0$ *and* $a_{11}a_{22} \geq \det(A) > 0$.

In the following, let $\delta = \det(A) > 0$ and $q_1, q_2 \in (0, 1]$, $h > 0$ arbitrarily fixed. With the aim of establishing easily applicable stability conditions, we introduce the following notations in the (a_{11}, a_{22})-plane:

- the line

$$l(\delta, q_1, q_2, h) \; : \quad a_{11}\left(\frac{h}{2}\right)^{q_1} + a_{22}\left(\frac{h}{2}\right)^{q_2} + \delta\left(\frac{h}{2}\right)^{q_1+q_2} + 1 = 0$$

- for $q_1 \neq q_2$, the smooth parametric curve

$$\Gamma(\delta, q_1, q_2, h) \; : \quad \begin{array}{l} a_{11} = \rho_2(q_1, q_2, \theta)u(\theta, h)^{q_1} - \delta\rho_1(q_1, q_2, \theta)u(\theta, h)^{-q_2} \\ a_{22} = \delta\rho_2(q_1, q_2, \theta)u(\theta, h)^{-q_1} - \rho_1(q_1, q_2, \theta)u(\theta, h)^{q_2} \end{array},$$

where $\theta \in \left(0, \frac{\pi}{2}\right)$, $\rho_k(q_1, q_2, \theta) = \frac{\sin(2-q_k)\theta}{\sin(q_2-q_1)\theta}$ for $k = \overline{1, 2}$, $u(\theta, h) = \frac{2}{h}\cos\theta$.

– for $q_1 = q_2 =: q$, the line

$$\Lambda(\delta, q, h): \quad a_{11} + a_{22} = -2\sqrt{\delta}\cos\left[(2-q)\arccos\left(\frac{h}{2}\delta^{1/2q}\right)\right].$$

With these notations, the following stability result is obtained:

Theorem 3 (Fractional-Order-Dependent Stability Results)
Let $q_1, q_2 \in (0, 1]$, $h > 0$ *arbitrarily fixed and* $0 < \delta = \det(A) < \left(\frac{2}{h}\right)^{q_1+q_2}$.

1. *If* $q_1 \neq q_2$, *system (2) is* $\mathcal{O}(n^{-q})$*-asymptotically stable* ($q = \min\{q_1, q_2\}$) *if and only if* (a_{11}, a_{22}) *are in the domain above the line* $l(\delta, q_1, q_2, h)$ *and below the curve* $\Gamma(\delta, q_1, q_2, h)$.
2. *If* $q_1 = q_2 := q$, *system (2) is* $\mathcal{O}(n^{-q})$*-asymptotically stable if and only if*

$$-\delta\left(\frac{h}{2}\right)^{q} - \left(\frac{h}{2}\right)^{-q} < a_{11} + a_{22} < -2\sqrt{\delta}\cos\left[(2-q)\arccos\left(\frac{h}{2}\delta^{1/2q}\right)\right].$$

It is worth noting that if $(a_{11}, a_{22}) \in \Gamma(\delta, q_1, q_2, h)$, the characteristic function $\Delta_A(z)$ has a pair of complex conjugated roots on the unit circle, which is a requirement for the discrete Hopf bifurcation in the classical theory of discrete dynamical systems. On the other hand, if $(a_{11}, a_{22}) \in l(\delta, q_1, q_2, h)$, the characteristic function $\Delta_A(z)$ has a root $z = -1$, which is associated with the occurrence of a flip bifurcation in discrete dynamical systems. However, it is important to emphasize that the bifurcation theory of systems of fractional-order difference equations remains yet to be investigated.

The results obtained in Theorem 3 are exemplified in Fig. 1 for different values of the discretization step h, while fixing $\delta = \det(A) = 5$ and the fractional orders $q_1 = 0.8$ and $q_2 = 0.4$.

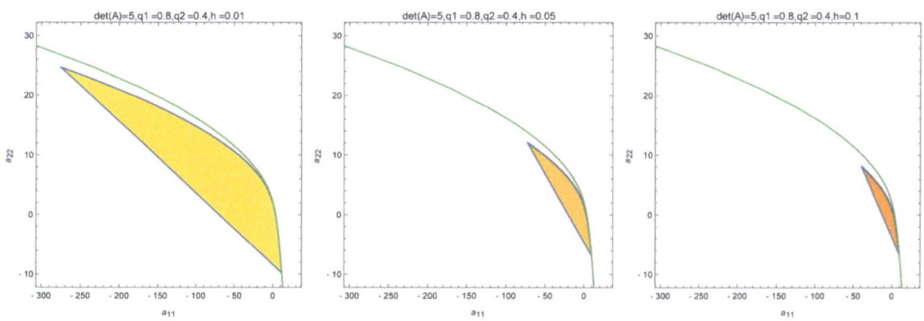

Fig. 1 Stability regions in the (a_{11}, a_{22})-plane, for different values of the discretization step h. The green curve represents the limiting case when $h \to 0$, i.e., it represents the boundary of the stability region corresponding to the continuous-time counterpart [3]

4 Applications to a Fractional-Order Rulkov-Type Neuronal Model

As an application to the theoretical results presented above, we analyze the following discrete-time fractional-order Rulkov-type model, describing the spiking behavior of a biological neuron:

$$\begin{aligned} {}^c\Delta^{q_1} x(n) &= \frac{\alpha}{1 + x(n)^2} - x(n) + y(n) \\ {}^c\Delta^{q_2} y(n) &= -\mu(x(n) - \sigma) \end{aligned}, \tag{3}$$

where x represents the membrane potential, y is a gating variable, with $0 < \mu \ll 1$, σ acts as an external current applied to the neuron, and $\alpha > 0$ is a nonlinearity parameter. We will assume for simplicity that $0 < q_1 < q_2 \leq 1$.

The unique fixed point of system (3) is $(x^*, y^*) = \left(\sigma, \sigma - \frac{\alpha}{1 + \sigma^2}\right)$. The Jacobian matrix of system (3) at (x^*, y^*) is

$$J = \begin{pmatrix} -1 - \frac{2\alpha\sigma}{(1+\sigma^2)^2} & 1 \\ -\mu & 0 \end{pmatrix}.$$

With the notations from the previous section, we observe that

$$a_{11} = -1 - \frac{2\alpha\sigma}{(1+\sigma^2)^2} < 0, \quad a_{22} = 0, \quad 0 < \delta = \mu \ll 1 < 2^{q_1+q_2}, \quad h = 1.$$

Theorem 3 implies that the fixed point (x^*, y^*) is asymptotically stable if and only if

$$\begin{aligned} -2^{-q_2}\mu - 2^{q_1} < -1 - \frac{2\alpha\sigma}{(1+\sigma^2)^2} &< \rho_2(q_1, q_2, \theta^*)(2\cos\theta^*)^{q_1} \\ &- \mu\rho_1(q_1, q_2, \theta^*)(2\cos\theta^*)^{-q_2}, \end{aligned}$$

where $\theta^* = \theta^*(q_1, q_2) \in \left(0, \frac{\pi}{2}\right)$ is the unique solution of the equation

$$\frac{\sin(2 - q_1)\theta^*}{\sin(2 - q_2)\theta^*} \cdot (2\cos\theta^*)^{q_1+q_2} = \mu. \tag{4}$$

Based on the above inequalities, we deduce that a sufficient condition for the instability of the fixed point (x^*, y^*), regardless of the choice of fractional orders q_1 and q_2, is the fulfillment of one of the following inequalities:

$$-\frac{2\alpha\sigma}{(1+\sigma^2)^2} < -\frac{\mu}{2} - 1 \qquad \text{or} \qquad -\frac{2\alpha\sigma}{(1+\sigma^2)^2} > M(\mu) + 1$$

where

$$M(\mu) = \sup_{0<q_1<q_2<1} \left[\rho_2(q_1, q_2, \theta^*)(2\cos\theta^*)^{q_1} - \mu\rho_1(q_1, q_2, \theta^*)(2\cos\theta^*)^{-q_2}\right],$$

with $\theta^* = \theta^*(q_1, q_2)$ given by (4).

For the numerical simulations, $\mu = 0.001$ is fixed. Figure 2 shows the stability region in the (α, σ)-plane for system (3) with fixed values of the fractional orders $q_1 = 0.6$ and $q_2 = 1$ (Fig. 3). Furthermore, fixing $\sigma = -1$, we find the critical value $\alpha^* = 2.09278$ where a Hopf bifurcation is expected to occur (see Fig. 4).

On the other hand, as our aim is observing neuronal bursting behavior in the fractional-order model, we search for values of the parameters α and σ for which the unique equilibrium of system (3) is guaranteed to be unstable, for every choice of the fractional orders q_1 and q_2. Numerical evaluation of $M(\mu)$ given above using differential evolution or random search returns the value 1.0083. Therefore, the fractional-order independent instability region plotted in Fig. 3 is obtained. Different types of bursting behavior can be observed in Fig. 5 for $\alpha = 4.5$, $\sigma = -1$ belonging to the instability region from Fig. 3, by varying q_1 between 0.01 and 1, for fixed $q_2 = 1$.

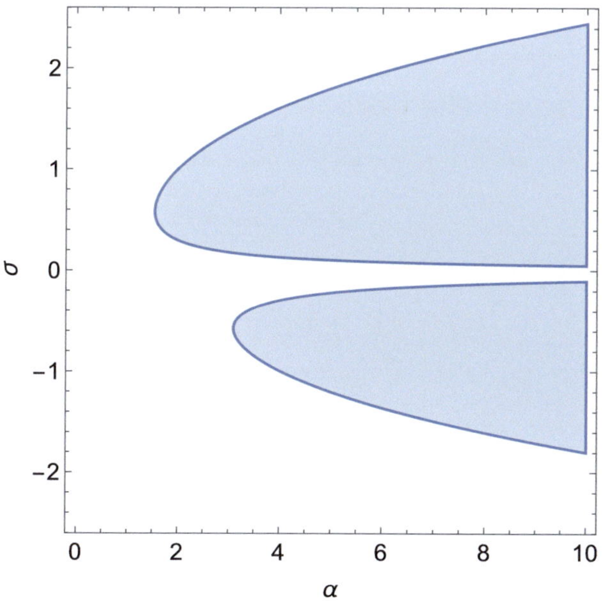

Fig. 2 Stability region in the (α, σ)-plane for system (3) with $\mu = 0.001$, $q_1 = 0.6$, and $q_2 = 1$

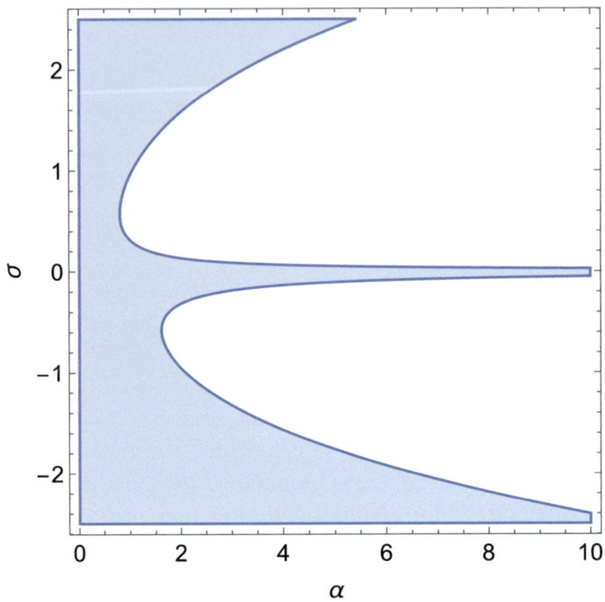

Fig. 3 Fractional-order independent instability region in the (α, σ)-plane for system (3) with $\mu = 0.001$

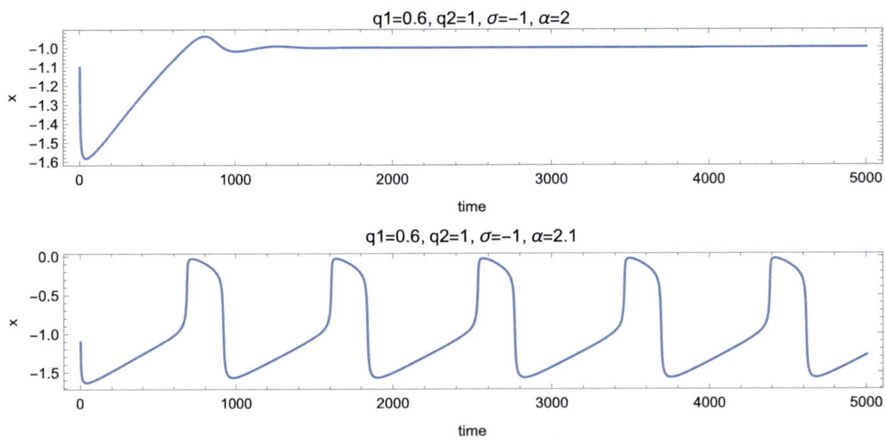

Fig. 4 A Hopf bifurcation occurs in the Rulkov model (3) with $\mu = 0.001$, $\sigma = -1$, $q_1 = 0.6$, $q_2 = 1$ at the critical value $\alpha^* = 2.09278$. The equilibrium state (x^*, y^*) is asymptotically stable for $\alpha < \alpha^*$ and it loses its stability for $\alpha > \alpha^*$. An asymptotically stable limit cycle appears for $\alpha > \alpha^*$, suggesting the occurrence of a supercritical Hopf bifurcation

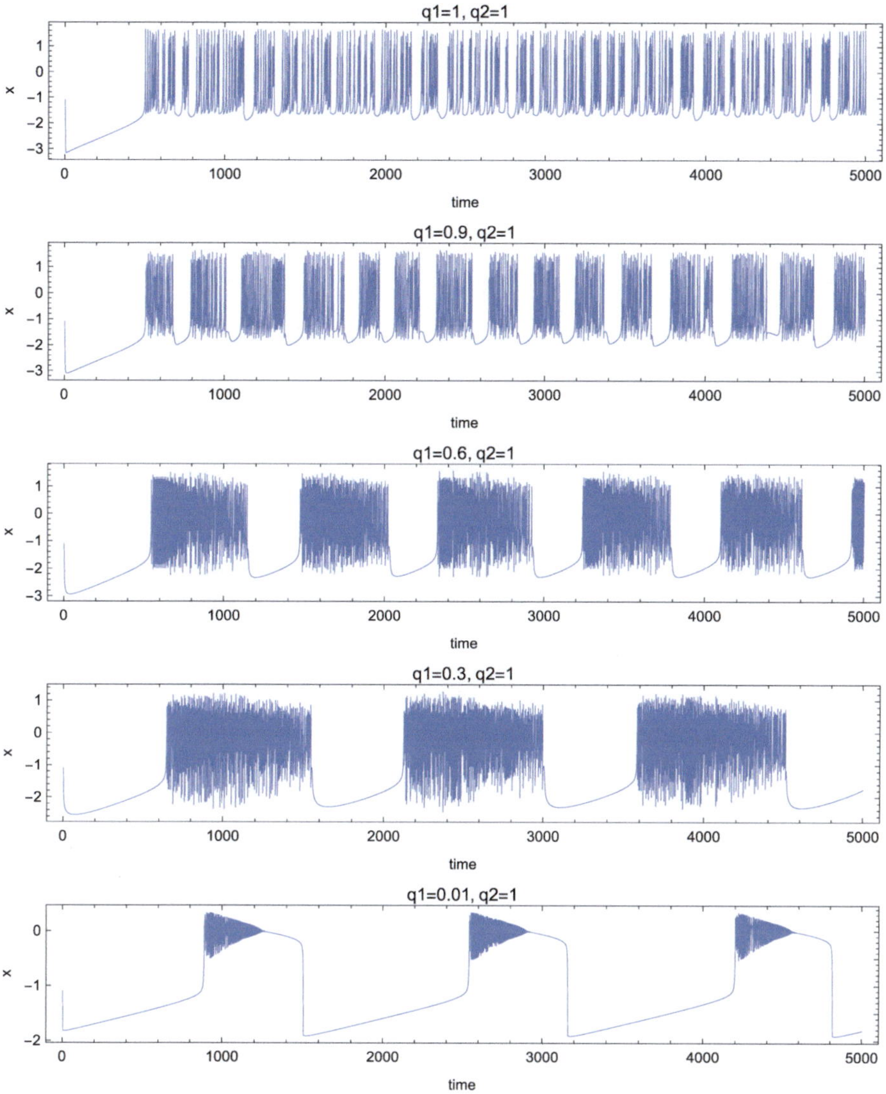

Fig. 5 Bursting behavior in the Rulkov model (3) with $\mu = 0.001$, $\alpha = 4.5$, $\sigma = -1$ and the fractional orders $q_2 = 1$ and q_1 varying from 0.01 to 1

5 Conclusions

Theoretical results concerning the asymptotic stability and instability of two-dimensional incommensurate systems of autonomous linear fractional-order Caputo difference equations has been presented and applied in the investigation of a fractional-order Rulkov neuronal model. Numerical simulations reveal that chaotic bursting can be modulated by varying the fractional orders of the system.

Acknowledgements This article is based upon work from COST Action CA15225, a network supported by COST (European Cooperation in Science and Technology). The work was supported by Polish funds of National Science Center, granted on the basis of decision DEC-2016/23/B/ST7/03686.

References

1. Anastasio, T.J.: The fractional-order dynamics of brainstem vestibulo-oculomotor neurons. Biol. Cybern. **72**(1), 69–79 (1994)
2. Brandibur, O., Kaslik, E.: Stability properties of a two-dimensional system involving one Caputo derivative and applications to the investigation of a fractional-order Morris-Lecar neuronal model. Nonlinear Dyn. **90**(4), 2371–2386 (2017)
3. Brandibur, O., Kaslik, E.: Stability of two-component incommensurate fractional-order systems and applications to the investigation of a Fitzhugh–Nagumo neuronal model. Math. Methods Appl. Sci. **41**(17), 7182–7194 (2018)
4. Du, M., Wang, Z., Hu, H.: Measuring memory with the order of fractional derivative. Sci. Rep. **3**, 3431 (2013)
5. Lundstrom, B.N., Higgs, M.H., Spain, W.J., Fairhall, A.L.: Fractional differentiation by neocortical pyramidal neurons. Nat. Neurosci. **11**(11), 1335–1342 (2008)
6. Mozyrska, D., Wyrwas, M.: Explicit criteria for stability of fractional h-difference two-dimensional systems. Int. J. Dyn. Control **5**(1), 4–9 (2017)
7. Mozyrska, D., Wyrwas, M.: Stability by linear approximation and the relation between the stability of difference and differential fractional systems. Math. Methods Appl. Sci. **40**(11), 4080–4091 (2017)
8. Rulkov, N.F.: Regularization of synchronized chaotic bursts. Phys. Rev. Lett. **86**(1), 183–186 (2001)
9. Rulkov, N.F.: Modeling of spiking-bursting neural behavior using two-dimensional map. Phys. Rev. E **65**(4), 041922 (2002)
10. Weinberg, S.H.: Membrane capacitive memory alters spiking in neurons described by the fractional-order Hodgkin–Huxley model. PLoS One **10**(5), e0126,629 (2015)

Independent Fractional Type Modes of Free and Forced Vibrations of Discrete Continuum Hybrid Systems of Fractional Type with Multi-Deformable Bodies

Katica R. (Stevanović) Hedrih

Abstract Hybrid systems of fractional type with multi-deformable bodies exhibit very complex dynamics. In this chapter, an advanced analysis of independent fractional type modes of free and hybrid modes of forced vibrations of these hybrid systems is performed. The analyses are based on author's previous published results of the analytical dynamics and vibrations of the fractional type discrete and discrete continuum hybrid systems. A hybrid discrete continuum system of fractional type with multi-deformable bodies consists of ideal elastic bodies (belts, beams, membranes or plates) with equal boundary conditions, coupled by standard light fractional type discrete continuum layers in transversal direction permitting the system transversal vibrations. Independent fractional type hybrid forced modes appear in each of the eigen time functions of each of the eigen amplitude forms in both, free and forced regimes of transversal vibrations of discrete continuum system. Number of the independent fractional type modes in each of the eigen time function of each of the eigen amplitude shape is equal to number of the coupled deformable bodies in hybrid fractional type system (HFS) plus number of the external forced excitation frequencies.

The main new result, presented in this chapter, is the description of the hybrid forced fractional type vibration modes with corresponding analytical approximate solution expressed by integral of the convolution.

On the basis of this result it is possible to define independent fractional type hybrid modes in HFS with two characteristic numbers and a corresponding number of external excitation forced frequencies. One of these characteristic numbers corresponds to the square of eigen frequencies. The other characteristic number corresponds to fractional type system properties.

K. R. (Stevanović) Hedrih (✉)
Department of Mechanics, Mathematical Institute of the Serbian Academy of Science and Arts, Belgrade, Serbia
e-mail: khedrih@sbb.rs; katicah@mi.sanu.ac.rs

W. Lacarbonara et al. (eds.), *New Trends in Nonlinear Dynamics*,
https://doi.org/10.1007/978-3-030-34724-6_32

Keywords Discrete continuum hybrid systems of multi-deformable bodies · Fractional type hybrid forced modes · Generalized function of fractional type dissipation of the energy · Functions in convoluting · Theorem

1 A Brief Historical Introduction to Fractional Calculus

Fractional calculus was initiated by Leibniz and L'Hopital as a result of a correspondence which lasted several months during 1695. That year, Leibniz wrote a letter to L'Hopital raising the following question [1]: "*Can the meaning of derivatives with integer order be generalized to derivatives with non-integer orders*?" L'Hopital was somewhat curious about the above question and replied by another simple one to Leibniz: "*What if the order will be* 1/2?". Leibniz in a letter dated September 30, 1695, replied: "*It will lead to a paradox, from which one day useful consequences will be drawn*."

Since the nineteenth century, the theory of fractional calculus has developed rapidly, mostly as a foundation for a number of applied disciplines, especially for applications in rheology and viscoelasticity, to define constitutive stress-strain relations of the material with dissipation. All fractional order differential operators in fractional calculus are therefore an excellent set of tools for describing the memory and hereditary properties of various materials and viscoelastic processes [2–5].

Between numerous definitions and forms of the fractional order derivatives, in this chapter, a differential fractional order operator denoted by $\mathfrak{D}_t^{\alpha}[\bullet]$ is used. This operator is fractional order differential operator of the αth derivative, with respect to time in the following form (see [1–11]):

$$\mathfrak{D}_t^{\alpha}[x(t)]=\frac{d^{\alpha}x(t)}{dt^{\alpha}}=x^{(\alpha)}(t)=\frac{1}{\Gamma(1-\alpha)}\frac{d}{dt}\int_0^t\frac{x(\tau)}{(t-\tau)^{\alpha}}d\tau,\quad \alpha\in(0.1),\ t\in(0.b) \tag{1}$$

where α is a rational number between 0 and 1, $0<\alpha<1$, determined experimentally and $\Gamma(1-\alpha)$ is the Gamma function. In this chapter, independent fractional type modes are based on the known two ordinary fractional order differential equations in the following forms:

$$\ddot{\xi}+\omega_{(\alpha)}^2\mathfrak{D}_t^{\alpha}[\xi]+\omega_0^2\xi=\begin{matrix} 0, & \text{for free vibrations} \\ h_0\sin(\Omega_0 t+\vartheta_0), & \text{for forced vibrations} \end{matrix},$$

$$\alpha\in(0.1),\ t\in(0.b) \tag{2}$$

where ξ is coordinate-function of time, ω_0^* and $\omega_{(\alpha)}$ are constant coefficients of ordinary fractional order differential equation, and α is a rational number between 0 and 1, $0 < \alpha < 1$. Analytical approximate solution of the first ordinary fractional order homogeneous differential equation in (2) is known from literature [2–4, 9]. Analytical approximate particular solution of the second ordinary fractional order non-homogeneous differential equation in (2) can be obtained by use Laplace transform and inverse of Laplace transform and properties of Laplace transform of the two functions in convolution.

2 Fractional Type Oscillator with One Degree of Freedom and Generalized Function of Fractional-Type Dissipation of the Energy

Consider a mechanical fractional-type system, with one degree of freedom, presented in Fig. 1a*. A simple mechanical fractional type oscillator contains a mass particle, mass m, moving translator along a line trace and coupled by a standard light fractional type element at one fixed point in same direction of mass particle displacement ξ. Constitutive relation between reactive force $P(\xi, \mathfrak{D}[\xi])$ and axial extension deformation–elongation ξ of the standard light fractional type element is in the form [1, 6–11]:

$$P(\xi, \mathfrak{D}[\xi]) = P(t) = -\left\{c_0\xi(t) + c_\alpha \mathfrak{D}_t^\alpha [\xi(t)]\right\} \tag{3}$$

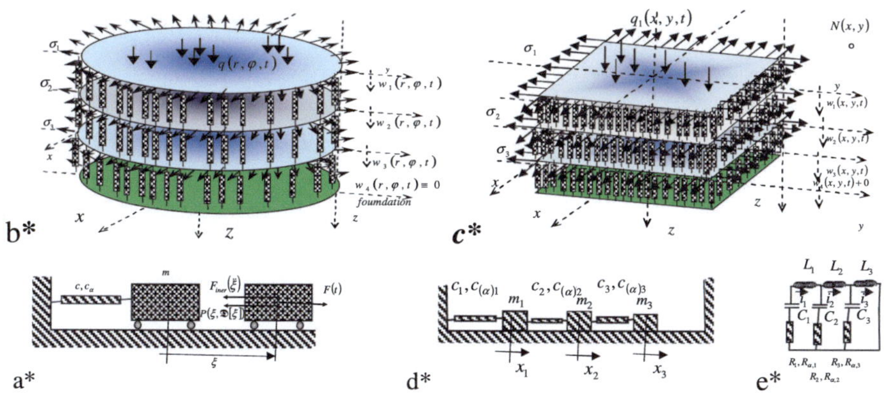

Fig. 1 (**a***) Fractional order oscillator with one degree of freedom. Qualitative and mathematical analogous fractional-type chains: (**d***) mechanical fractional-type chain system and (**e***) electrical fractional-type chain system; (**b***) hybrid multi-circular membrane fractional-type system and (**c***) hybrid multi-rectangular membrane fractional-type system on the fractional-type discrete continuum foundation

where $\mathfrak{D}_t^\alpha[\bullet]$ is fractional order differential operator of the α derivative with respect to time t in the form (1), and where c_0 and c_α are rigidity coefficients—the momentary and the prolonged one, α is a rational number between 0 and 1, $0<\alpha<1$, determined experimentally. All these parameters depend of the material properties, whose theoretical abstraction is the standard light fractional type element [1, 9]. Expressions for the kinetic $\mathbf{E_k}$ and potential $\mathbf{E}_p$ energies and generalized function $\Phi_{\alpha,\,0<\alpha<1}$ of fractional type dissipation of system energy of the mechanical fractional type oscillator with one degree of freedom, presented in Fig. 1, are in the following forms: $\mathbf{E_k}=\frac{1}{2}m\dot{\xi}^2$, $\mathbf{E_k}=\frac{1}{2}c_0\xi^2$ and $\Phi_{\alpha,0<\alpha<1}=\frac{1}{2}c_\alpha\left\{\mathfrak{D}_t^\alpha[\xi]\right\}^2$ [1, 6]. Mass particle in the considered oscillator is loaded by the following active and fictive forces: $P(\xi, D[\xi])$ —reactive force in the form (3), $F(t)=F_0\sin(\Omega_0 t+\vartheta_0)$ —external active force with amplitude F_0 and circular frequency Ω_0, and phase ϑ_0, and $F_{\text{iner}}\left(\ddot{\xi}\right)=-m\ddot{\xi}$—fictive force of inertia, and on the basis by principle of dynamical equilibrium of mass particle, we can write the ordinary fractional order differential equation in the form (2), where the following denotation $\omega_0^2=\frac{c_0}{m}$ and $\omega_{(\alpha)}^2=\frac{c_{(\alpha)}}{m}$, are introduced. Analytical approximate solution for free regimes of oscillations is solution of the ordinary fractional order differential equation, the first one in (2), and known from literature, see [1, 2, 5, 10, 11]. For obtaining analytical approximate solution for forced regimes of oscillations it is necessary to determine particular solution of the ordinary fractional order differential equation, the second one in Eq. (2). For that reason, Laplace transform noted by operator $\mathfrak{L}\{\bullet\}$, is applied to the ordinary fractional order differential equation, second in Eq. (2). Taking into account that necessary Laplace transformations are:

$$\mathfrak{L}\left\{\ddot{\xi}\right\}=p^2\mathfrak{L}\{\xi\}-\left[p\xi(0)+\dot{\xi}(0)\right],$$

$$\mathfrak{L}\left\{\mathfrak{D}_t^\alpha[\xi]\right\}=p^\alpha\mathfrak{L}\{\xi\}-\frac{d^{\alpha-1}}{dt^{\alpha-1}}\xi(0),\quad \text{for}\quad \frac{d^{\alpha-1}}{dt^{\alpha-1}}\xi(0)=0$$

it follows $\mathfrak{L}\left\{\mathfrak{D}_t^\alpha[\xi]\right\}=p^\alpha\mathfrak{L}\{\xi\}$ and $\mathfrak{L}\{\sin(\Omega_0 t)\}=\frac{\Omega_0}{p^2+\Omega_0^2}$, where p is a complex variable, with real and imaginary part, and the solution for $\mathfrak{L}\{\xi\}$ is in the following form:

$$\mathfrak{L}\{\xi\}=\frac{\left[p\xi(0)+\dot{\xi}(0)\right]}{\left\langle p^2+\omega_{(\alpha)}^2p^\alpha+\omega_0^2\right\rangle}+\frac{\Omega_0 h_0}{\left(p^2+\Omega_0^2\right)\left\langle p^2+\omega_{(\alpha)}^2p^\alpha+\omega_0^2\right\rangle} \tag{4}$$

Then, it is necessary to apply the inverse Laplace transform $\mathfrak{L}^{-1}\mathfrak{L}\{\bullet\}$ to determine the general solution $\xi(t)=\mathfrak{L}^{-1}\mathfrak{L}\{\xi\}\{\xi\}$ of ordinary fractional order differential equation (2).

The obtained solution along $\mathfrak{L}\{\xi\}$ in the form (4) can be separated in two terms:

* first term in solution (4) corresponds to solution $\xi_{\text{free}}(t,\alpha,\omega_0,\omega_{(\alpha)})$ of ordinary fractional order differential equation, first in (2) and describes free fractional type vibrations. Analytical approximate solution $\xi_{\text{free}}(t,\alpha,\omega_0,\omega_{(\alpha)})$ is in the form:

$$\xi_{\text{free}}\left(t,\alpha,\omega_0,\omega_{(\alpha)}\right)=\xi(0)\sum_{k=0}^{\infty}(-1)^k\omega_{(\alpha)}^{2k}t^{2k}\sum_{j=0}^{k}\binom{k}{j}\frac{(\mp 1)^j\omega_{(\alpha)}^{-2j}t^{-\alpha j}}{\omega_0^{2j}\Gamma(2k+1-\alpha j)}$$
$$+\dot{\xi}(0)\sum_{k=0}^{\infty}(-1)^k\omega_{(\alpha)}^{2k}t^{2k+1}\sum_{j=0}^{k}\binom{k}{j}\frac{(\mp 1)^j\omega_{(\alpha)}^{-2j}t^{-\alpha j}}{\omega_0^{2j}\Gamma(2k+2-\alpha j)},$$
$$\alpha\in(0.1),\quad t\in(0.b) \tag{5}$$

where $\xi(0)$ and $\dot{\xi}(0)$ are integral constants. In previous approximate solution (5) of the ordinary fractional order differential equation, first in (2), it is possible to separate two particular approximate solutions, as fractional type free modes, which look like cosine $\xi_{\cos}(t,\alpha,\omega_0,\omega_{(\alpha)})$ and look like sine $\xi_{\sin}(t,\alpha,\omega_0,\omega_{(\alpha)})$, (for details see [1, 9, 10]):

* second term in solution (4) corresponds to Laplace transform $\mathfrak{L}\{\xi_{p,\sin}\}$ of particular solution $\xi_{p,\sin}$ of the ordinary fractional order differential equation, second in (2), and describes one forced mode of a fractional type forced vibrations. Let us start by components of two Laplace transforms of two functions in the forms:

$$\mathfrak{L}\{f_1(t)\}=\frac{\Omega_0 h_0}{\left(p^2+\Omega_0^2\right)}=h_0\mathfrak{L}\{\sin\Omega_0 t\},\quad \mathfrak{L}\{f_2(t)\}=\frac{1}{\left(p^2+\omega_{(\alpha)}^2p^\alpha+\omega_0^2\right)}=\mathfrak{L}\{\xi_{\sin}(t)\}$$
$$\mathfrak{L}\{f_2(t)\}=\mathfrak{L}\left\{\sum_{k=0}^{\infty}(-1)^k\omega_{(\alpha)}^{2k}t^{2k+1}\sum_{m=0}^{k}\binom{k}{m}\frac{\omega_{(\alpha)}^{-2m}t^{-\alpha m}}{\omega_0^{2m}\Gamma(2k+2-\alpha m)}\right\} \tag{6}$$

The particular solution corresponds to one of the modes of forced vibrations which is in the function $f_3(t)$ whose Laplace transform $\mathfrak{L}\{f_3(t)\}$ is in the product of Laplace transforms of the two previous terms $\mathfrak{L}\{f_1(t)\}$ and $\mathfrak{L}\{f_2(t)\}:\mathfrak{L}\{f_3(t)\}=\mathfrak{L}\{f_1(t)\}*\mathfrak{L}\{f_2(t)\}$. Three functions $f_3(t)$, $f_1(t)$, and $f_2(t)$ are in convolution (denoted by *). Then, the particular solution $\xi_{p,\text{sub}}(t,\alpha,\omega_0,\omega_{(\alpha)},\Omega_0)$ of the ordinary fractional order differential equation, second in (2), which corresponds to one forced mode of fractional-type vibrations under the external sinusoidal single frequency excitation $h_0\sin(\Omega_0 t)$, is given by

$$\xi_{p,\sin}\left(t,\alpha,\omega_0,\omega_{(\alpha)},\Omega_0\right)$$
$$=\langle h_0\sin(\Omega_0 t)\rangle*\left\langle\sum_{k=0}^{\infty}(-1)^k\omega_{(\alpha)}^{2k}t^{2k+1}\sum_{m=0}^{k}\binom{k}{m}\frac{\omega_{(\alpha)}^{-2m}t^{-\alpha m}}{\omega_0^{2m}\Gamma(2k+2-\alpha m)}\right\rangle$$

$$\xi_{p,\mathrm{sub}}\left(t,\alpha,\omega_0,\omega_{(\alpha)},\Omega_0\right) = \int_0^t \langle h_0 \sin(\Omega_0\,(t-\tau))\rangle \left\langle \sum_{k=0}^{\infty} (-1)^k \omega_{(\alpha)}^{2k} \tau^{2k+1} \sum_{m=0}^{k} \binom{k}{m} \frac{\omega_{(\alpha)}^{-2m} \tau^{-\alpha m}}{\omega_0^{2m} \Gamma\,(2k+2-\alpha m)} \right\rangle d\tau \tag{7}$$

General approximate solution $\xi(t,\alpha,\omega_0,\omega_{(\alpha)})$ for forced vibrations of a fractional-type oscillator, with one degree of freedom, loaded by a single-frequency external excitation is in the form of sum of analytical approximate solutions for free vibrations modes $\xi_{\mathrm{free}}(t,\alpha,\omega_0,\omega_{(\alpha)})$ and the particular solution for a forced mode $\xi_{p,\,\mathrm{sub}}(t,\alpha,\omega_0,\omega_{(\alpha)},\Omega_0)$, and for $\alpha \in (0.1)$, $t \in (0.\ b)$, is in the following form:

$$\begin{aligned} \xi\left(t,\alpha,\omega_0,\omega_{(\alpha)}\right) &= \xi(0) \sum_{k=0}^{\infty} (-1)^k \omega_{(\alpha)}^{2k} t^{2k} \sum_{j=0}^{k} \binom{k}{j} \frac{(\mp 1)^j \omega_{(\alpha)}^{-2j} t^{-\alpha j}}{\omega_0^{2j} \Gamma\,(2k+1-\alpha j)} + \\ &+ \dot{\xi}(0) \sum_{k=0}^{\infty} (-1)^k \omega_{(\alpha)}^{2k} t^{2k+1} \sum_{j=0}^{k} (kj) \frac{(\mp 1)^j \omega_{(\alpha)}^{-2j} t^{-\alpha j}}{\omega_0^{2j} \Gamma\,(2k+2-\alpha j)} + \\ &+ \int_0^t \langle h_0 \sin\left(\Omega_0\,(t-\tau)\right)\rangle \\ &\left\langle \sum_{k=0}^{\infty} (-1)^k \omega_{(\alpha)}^{2k} \tau^{2k+1} \sum_{m=0}^{k} \binom{k}{m} \frac{\omega_{(\alpha)}^{-2m} \tau^{-\alpha m}}{\omega_0^{2m} \Gamma\,(2k+2-\alpha m)} \right\rangle d\tau \end{aligned} \tag{8}$$

3 Description of the Hybrid Discrete Continuum Systems of Fractional Type with Multi-Deformable Bodies

The hybrid discrete continuum systems of fractional type with multi-deformable bodies are composed by finite number of deformable ideally elastic bodies (strings, belts, beams, membranes, or plates) with equal contours and equal boundary conditions, see Fig. 1b*, c*. Deformable bodies are coupled by standard light fractional type discrete continuum layers in transversal direction permitting the system transversal vibrations. A fractional type discrete continuum layer contains homogeneous distributed standard light fractional type element determined by constitutive stress-strain relation expressed by fractional order differential operator $\mathfrak{D}_t^{\alpha}\,[\bullet]$. The force which defines the interactions between two deformable bodies is in the form:

$$\begin{aligned} Q_{\alpha,k}(t) = &-\left\{c_{0(k,k+1)}\left[w_{k+1}(x,y,t) - w_k(x,y,t)\right]\right\} \\ &-\left\{c_{\alpha(k,k+1)}\mathfrak{D}_t^{\alpha}\left[w_{k+1}(x,y,t) - w_k(x,y,t)\right]\right\} = -Q_{\alpha,k+1}(t) \end{aligned} \tag{9}$$

In the previous expression, $w_k(x, y, t)$ and $w_{k+1}(x, y, t)$ are transversal displacements of the middle surface points of thin ideal elastic plates (or membranes, see Fig. 1b*, c*), where x and y are the coordinates of plate (or membrane) middle surface points, and k and $k+1$ are order of the plates (or membranes) in hybrid fractional type system (for details see [8, 9]). Taking into account the previously published system of coupled partial fractional order differential equations of transversal vibrations of the hybrid discrete continuum systems of fractional type with multi-deformable bodies containing, in the considered case, in each particular case three deformable bodies (strings, belts, beams, membranes, or plates) in listed [8, 9] with corresponding solutions for free vibrations, it can be seen that each eigen amplitude form $\mathbf{W}_{nm}(x, y)$, $n, m = 1, 2, 3, 4, \ldots, \infty$ contains three eigen time functions $T_{k(nm)}(t)$, $k = 1, 2, 3$, $n, m = 1, 2, 3, 4, \ldots, \infty$ with three eigen frequencies and three eigen characteristic numbers expressing fractional type of eigen time functions, for free vibrations. On the basis of these mathematical results, it is visible and *it is possible to produce new additional frequency analysis and to conclude/determine according properties for free as well as for forced transversal vibrations of considered hybrid multi-deformable body system dynamics.*

Each subsystem of the obtained series, contains three independent ordinary fractional order differential equations, along eigen main coordinates $\xi_{(nm)s}(t)$, $s = 1, 2, 3$, $n, m = 1, 2, 3, 4, \ldots, \infty$ for each of eigen time functions, for free or forced vibrations, independent hybrid modes of three plate- or three-membrane system. One of that independent ordinary fractional order differential equation, describing independent fractional type hybrid mode of forced vibrations is in the following form (see [9]):

$$\left\langle \ddot{\xi}_{(nm)1}(t) + \tilde{\omega}^2_{nm(1)}\xi_{(nm)1}(t) + \tilde{\omega}^2_{\alpha(nm)(1)}\mathfrak{D}^{\alpha}_{t}\left[\xi_{(nm)1}(t)\right]\right\rangle$$
$$= \frac{\begin{vmatrix} h_{01,nm}\sin\left(\Omega_{1,nm}t + \vartheta_{1,nm}\right) & K^{(2)}_{(nm)31} & K^{(3)}_{(nm)31} \\ h_{02,nm}\sin\left(\Omega_{2,nm}t + \vartheta_{2,nm}\right) & K^{(2)}_{(nm)32} & K^{(3)}_{(nm)32} \\ h_{03,nm}\sin\left(\Omega_{3,nm}t + \vartheta_{3,nm}\right) & K^{(2)}_{(nm)33} & K^{(3)}_{(nm)33} \end{vmatrix}}{\begin{vmatrix} K^{(1)}_{(nm)31} & K^{(2)}_{(nm)31} & K^{(3)}_{(nm)31} \\ K^{(1)}_{(nm)32} & K^{(2)}_{(nm)32} & K^{(3)}_{(nm)32} \\ K^{(1)}_{(nm)33} & K^{(2)}_{(nm)33} & K^{(3)}_{(nm)33} \end{vmatrix}} \tag{10}$$

and is a representation of other two OFODEs for forced transversal vibrations of a considered hybrid three plate- or three-membrane system.

New analytical approximate solution of the previous independent ordinary fractional order differential equation (10) on the basis of the expression (8) is in the form:

$$
\begin{aligned}
&\xi_{(nm)1}\left(t, \alpha, \omega_0, \omega_{(\alpha)}\right)\\
&= \xi_{(nm)1}(0) \sum_{k=0}^{\infty} (-1)^k \tilde{\omega}_{\alpha(nm)(1)}^{2k} t^{2k} \sum_{j=0}^{k} \binom{k}{j} \frac{(\mp 1)^j \tilde{\omega}_{\alpha(nm)(1)}^{-2j} t^{-\alpha j}}{\tilde{\omega}_{nm(1)}^{2j} \Gamma(2k+1-\alpha j)}\\
&+ \dot{\xi}_{(nm)1}(0) \sum_{k=0}^{\infty} (-1)^k \tilde{\omega}_{\alpha(nm)(1)}^{2k} t^{2k+1} \sum_{j=0}^{k} \binom{k}{j} \frac{(\mp 1)^j \tilde{\omega}_{\alpha(nm)(1)}^{-2j} t^{-\alpha j}}{\tilde{\omega}_{nm(1)}^{2j} \Gamma(2k+2-\alpha j)}\\
&+ \int_0^t \left\langle \tilde{h}_{01,nm} \sin\left(\Omega_{1,nm}(t-\tau) + \vartheta_{1,nm}\right)\right\rangle\\
&\left\langle \sum_{k=0}^{\infty} (-1)^k \tilde{\omega}_{\alpha(nm)(1)}^{2k} \tau^{2k+1} \sum_{m=0}^{k} \binom{k}{m} \frac{\tilde{\omega}_{\alpha(nm)(1)}^{-2m} \tau^{-\alpha m}}{\tilde{\omega}_{nm(1)}^{2m} \Gamma(2k+2-\alpha m)} \right\rangle d\tau\\
&+ \int_0^t \left\langle \tilde{h}_{02,nm} \sin\left(\Omega_{2,nm}(t-\tau) + \vartheta_{2,nm}\right)\right\rangle\\
&\left\langle \sum_{k=0}^{\infty} (-1)^k \tilde{\omega}_{\alpha(nm)(1)}^{2k} \tau^{2k+1} \sum_{m=0}^{k} \binom{k}{m} \frac{\tilde{\omega}_{\alpha(nm)(1)}^{-2m} \tau^{-\alpha m}}{\tilde{\omega}_{nm(1)}^{2m} \Gamma(2k+2-\alpha m)} \right\rangle d\tau\\
&+ \int_0^t \left\langle \tilde{h}_{03,nm} \sin\left(\Omega_{3,nm}(t-\tau) + \vartheta_{3,nm}\right)\right\rangle\\
&\left\langle \sum_{k=0}^{\infty} (-1)^k \tilde{\omega}_{\alpha(nm)(1)}^{2k} \tau^{2k+1} \sum_{m=0}^{k} \binom{k}{m} \frac{\tilde{\omega}_{\alpha(nm)(1)}^{-2m} \tau^{-\alpha m}}{\tilde{\omega}_{nm(1)}^{2m} \Gamma(2k+2-\alpha m)} \right\rangle d\tau
\end{aligned}
\tag{11}
$$

Previously *new derived approximate general solution* (11) is the analytical approximate description of one independent eigen fractional type hybrid forced mode, one of three forced hybrid modes of an eigen time function of transversal vibrations of a deformable body (plate or membrane) in the considered hybrid system. It is visible that this forced hybrid mode contains three frequencies of the forced external excitations applied as surface distributed three single frequency excitations along each of the three deformable bodies. Then this independent mode is like the fourth frequency mode of forced vibrations. Independent fractional type forced hybrid modes appear in each of eigen time functions of each of the infinite number of eigen amplitude modes in regimes of the forced vibrations of the forced transversal vibrations of the discrete continuous system. A number of the independent fractional type hybrid modes in each of the eigen time functions of each eigen amplitude mode is equal to the number of the coupled deformable bodies in hybrid fractional type system.

4 Concluding Remarks

Solutions (8) and (11) representing the new results obtained for fractional-type forced transversal vibrations for three coupled deformable bodies (beam, plate, or membrane) with the same contour and boundary conditions, in hybrid fractional-type system, can be generalized to hybrid systems with any finite number of coupled deformable bodies. This particularly applies to kinetic parameters of the eigen time functions and eigen hybrid forced modes. As final remark a theorem is formulated in the following form:

Theorem *Dynamics of hybrid fractional-type system, which contains N deformable bodies (beams, plates, or membranes), with equal boundary conditions, each excited by surface external single-frequency excitation with frequency* $\Omega_{k,\,nm}$ $k = 1, 2, \ldots, K$, *and coupled by discrete continuum fractional order layers, and with displacements* $w_k(x, y, t)$ *and* $w_{k+1}(x, y, t)$ *described by the corresponding system of coupled partial fractional order differential equations, and in each of the eigen amplitude form* $\mathbf{W}_{nm}(x, y)$, *from the set of infinite numbers is described by corresponding like* $N + K$ *frequency eigen time functions* $T_{k(nm)}(t) = \sum_{s=1}^{s=N} \mathbf{K}_{(nm)Nk}^{(s)} \xi_{(nm)s}(t)$, *where* $\xi_{(nm)s}(t)$ $s = 1, 2, 3 \ldots, N$, *are independent eigen main fractional-type time hybrid forced modes of corresponding subsystem in one eigen amplitude form* $\mathbf{W}_{nm}(x, y)$. *These eigen main fractional-type time hybrid modes* $\xi_{(nm)s}(t)$, $s = 1, 2, 3 \ldots, N$, *are described by the system of the solutions of independent ordinary fractional order differential equations in the forms (2) or (10) with approximate solution in the form (8) or (11), with two sets of characteristic numbers:* $\tilde{\omega}_{nm(s)}^{2}$ *and* $\tilde{\omega}_{\alpha(nm)(s)}^{2}$, *and one set with external excitation frequencies* $\Omega_{k,\,nm}$ $k = 1, 2, \ldots, K$. *The first set contains characteristic numbers* $\tilde{\omega}_{nm(s)}^{2}$ *which are the squares of eigen frequencies, the same as for corresponding linear system free vibrations, and the second set* $\tilde{\omega}_{\alpha(nm)(s)}^{2}$ *corresponds to eigen frequency fractional-type modes.*

In a fractional type system dynamics with independent fractional-type modes, no interactions between independent fractional type modes can occur. Also, transfer energy between independent fractional types modes does not appear. But, in each independent hybrid forced mode fractional-type dissipation of energy during the system oscillations occurs. In general, the dynamics of a fractional-type system is nonlinear and no independent fractional type modes for free as well as for forced vibrations exist. Knowledge about independent fractional-type modes can have an important role for investigation of nonlinear dynamics of general fractional-type system dynamics.

Acknowledgement Parts of this research were supported by the Ministry of Sciences of Republic Serbia trough Mathematical Institute SANU Belgrade Grants OI 174001 "Dynamics of hybrid systems with complex structures. Mechanics of materials".

References

1. Hedrih (Stevanović), K.: Analytical dynamics of fractional type discrete system, review paper. Adv. Theor. Appl. Mech. **11**(1), 15–47 (2018). https://doi.org/10.12988/atam.2018.883
2. Atanacković, T., Pilipovićc, S., Stanković, B., Zorica, D.: Fractional Calculus with Applications in Mechanics: Vibrations and Diffusion Processes. ISTE, London (2014)
3. Bačlić, B.S., Atanacković, T.M.: Stability and creep of a fractional derivative order viscoelastic Rod, Bulletin T, CXXI de L'Academie Serbe des Sciences st de Arts - 2000. Class Sci. Math. Nat. Sci. **25**, 115–131 (2000)
4. Gorenflo, R., Mainardi, F.: Fractional calculus: integral and differential equations of fractional order. In: Carpinteri, A., Mainardi, F. (eds.) Fractals and Fractional Calculus in Continuum Mechanics, vol. 1997, pp. 223–276. Springer Verlag, Berlin (2000)
5. Rossikhin, Y.A., Shitikova, M.V.: Application of fractional calculus for dynamic problems of solid mechanics: novel trends and recent results. Appl. Mech. Rev. **63**(1), 010801 (2010)
6. Hedrih (Stevanović), K.: Generalized function of fractional order dissipation of system energy and extended Lagrange differential Lagrange equation in matrix form, Dedicated to 86th Anniversary of Radu MIRON'S Birth. Tensor. **75**(1), 35–51 (2014)
7. Hedrih (Stevanović), K.: Partial fractional differential equations of creeping and vibrations of plate and their solutions (First Part). J. Mech. Behav. Mater. **16**(4–5), 305–314 (2005). ISSN 0334-8938. http://www.freundpublishing.com/JOURNALS/materials_science_and_engineering.htm
8. Hedrih (Stevanović), K.: Multi membrane fractional order system vibrations. Theor. Appl. Mech. **41**(S1), 43–61 (2014). https://doi.org/10.2298/TAM14S1043H.I. Series: Special Issue—Dedicated to memory of Anton D. Bilimovič (1879-1970). http://www.mi.sanu.ac.rs/projects/TAM-SpecialIssue41-2014-BILIMOVIC.pdf
9. Hedrih (Stevanović), K.: Elements of mathematical phenomenology in dynamics of multi-body system with fractional order discrete continuum layers, Dedicated to the 100th Anniversary of the Russian Academician Yury Rabotnov, v, Special issue. Int. J. Mech. **8**, 345–352 (2014). ISSN: 1998-4448. http://www.naun.org/main/NAUN/mechanics/2014/b042003-061.pdf
10. Hedrih (Stevanović), K., Filipovski, A., et al.: Facta Univ. Ser. Mech. Autom. Control Robot. **3**(12), 327–350 (2002). YU ISSN 0354-2009
11. Hedrih (Stevanović), K., Machado, J.M.T.: Discrete fractional order system vibrations. Int. J. Non-Linear Mech. **73**, 2–11 (2013). https://doi.org/10.1016/j.ijnonlinmec.2014.11.009. ISSN 0020-7462. http://authors.elsevier.com/authorforms/NLM2407/7c32b6b4f19f2471fb24556142da3cd1

Non-Smooth Bifurcation in Two Fractional-Order Memristive Circuits

Yajuan Yu and Zaihua Wang

Abstract Two fractional-order Chua's memristive circuits named Model 1 and Model 2 are proposed. Model 1 is a fractional-order memristive circuit with only the memristor described by a fractional-order derivative due to the memory loss observed experimentally, while Model 2 is a direct fractional-order generalization of integer-order Chua's memristive circuit without considering the physical background. Both models are non-smooth systems with a line equilibrium depending on the memristor's initial state. Numerical simulation shows that both models exhibit multi-stability and different steady states switch via "grazing bifurcation" or "tangent bifurcation," "intermittent chaos" is found in Model 1 as the fractional order is close to 0 or 1, but no "intermittent chaos" is found in Model 2 as the fractional order is between 0 and 1.

Keywords Tangent bifurcation · Fractional-order memristor · Stability · Non-smoothness

1 Introduction

Capacitor, inductor, and resistor are known as three traditional circuit elements for more than one century. As the fourth basic circuit element with an adjustable resistance or conductance, memristor was fabricated for the first time as a physical device without internal power supply in HP Laboratory [1] in 2008, 37 years after it was originally introduced by Lenon O Chua [2] in 1971. A memristive circuit stands for a circuit with a memristor. It is essentially a nonlinear system, thus, loss of

Y. Yu
School of Mathematics and Physics, Changzhou University, Changzhou, China
e-mail: yyjxye@cczu.edu.cn

Z. Wang (✉)
State Key Lab of Mechanics and Control of Mechanical Structures, Nanjing University of Aeronautics and Astronautics, Nanjing, China
e-mail: zhwang@nuaa.edu.cn

W. Lacarbonara et al. (eds.), *New Trends in Nonlinear Dynamics*,
https://doi.org/10.1007/978-3-030-34724-6_33

stability, various kinds of bifurcation and even chaos can be observed in memristive circuits. By introducing different nonlinearities, memristive circuits exhibit self-excited attractors [3, 4], hidden attractors [5, 6], coexisting multiple attractors [7], and coexisting infinitely many attractors [8]. In these cases, the circuit systems diversify in performance with unchanging parameters, and switch between different coexisting states. On the one hand, such complicated dynamics can lead to unexpected and even catastrophic behaviors in engineering applications, such as multiple attractors are the source of unpredictability [9]. On the other hand, chaotic systems with multiple attractors increase the confidentiality in the communications [10], and using memristor to construct chaotic circuits with multiple attractors is a new control strategy in such applications. Interestingly, memristor's stateful logic property and similarity to synapse can be potentially applied to neuromorphic computing [11]. Thus, it is highly required to study nonlinear dynamics of memristive circuits.

Memristor and memristive circuits can be modeled mainly in three different ways. Firstly, memristor and memristive circuits are described by using integer-order derivatives. Secondly, the work [1] shows that there is a memory loss of the HP TiO_2 linear model for memristor, the width of the doped layer of HP TiO_2 model cannot be equal to zero or the whole width [1], thus, HP TiO_2 memristor with the edge memory loss can be mathematically modeled by a fractional-order derivative with the order between 0 (no memory) and 1 (full memory), and the memristor in the series circuits has capacitive properties or inductive properties [12]. Thirdly as given in [13–15], some fractional-order memristive models are generalization of the traditional ones, simply replacing the integer-order derivative with the fractional-order derivative without considering the physical background. For fractional-order memristive models, the memory strength of the memristor can be adjusted by changing the order, as shown in [12].

Many features can be kept from integer-order memristive models to fractional-order ones. There are also differences in the nonlinear dynamics between the fractional-order memristive circuits and the corresponding integer-order systems. For example [16], both the fractional-order and its corresponding integer-order systems exhibit chaotic behavior, but the trajectory of the fractional-order system exhibits a single-scroll attractor while in the corresponding integer-order system the trajectory is a double-scroll attractor. Thus, there comes a question: what difference can be found between two kinds of fractional-order memristive circuits obtained in the second way and the third way above?

This paper studies the non-smooth bifurcation in two fractional-order Chua's memristive circuits, with a comparison between the one proposed in [17] and the other to be discussed below. It begins with the modeling of the memristive circuits in Sect. 2, then stability and bifurcation analysis of the two models are displayed in Sect. 3. Finally it ends with a summary in Sect. 4.

2 Models

2.1 Integer-Order Chua's Memristive Circuit

Figure 1 is a Chua's circuit including a flux-controlled memristor M, where the memristor's equivalent circuit is realized in Fig. 2 [18] with the state equations

$$\begin{aligned} i_M(t) &= W(v_0)v_M(t), \\ C_0\frac{\mathrm{d}v_0}{\mathrm{d}t} &= -\frac{v_1}{R_0}, \end{aligned} \tag{1}$$

where $W(v_0) = -a + b|v_0|$ is the memductance with $a = 0.6667\,\text{mS}$, $b = 0.403\,\text{mS/Wb}$. The memristor's flux $\varphi_M(t) = -C_0R_0v_0(t)$, derived from Fig. 2. The following set of nonlinear differential equations are state equations of circuit displayed in Fig. 1

$$\begin{aligned} C_1\frac{\mathrm{d}v_1}{\mathrm{d}t} &= \frac{v_2 - v_1}{R} - W(v_0)v_1, \\ C_2\frac{\mathrm{d}v_2}{\mathrm{d}t} &= \frac{v_1 - v_2}{R} - i_3, \\ L\frac{\mathrm{d}i_3}{\mathrm{d}t} &= v_2, \\ C_0\frac{\mathrm{d}v_0}{\mathrm{d}t} &= -\frac{v_1}{R_0}, \end{aligned} \tag{2}$$

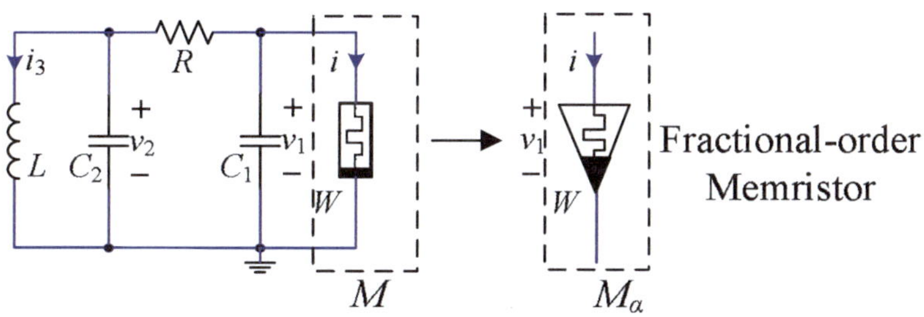

Fig. 1 The Chua's circuit with a fractional-order memristor

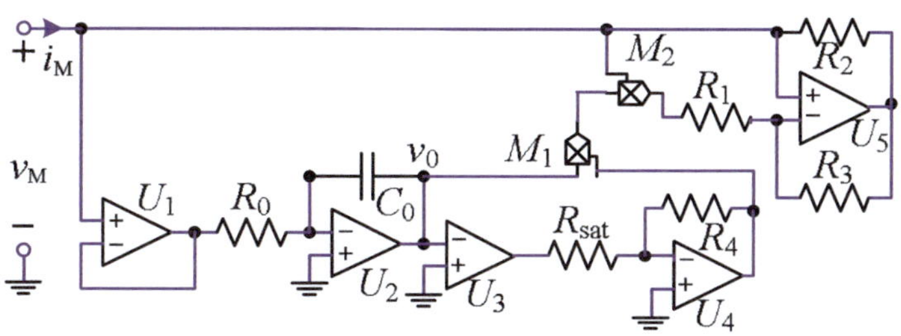

Fig. 2 Memristor equivalent circuit realization

where $R = 2\,\text{k}\Omega$, $C_1 = 6.8\,\text{nF}$, $C_2 = 68\,\text{nF}$, $L = 17.2\,\text{mH}$, $R_0 = 2\,\text{k}\Omega$, and $C_0 = 68\,\text{nF}$.

Let $t = RC_2\tau$, and

$$v_1(t) = v_1(RC_2\tau) = x_1(\tau), \quad v_2(t) = v_2(RC_2\tau) = x_2(\tau),$$
$$Ri_3(t) = Ri_3(RC_2\tau) = x_3(\tau), \quad v_0(t) = v_0(RC_2\tau) = x_4(\tau),$$
$$m = \frac{C_2}{C_1}, \quad n = \frac{R^2C_2}{L}, \quad p = \frac{RC_2}{R_0C_0} = 1, \quad A = aR, \quad B = bR,$$

then Eq. (2) is changed into

$$\begin{aligned} \frac{\mathrm{d}x_1}{\mathrm{d}\tau} &= m\,(x_2 - x_1) - m\,(B|x_4| - A)x_1, \\ \frac{\mathrm{d}x_2}{\mathrm{d}\tau} &= x_1 - x_2 - x_3, \\ \frac{\mathrm{d}x_3}{\mathrm{d}\tau} &= n\,x_2, \\ \frac{\mathrm{d}x_4}{\mathrm{d}\tau} &= -p\,x_1, \end{aligned} \tag{3}$$

where $A = 1.3334$, $B = 0.8060$, $m = 10$, $n = 15.814$, $p = 1$. For Eq. (2), the phenomenon of extreme multi-stability revealing the coexistence of infinitely many attractors was shown in [18], where the emergence of extreme multi-stability depended on the initial state of the memristor.

2.2 Fractional Model 1

With the memory loss taken into consideration, the state equations of α-order ($0 < \alpha < 1$) flux-controlled memristor are described as below

$$\begin{aligned} i_M(t) &= W(v_0)v_M(t), \\ C_0({}_0^C D_t^\alpha v_0) &= -\frac{v_1}{R_0}, \quad \alpha \in (0, 1), \end{aligned} \tag{4}$$

where the α-order derivative of $v_0(t)$ in the sense of Caputo's definition is given by [19]

$${}_0^C D_t^\alpha v_0(t) = \frac{1}{\Gamma(1-\alpha)} \int_0^t \frac{\dot{v}_0(\tau)}{(t-\tau)^\alpha} \mathrm{d}\tau. \tag{5}$$

Replacing the memristor in Eq. (3) by fractional-order memristor Eq. (4), one has Model 1 [17]

$$
\begin{aligned}
\frac{\mathrm{d}x_1}{\mathrm{d}\tau} &= m\,(x_2 - x_1) - m\,(B|x_4| - A)x_1,\\
\frac{\mathrm{d}x_2}{\mathrm{d}\tau} &= x_1 - x_2 - x_3,\\
\frac{\mathrm{d}x_3}{\mathrm{d}\tau} &= n\,x_2,\\
{}_0^C D_\tau^\alpha x_4 &= -p_\alpha\, x_1, \quad \alpha \in (0, 1),
\end{aligned}
\tag{6}
$$

where $p_\alpha = (RC_2)^\alpha / R_0 C_0$. Equation (6) has line equilibrium $\boldsymbol{x}^* = (0, 0, 0, c)^{\mathrm{T}}$ and c is any real number.

2.3 *Fractional Model 2*

A mathematical or logical generalization of the model (3) is to replace all the first-order derivatives with fractional derivatives of the same order, namely not only the capacitor, but also the inductor and the resistor in Fig. 2 are governed by a set of fractional differential equations of the same order. That is Model 2, described by

$$
\begin{aligned}
{}_0^C D_\tau^\alpha x_1 &= m\,(x_2 - x_1) - m\,(B|x_4| - A)x_1,\\
{}_0^C D_\tau^\alpha x_2 &= x_1 - x_2 - x_3,\\
{}_0^C D_\tau^\alpha x_3 &= n\,x_2,\\
{}_0^C D_\tau^\alpha x_4 &= -\,x_1,
\end{aligned}
\tag{7}
$$

where $A = 1.3334$, $B = 0.8060$, $m = 10$, $n = 15.814$, $0 < \alpha < 1$. Again, Eq. (7) has also line equilibrium $\boldsymbol{x}^* = (0, 0, 0, c)^{\mathrm{T}}$ and c is any real number.

3 Analysis of Stability and Bifurcation

The fractional Model 1 and Model 2 have the same line equilibrium $\boldsymbol{x}^* = (0, 0, 0, c)^{\mathrm{T}}$, which exhibits stability and non-smooth bifurcation. Below the numerical simulation is made by using the predictor-corrector algorithm developed in [20–22]. The sum of the orders of Model 2 is smaller than that of Model 1, thus it is expected that Model 2 has simpler nonlinear dynamics than Model 1.

3.1 The Characteristic Function

Let $f(x_1, x_2, x_3, x_4) = m\,(x_2 - x_1) - m\,(B|x_4| - A)x_1$, then

$$\frac{\partial f}{\partial x_1} = -m - m\,(B|x_4| - A), \quad \frac{\partial f}{\partial x_2} = m, \quad \frac{\partial f}{\partial x_3} = 0$$

At $x_4 \neq 0$, it holds $\frac{\partial f}{\partial x_4} = -mBx_1\mathrm{sgn}(x_4)$. At $x_4 = 0$, one has

$$\frac{\partial f}{\partial x_4}(0,0,0,0) = \lim_{\Delta x \to 0} \frac{f(0,0,0,\Delta x) - f(0,0,0,0)}{\Delta x} = 0.$$

Thus, the Jacobian matrix of the vector composed of the right-hand functions of Eq. (7) with respect to $x_1,\ x_2,\ x_3,\ x_4$ at $\boldsymbol{x}^*$ is a lower triangle matrix given by

$$\boldsymbol{J}(\boldsymbol{x}^*) = \begin{pmatrix} m(A - B \mid c \mid) - m & m & 0 & 0 \\ 1 & -1 & -1 & 0 \\ 0 & n & 0 & 0 \\ -1 & 0 & 0 & 0 \end{pmatrix}, \tag{8}$$

where $m(A - B \mid c \mid) - m$ is non-smooth at $c = 0$. The characteristic function of the fractional Model 2 is the determinant of the matrix $\boldsymbol{\Delta}(\lambda) = \mathbf{diag}(\lambda^\alpha,\ \lambda^\alpha,\ \lambda^\alpha,\ \lambda^\alpha) - \boldsymbol{J}(\boldsymbol{x}^*)$ given by

$$\det(\boldsymbol{\Delta}(\lambda)) = \lambda^\alpha(\lambda^{3\alpha} + d_2\lambda^{2\alpha} + d_1\lambda^\alpha + d_0) = \lambda^\alpha\, \Pi(\lambda, \alpha), \tag{9}$$

where $d_0 = n(m - m(A - B \mid c \mid) + 1) - n$, $d_1 = n - m(A - B \mid c \mid)$, $d_2 = m - m(A - B \mid c \mid) + 1$. The characteristic function of the fractional Model 1 is

$$\det(\boldsymbol{\Delta}(\lambda)) = \lambda^\alpha\, \Pi(\lambda, 1) \tag{10}$$

which is defined by the determinant of the matrix $\boldsymbol{\Delta}_1(\lambda) = \mathbf{diag}(\lambda,\ \lambda,\ \lambda,\ \lambda^\alpha) - \boldsymbol{J}_1(\boldsymbol{x}^*)$, where $\boldsymbol{\Delta}_1(\lambda)$ is almost the same as $\boldsymbol{\Delta}(\lambda)$ except for the first element in the fourth row of $\boldsymbol{\Delta}(\lambda)$. Here λ^α is a multi-valued function when α is a fractional number. It becomes a single-valued function if a cut on the negative real axis with the real part $\Re(\lambda) < 0$ is made, so as to have $\arg(\lambda) \in (-\pi, \pi]$ which denotes the principal argument of λ. In this case, $\lambda = 0$ is the unique root of $\lambda^\alpha = 0$. The root $\lambda = 0$ comes from the steady state of the fourth differential equation, which gives $x_4 = c$ for some real number c if the sub-state $(x_1, x_2, x_3)^{\mathrm{T}} = (0, 0, 0)^{\mathrm{T}}$ is asymptotically stable, a case when all the roots of

$$\Pi(\lambda, \alpha) = \lambda^{3\alpha} + d_2\lambda^{2\alpha} + d_1\lambda^\alpha + d_0$$

have negative real parts only, for $\alpha = 1$ or $0 < \alpha < 1$, respectively. The sub-state $(x_1, x_2, x_3)^{\mathrm{T}} = (0, 0, 0)^{\mathrm{T}}$ is not stable if there is at least one root of $\Pi(\lambda, \alpha)$ has positive real part.

3.2 *Fractional Model 1*

When $\alpha = 1$, the critical stable condition $\Pi(\mathrm{i}\omega, 1) = 0$ gives that there are $\omega \neq 0$ at $c = \pm 1.4433$, $c = \pm 0.5006$ referring to Fig. 3, where "s," "us" are pointed to stable and unstable, respectively. After its stability loss, a non-smooth bifurcation of the fractional Model 1 can occur and it is studied with details in [17]. Below is the summary of the main conclusions given in [17].

1. As c changes in $(-1.5, 1.5)$, there are four stable regions: $(-1.5, -1.4433)$, $(-0.5006, -0.4136)$, $(0.4136, 0.5006)$, and $(1.4433, 1.5)$. In the bifurcation diagrams of $x_4(t)$ shown in Fig. 4a, the four overlaps denoted by P_1, P_2, P_3, P_4 between the bifurcation diagram and the red line $x_4 = c$ correspond to these four stable regions. As long as c is in the four stable regions, $\boldsymbol{x}^* = (0, 0, 0, c)^{\mathrm{T}}$ is stable.
2. Figure 4a shows that three kinds of steady states, namely chaotic attractor, limit cycle, and stable equilibrium, switch via tangent bifurcation, rather than period-doubling bifurcation occurred in the integer-order memrisive circuit system [18].
3. For fixed initial value taken in unstable regions, chaos occurs only when the fractional-order is close to 0 or 1, refer to Fig. 4c. When the fixed fractional-order is close to 1, the circuit system exhibits "intermittent chaos" (Fig. 4b) resulted

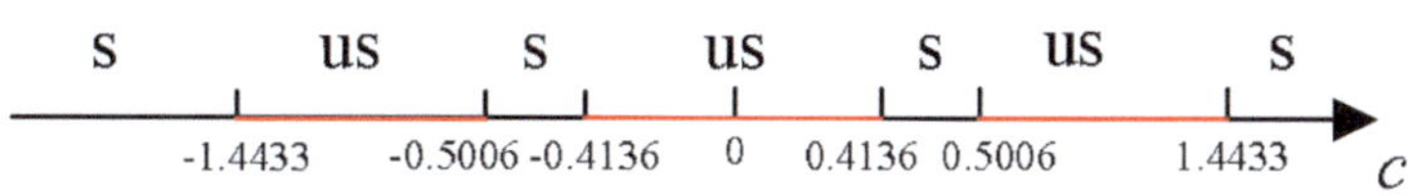

Fig. 3 The stable region of c

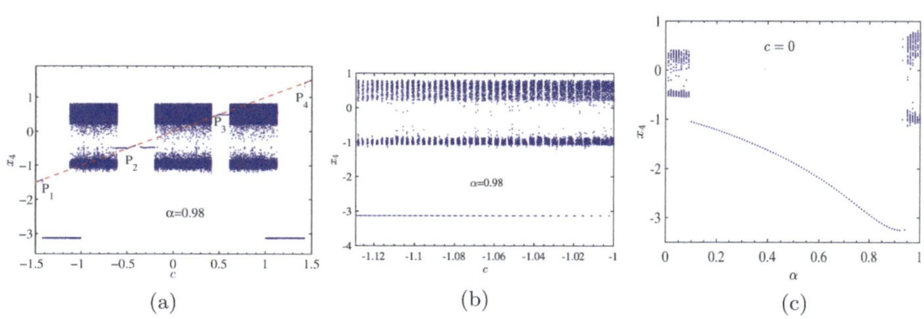

Fig. 4 Bifurcation diagrams of $x_4(t)$ with respect to c or α. **(a)** $\alpha = 0.98$, red dashed line: $x_4 = c$. **(b)** $c \in [-1.129, -1]$. **(c)** $c = 0$

from "tangent bifurcation" or "grazing bifurcation," for which limit cycle and chaotic attractor switch with very high frequency. When the fractional-order is not close to 0 or 1, the fractional-order memristive circuit has simple dynamical behaviors compared with the integer-order circuit.

4. With fixed initial states $(10^{-9}, 0, 0, 0)$, as α changes in $(0, 1)$, two kinds of steady states are found: namely chaotic attractor and limit cycle, which switch via tangent bifurcations, shown in Fig. 4c.

3.3 *Fractional Model 2*

When $0 < \alpha < 1$, the critical stable condition $\Pi(\mathrm{i}\omega, \alpha) = 0$ gives

$$(\mathrm{i}\omega)^{3\alpha} + d_2(\mathrm{i}\omega)^{2\alpha} + d_1(\mathrm{i}\omega)^{\alpha} + d_0 = 0.$$

Using Euler's formula $\mathrm{i}\omega = \omega \mathrm{e}^{\mathrm{i}\pi/2}$, separating the real part and the imaginary part yields

$$\begin{aligned} &\omega^{3\alpha}\cos(3\alpha\pi/2) + d_2\omega^{2\alpha}\cos(3\alpha\pi/2) + d_1\omega^{\alpha}\cos(3\alpha\pi/2) + d_0 = 0 \\ &\omega^{3\alpha}\sin(3\alpha\pi/2) + d_2\omega^{2\alpha}\sin(3\alpha\pi/2) + d_1\omega^{\alpha}\sin(3\alpha\pi/2) = 0. \end{aligned} \tag{11}$$

By solving Eqs. (11) with Matlab, the critical values c for different α are displayed in Table 1. When $\alpha \leq 0.95$, there are only two critical values $c = \pm 0.4136$.

Alternatively, let $\gamma = \lambda^{\alpha}$, $\Pi(\lambda, \alpha) = 0$ is changed into

$$\Pi(\gamma) = \gamma^3 + d_2\gamma^2 + d_1\gamma + d_0 = 0. \tag{12}$$

Then the sub-state $(x_1, x_2, x_3)^{\mathrm{T}} = (0, 0, 0)^{\mathrm{T}}$ is asymptotically stable if and only if $|\arg(\gamma)| > \alpha\pi/2$ of all roots γ of Eq. (12) or $Z = \min(|\arg(\gamma)|) - \alpha\pi/2 > 0$ [23], where $\arg(\gamma) \in (-\pi, \pi]$ is the principal argument of λ. Figure 5 shows three plots of $Z = \min(|\arg(\gamma)|) - \alpha\pi/2$ for $\alpha = 0.98$, $\alpha = 0.95$, and $\alpha = 0.7$. These plots are in accordance with Table 1. The order α in the model changes the stable region of $\Pi(\lambda, \alpha) = 0$. The smaller the order α is, the larger the stable region satisfying $\arg(\gamma) > \alpha\pi/2$ is. Thus, Model 2 has a larger stable region than Model 1.

Table 1 Critical value c_i for different order α

α	c_{-3}	c_{-2}	c_{-1}	c_1	c_2	c_3
0.99	−1.2653	−0.5144	−0.4136	0.4136	0.5144	1.2653
0.98	−1.1106	−0.5326	−0.4136	0.4136	0.5326	1.1106
0.97	−0.9680	−0.5589	−0.4136	0.4136	0.5589	1.9680
0.96	−0.8179	−0.6081	−0.4136	0.4136	0.6081	0.8179
0.95	/	/	−0.4136	0.4136	/	/

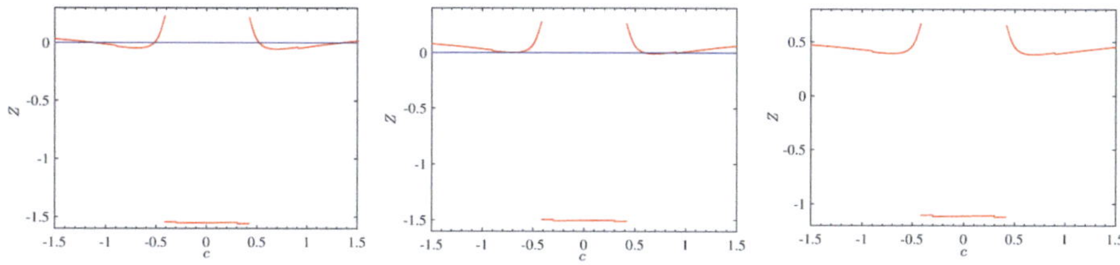

(a) (b) (c)

Fig. 5 The plots of $Z = \min(\arg(\gamma)) - \alpha\pi/2$ with respect to c, when α is fixed. (**a**) $\alpha = 0.98$. (**b**) $\alpha = 0.95$. (**c**) $\alpha = 0.7$

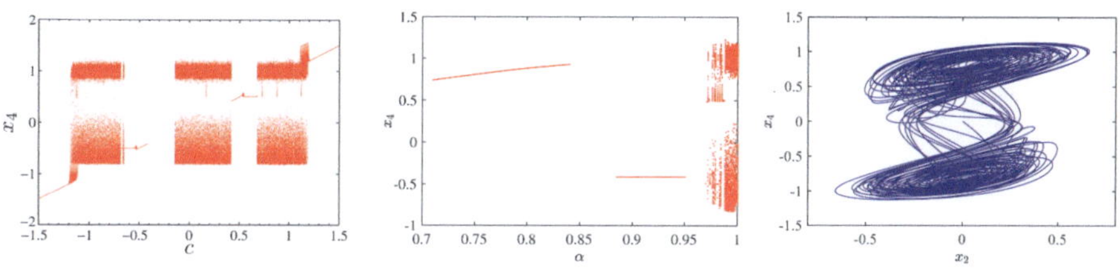

(a) (b) (c)

Fig. 6 (**a**) Bifurcation diagram of the local maxima of the state x_4 with respect to c, when $\alpha = 0.98$; (**b**) Bifurcation diagram of the local maxima of the state x_4 with respect to α, when $c = 0$; (**c**) Phase-portrait in the (x_2, x_4)-plane, when $\alpha = 0.98$ and $c = 0$

Let $c \in (-1.5, 1.5)$, as $0.95 < \alpha < 1$, there are four stable regions: $c \in (-1.5, c_{-3}), c \in (c_{-2}, c_{-1}), c \in (c_1, c_2)$, and $c \in (c_3, 1.5)$; as $0 < \alpha \leq 0.95$, there are two stable regions: $c \in (-1.5, -0.4136)$ and $c \in (0.4136, 1.5)$.

Figure 6a, b are the bifurcation diagrams of the local maxima of the variable x_4. In Fig. 6a, $\alpha = 0.98$ and $x_1(0) = 10^{-9}, x_2(0) = x_3(0) = 0$. As c increases from -1.5 to 1.5, two steady states of chaotic attractor and stable equilibrium switch via tangent bifurcation at $c_{\pm 3} = \pm 1.1106$, $c_{\pm 2} = \pm 0.5326$, and $c_{\pm 1} = \pm 0.4136$, respectively. As long as c is in four stable regions: $c \in (-1.5, -1.1106), c \in (-0.5326, -0.4136), c \in (0.4136, 0.5326)$, and $c \in (1.1106, 1.5)$, the equilibrium $(0, 0, 0, c)$ is stable. While c changes in unstable regions, unstable equilibrium $(0, 0, 0, c)$ stabilizes at two steady states of chaotic attractor and a stable point. For example, as $c \in (-1.1106, -0.5325)$, unstable equilibrium $(0, 0, 0, c)$ with $-1.1106 < c < -0.7$ stabilizes at chaotic attractor, and unstable equilibrium $(0, 0, 0, c)$ with $-0.6 < c < -0.5325$ stabilizes at a stable equilibrium. Figure 7a shows Eq. (7) with initial state $\boldsymbol{x}^* = (10^{-9}, 0, 0, -0.54)$ stabilizes at a stable equilibrium. Furthermore, Fig. 7b shows Eq. (7) with initial state $\boldsymbol{x}^* = (10^{-9}, 0, 0, c)$ stabilizes at a stable equilibrium, where $c = \pm 1.4433$ and $c = \pm 0.51$.

In Fig. 6b, the initial state is $\boldsymbol{x}^* = (10^{-9}, 0, 0, 0)$. As α changes in (0.7, 1), there are three steady states of chaotic attractor, limit cycle, and stable equilibrium. Figure 6c is a chaotic attractor resulted from the initial states $\boldsymbol{x}^* = (10^{-9}, 0, 0, 0)$. Figure 7c is a limit cycle as order $\alpha = 0.8$ and a stable equilibrium as order $\alpha = 0.9$.

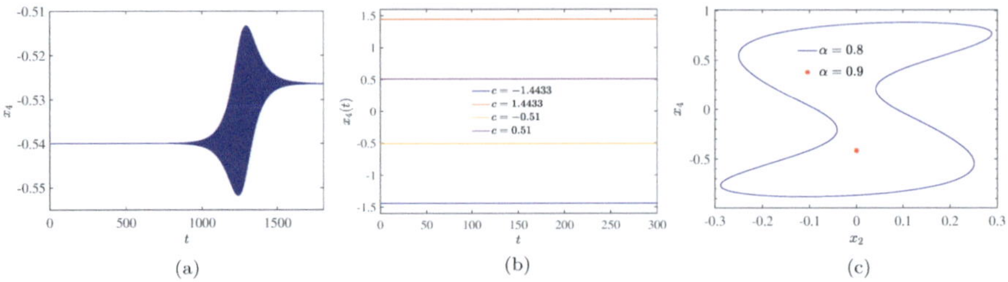

Fig. 7 (**a**) Time-history of the state x_4 when $\alpha = 0.98$ and $c = -0.54$; (**b**) Time-history of the state x_4 with $\alpha = 0.98$ and different c; (**c**) Phase-portrait in the (x_2, x_4)-plane when $c = 0$

In addition, Model 2 has two kinds of steady states: chaotic attractor and stable equilibrium (when $0.95 < \alpha < 1$) or limit cycle and stable equilibrium (when $0 < \alpha \leq 0.95$). The numerical simulation of the fractional Model 2 shows "tangent bifurcations" still exist due to the non-smoothness of the absolute function of c, but "intermittent chaos" is not observed.

4 Summary

Two fractional-order models of Chua's memristive circuits are generalized from the integer-order Chua's memristive circuit, and they are non-smooth systems with a line equilibrium depending on the memristor's initial state. It is found numerically that both models switch between steady states according to "grazing bifurcation" or "tangent bifurcation" resulted from the non-smoothness of the absolute function of c and the fractional order, the stable region of Model 2 increases as the order decreases. For fixed order, Model 2 has a lager stable region and has simpler nonlinear dynamics than Model 1. Three steady states of chaotic attractor, limit cycle, and stable equilibrium coexist for Model 1, and "intermittent chaos" occurs. However, only two kinds of steady states of chaotic attractor and stable equilibrium coexist in Model 2, and no "intermittent chaos" is found. The order of fractional-order derivative in the memristive Chua's circuits can be used to create complexity or to eliminate complexity of nonlinear dynamics of the two fractional-order models.

Acknowledgement This work was supported by Natural Science Foundation of China under Grants 11602035 and 11372354.

References

1. Strukov, D.B., Snider, G.S., Stewart, D.R. Williams, R.S.: The missing memristor found. Nature **453**, 80–83 (2008)
2. Chua, L.O.: Memristor-the missing circuit element. IEEE Trans. Circuit Theory **18**(5), 507–519 (1971)
3. Iton, M., Chua, L.O.: Memristor oscillators. Int. J. Bifurc. Chaos **18**, 3183–3206 (2008)
4. Iton M, Chua L.O.: Duality of memristor circuits. Int. J. Bifurcation Chaos **23**, 1330001 (2013)
5. Pham, V.T., Volos, C., Jafari, S., Wang, X., Vaidynathan, S.: Hidden hyperchaotic attractor in a novel simple memristive neural network. J. Optoelectron. Adv. Mater. **8**, 11–12 (2014).
6. Bao, B.C., Bao, H., Wang, N., Chen, M., Xu, Q.: Hidden extreme multistability in memristive hyperchaotic system. Chaos, Solitons Fractals **94**, 102–111 (2017)
7. Njitacke, Z.T., Kengne, J., Fotsin, H.B., Negou, A.N. Tchiotsop, D.: Coexistence of multiple attractors and crisis route to chaos in a novel memristive diode bridge-based Jerk circuit. Chaos, Solitons Fractals **91**, 180–197 (2016)
8. Li, Q.D., Zeng, H.Z. Li, J.: Hyperchaos in a 4D memristive circuit with infinitely many stable equilibria. Nonlinear Dyn. **79**, 2295–2308 (2015)
9. Dutta, M., Nusse, H.E., Ott, E., Yorke, J.A., Yuan, G.: Multiple attractor bifurcations: a source of unpredictability in piecewise smooth systems. Phys. Rev. Lett. **83**, 4281–4284 (1999)
10. Carroll, T., Pecora, L.: Using multiple attractor chaotic systems for communication. Chaos: Interdiscip. J. Nonlinear Sci. **9**, 445–451 (1999)
11. Ascoli, A., Tetzlaff, R., Menzel, S.: Exploring the dynamics of real-world memristors on the basis of circuit theoretic model predictions. IEEE Circuits Syst. Mag. **18**, 48–76 (2018)
12. Yu, Y.J., Wang, Z.H.: A fractional-order memristor model and the fingerprint of the simple series circuits including a fractional-order memristor. Acta Phys. Sin. **64**, 0238401 (2015) (in Chinese)
13. Fouda, M.E., Radwan, A.G.: On the fractional-order memristor model. JFCA **4**, 1–7 (2013)
14. Fouda, M.E., Radwan, A.G.: Fractional-order memristor response under DC and periodic signals. Circuits Syst. Signal Process. **34**, 961–970 (2015)
15. Cafagna, D., Grassi, G.: On the simplest fractional-order memristor-based chaotic system. Nonlinear Dyn. **70**, 1185–1197 (2012)
16. Yang, N.N., Xu, C., Wu, C.J., Jia, R., Liu, C.X.: Modeling and analysis of a fractional-order generalized memristor-based chaotic system and circuit implementation. Int. J. Bifurcation Chaos **27**, 1750199 (2017)
17. Yu, Y.J., Wang, Z.H.: Initial state dependent nonsmooth bifurcations in a fractional-order memristive circuit. Int. J. Bifurcation Chaos **28**, 1850091 (2018)
18. Bao, B.C., Xu, Q., Bao, H., Chen, M.: Extreme multistability in a memristive circuit. Electron. Lett. **52**, 1008–1010 (2016)
19. Podlubny, I.: Fractional Differential Equations. Academic Press, San Diego (1999)
20. Diethelm, K., Ford, N.J., Freed, A.D.: A predictor-corrector approach for the numerical solution of fractional differential equations. Nonlinear Dyn. **29**, 3–22 (2002)
21. Diethelm, K., Ford, N.J.: Analysis of fractional differential equations. J. Math. Anal. Appl. **265**, 229–248 (2002)
22. Danca, M.F., Kuznetsov, N.: Matlab code for Lyapunov exponents of fractional-order systems. Int. J. Bifurcation Chaos **28**, 1850067 (2018)
23. Matignon, D.: Stability properties for generalized fractional differential systems. ESAIM Proc. **5**, 145–158 (1998)

Author Index

W. Lacarbonara et al. (eds.), *New Trends in Nonlinear Dynamics*,
https://doi.org/10.1007/978-3-030-34724-6

C

L

Subject Index

W. Lacarbonara et al. (eds.), *New Trends in Nonlinear Dynamics*,
https://doi.org/10.1007/978-3-030-34724-6

MIX
Papier aus verantwortungsvollen Quellen
Paper from responsible sources
FSC® C105338

If you have any concerns about our products,
you can contact us on
ProductSafety@springernature.com

In case Publisher is established outside the EU,
the EU authorized representative is:
Springer Nature Customer Service Center GmbH
Europaplatz 3, 69115 Heidelberg, Germany

Printed by Libri Plureos GmbH
in Hamburg, Germany